Turning Wind into Power

European University Studies
Europäische Hochschulschriften
Publications Universitaires Européennes

Series XXXI
Political Science

Reihe XXXI Série XXXI
Politikwissenschaft
Sciences politiques

Vol./Bd. 601

PETER LANG
Frankfurt am Main · Berlin · Bern · Bruxelles · New York · Oxford · Wien

Gudrun Elisabeth Benecke

Turning Wind into Power

Effects of Stakeholder Networks on Renewable Energy Governance in India

PETER LANG
Internationaler Verlag der Wissenschaften

Bibliographic Information published by the Deutsche Nationalbibliothek
The Deutsche Nationalbibliothek lists this publication in the Deutsche Nationalbibliografie; detailed bibliographic data is available in the internet at http://dnb.d-nb.de.

Zugl.: Potsdam, Univ., Diss., 2010

517
ISSN 0721-3654
ISBN 978-3-631-61570-6

www.peterlang.de

Acknowledgements

Ultimately, shortly before the submission of the thesis, remains the most difficult thing to do as words are insufficient to express my gratitude and thanks to all the people who have been with me in this venture.

Having emerged quietly in the course of my PhD and turning out as the most valuable support particularly during this last phase is my partner Torsten Eichhorn. Without his unshakable faith in me and his visible, invisible and undemanding presence, I would have lost the strength to keep up until the very end.

For his strategic advice I would like to thank my first supervisor and project leader Prof. Dr. Harald Fuhr who encouraged my search for new perspectives while retaining the big picture. I am very grateful to the concise and instructive recommendations by Prof. Dr. Frank Biermann as well as personal advice that kept me in good spirits during my field research in India. Last but not least, I owe a lot to Dr. Axel Michaelowa who consciously followed and guided my research right from the beginning. At this stage, I would also like to thank Prof. Dr. Johannes Hamhaber who strongly encouraged me to pursue the New Economic Sociology track.

Without the financial support and the community network of the SFB 700 I would not have been able to conduct my research this independently. In addition to conceptual guidance, I appreciate highly the opportunities to discuss ideas and share lessons-learnt regarding research and personal experiences. My Indian field research will remain an unforgettable period of my life and I owe many people who assisted me. Most and foremost, I do thank TERI, and in particular my supervisors Amit Kumar and Dr. Vivek Kumar for providing the opportunity to use their facilities and offered guidance. Furthermore, I extend my gratitude to the GTZ office and Ms. Pamposh Bhat. Finally, I am extremely grateful for the openness of all my interview partners that unquestioningly opened new avenues for my research. Finally, I do appreciate the silent understanding of my colleagues at the Heinrich-Böll-Stiftung in these last few months of my PhD.

In order for the acknowledgements not to turn into a thesis, I would like to end with some very personal and deeply heartfelt thanks to all my friends who stayed with me during this time. Thanks to Miriam, Lars, Markus, Hendrikje and Melanie that have accompanied my struggles – I shall never forget you backing me up during my sickness leave period. Thanks to Anke – without words. Many thanks also to Anni, Giz and particularly Gunti – this thesis would not go into print without you being there.

Contents

Figures & Tables

Maps

Figures

Tables

Acronyms

ANERT: Agency for Non-Conventional Energy Resources and Technology
C-WET: Centre for Wind Energy Technology
CDM: Clean Development Mechanism
CERC: Central Electricity Regulatory Commission
CPI – M: Communist Party India - Marxist
CS: Civil Society
CSE: Centre for Science and Environment
CPI: Corruption Perception Index
DANIDA: Danish International Development Agency
DMK: Dravida Munnetra Kazagha (National Party Tamil Nadu)
DNES: Department for Non-Conventional Energy Sources
DOE: Designated Operational Entity
GDP: Gross Domestic Product
GHG: Greenhouse Gas
HDI: Human Development Index
HPI: Human Poverty Index
I: Investor
IAEA: International Atomic Energy Agency
IAS: Indian Administrative Service
IEA: International Energy Agency
INC: India National Congress
InWEA: Indian Wind Energy Association
IPCC: International Panel on Climate Change
IREDA: Indian Renewable Energy Development Agency
IT: Information Technology
FDI: Foreign Direct Investment
IPP: Independent Power Producer
JI: Joint Implementation
KSEB: Kerala State Electricity Board
KSERC: Kerala State Electricity Regulatory Commission
LP: Local People
M: Manufacturer
MAT: Minimum Alternate Tax
MDG: Millennium Development Goals
MNRE: Ministry of New and Renewable Energy Sources (India)
MoP: Ministry of Power
MW: Megawatt
NEDA. Uttar Pradesh Non-Conventional Energy Development Agency
NGO: Non-Governmental Organisations
NEPC: National Electricity Petroleum Corporation
ODA: Official Development Assistance
OECD: Organisation for Economic Co-operation and Development
PDD: Project Design Documents
PPA: Power Purchase Agreement
PPP: Public-Private Partnerships
PLF: Plant Load Factor
PSR: Power Sector Rating

RPS: Renewable Portfolio Standard
REN: Renewable Energy
R&D: Research and Development
RRB: Indian Wind Company
Rs: Rupees
SDP: State Domestic Product
SEB: State Electricity Board
SERC: State Electricity Regulatory Commission
SNA: State Nodal Agency (of MNRE)
SP: Service Provider
Suzlon: Wind Company
T&D: Transmission and Distribution
TEDA: Tamil Nadu Environmental Development Agency
TERI: The Energy Resources Institute
TNEB: Tamil Nadu Electricity Board
TNERC: Tamil Nadu Electricity Regulatory Commission
TNSG: Tamil Nadu State Government
UN: United Nations
UNDP: United Nations Development Programme
UNEP: United Nations Environment Programme
USA: United States of America
Vestas: Wind Company
WCED: World Commission on Environment and Development
WWEA: World Wind Energy Association

CHAPTER I

Governance Failure – Market Failure? Spatial Variations in Renewable Energy Deployment

CHAPTER I
Governance Failure – Market Failure? - Spatial Variations in Renewable Energy Deployment

"Of all the forces of nature, I should think that wind contains the largest amount of motive power—that is, power to move things.... As yet, the wind is an untamed and unharnessed force; and quite possibly one of the greatest discoveries hereafter to be made, will be the taming, and harnessing of it."
(Abraham Lincoln 1860)

In a climate-constrained world, renewable energy is paramount to sustainable development particularly for developing countries like India. Spatial variations in deployment call into question the potential of renewable energies to meet development imperatives while protecting the climate. This dissertation will explore conditions for effective renewable energy governance due to the demand for understanding what successful deployment requires and derives at the relevance of local stakeholder networks. This first chapter will introduce the contextual background and provide the methodological roadmap to research on renewable energy governance taking the case of wind power deployment in India.

I.1 Setting the Context for Renewable Energy Governance

Gloomy climate change scenarios and the development dilemmas of the 21st century create the demand for clean development alternatives and give justice to Lincoln's thesis.
This section will proceed firstly to stress the importance to examine renewable energy governance considering the realities of climate change and developmental impasses. Secondly, the role of renewable energies in pursuing sustainable development paths is discussed.

I.1.1 Climate Change and Development Dilemmas

Realities and impacts of climate change pose threats to human livelihood and development worldwide. Developing countries are hit hardest as they face the dilemma of how to achieve development objectives in a climate-constrained world (Barnett 2000). The quest for energy to fuel socio-economic development and growth stimulates debates about sustainable development. This section will discuss the argument that the global re-birth of renewable energy for power generation responds to the complex interdependencies between multilevel environmental and developmental challenges.

Despite increasing concerns about the sustainability of the energy system, the present global energy scenario is characterised by a heavy dependence on fossil fuels. In 2002, oil, coal and gas contributed roughly to about 78 % of primary energy consumption (Goldemberg 2006: 5). Global economic, demographic and geopolitical developments as well as climate change create an international demand for renewable energy (Dincer 2000: 157). As a result, energy policies are a high-priority issue on governments' policy agendas worldwide (Banerjee 2005).

From the perspective of the industrialised world, climate change and environmental concerns are the main drivers for renewable energies. Initially, renewable energies were crucial to reduce the one-sided dependence on fossil fuels and to circumvent critical geopolitical constellations, e.g. the 1970s oil crisis or recent gas disputes with Russia. Lately, the potential of renewable energy sources to mitigate climate change has been recognised. Fossil fuels contribute a large extent to global carbon dioxide emissions that increase at about 0.5 % annually (Goldemberg 2006:5). Particularly localised environmental impacts such as acid rain, ozone and forest depletion created the awareness that alternative development pathways are needed (Dincer 2000: 158). Increasing the share of renewable energies emerges as a crucial, viable and sustainable alternative (Wilkins 2003: 204) especially as the costs of renewable energy technologies have declined[1].

With increasing economic growth in the developed world, the demand for energy is soaring and energy security concerns are of central importance (Goldemberg 2006: 6). Renewable energy provides solutions to the quest for additional energy sources and mitigates environmental damage. The example of India reflects the past and recent trends that on the whole have pushed for alternative energy systems. According to the energy scenario projections, India will represent a third of global energy demand by 2050 which will place immense stress on existing energy sources (Greenwood et al 2007: 45). This is triggered by sustained economic growth and rising income levels, which generates energy and electricity demand for consumptive and productive purposes. In order to meet Government's targets of energy security by 2020 and energy independence by 2030 (Greenwood et al 2007: 45), renewable energy has become one of the priority sector. Coal, oil and gas are still the three primary constituents of India's energy mix despite the rich natural resource endowment (TERI 2004). The Indian energy mix was traditionally dominated by a large share of renewable energy sources such as firewood, animal dung and agricultural residues[2] (Veda 2002: 2) that progressively declined with the shift to commercial and modern fuels. In 2007, renewable energy sources held a share of about 7.74% in the Indian energy mix (PowerLine 2007: 12) which Government aims to increase to 10% by 2012 (PowerLine 2007: 124).

The quest for alternative energy sources plays a major role (Ramana et al 2001: 5) since power generation is inadequate to meet increasing demands for energy and electricity services. Peaking power shortages about 20 % (PowerLine 2007: 12) characterise the poor shape of the Indian power sector. In the light of governance debates[3], the insufficient provision of essential electricity services to secure human welfare reflects governance failures. Renewable energies might not only constitute an option for clean development but also give rise to alternative ways to govern sustainable energy solutions. Given present and prospective economic growth and energy demand scenarios, the main question really is how India might tackle its energy problem.

---

[1] This depends of course on the type of renewable energy technology.

[2] Particularly in the rural sector, non-commercial and traditional fuels still play an important role in the energy portfolio.

[3] These refer to governance debates in the context of the research centre SFB 700 ("Governance in areas of limited statehood") which will be elaborated in Chapter IV as later in this section.

In summary, despite distinct perspectives on renewable energy debates (Wilkins 2003: 203) the common global challenge boils down to sustainable development, i.e. the question how to pursue an energy pathway that caters for the present demand without compromising on future developments.

I.1.2 Renewable Energies for Sustainable Development

The introductory quote depicted wind as the "*power to move things*" that, literally speaking, implies the potential of wind energy to provide an alternative sustainable development path. This section will highlight the advantages of alternative energy sources to respond to global energy and climate challenges particularly in developing countries. The following discussion will demonstrate that the relevance of renewable energy for sustainable development creates the demand for research on pre-conditions required in order to harness this potential.

As point of departure, the quest for renewable energy is driven by the demand to mitigate climate change without sacrificing energy security in order to cater for the basic needs of the population and secure sustainable livelihoods (Gupta 2003: 159). Renewable energy emerges as a viable option for a global sustainable transformation of social, economic and political systems.
The sustainable development discourse (Hopwood et al 2005) results from a growing international awareness of global linkages between mounting environmental problems, socio-economic issues and concerns about the future of humanity. The idea of sustainable development has gained momentum with the Brundtland report (WCED 1987) that explicitly recognises the interconnection between humans and the environment. Sustainable development is defined as development that "*meets the needs of the present without compromising the ability of future generations to meet their own needs*" (Smith & Rees 1998: 3). Since present growth models have not only led to environmental damage but also failed to eradicate poverty (Hopwood et al 2005: 39), alternative development pathways are envisioned that tackle these challenges. The ambition of sustainable development is to create long term ecologically stable, socially viable and economically productive systems (Gupta 2003: 157) in order to achieve sustainable livelihood.

Social Sustainability Dimension – Meeting Energy Demands
The first important linkage between renewable energy and sustainable development regards the social dimension. Sustainable development in this sense implies that all societal groups have the opportunity to meet their needs without excluding other actors (Wienges 2008).
One key challenge to sustainable development is the availability of energy services which has a critical bearing on the people's livelihoods. Particularly developing nations often lack energy services for vital consumptive and productive purposes such as water purification, cooking, heating and cooling, lighting to name but a few examples. The lack of access to available energy services restrains economic development and living conditions (Goldemberg 2006: 9). Not only the United Nation's poverty alleviation goals (Sherman 2005: 129) but also the World Summit on Sustainable Development recognise universal access to energy as essential for development (Srivastava & Rehman 2006: 643). Without the effective delivery of energy services people are impeded in engaging in productive efforts or improving their quality of life (Barnes & Floor 1996: 497). Another key challenge to social sustainability concerns

the affordability of energy services (McDade et al 2006: 217). Developing country households have to spent relatively more income on energy and also pay more in absolute terms (IAEA 2005: 17). This has direct impacts on poverty, employment, education, demography, local pollution and health (IAEA 2005: 16). Furthermore, present energy systems in developing countries are gendered meaning that the burden of acquiring traditional energy resources, e.g. collecting wood, is often born by women (Wilkins 2003: 203).

In summary, the demand for renewable energy to augment power generation in developing countries arises out of the increasingly dismal energy demand scenario that bears implications on social equity issues such as energy access, availability and affordability.

Environmental Sustainability Dimension – Integrating Environmental Concerns

The second linkage between renewable energy and sustainable development touches upon the environmental dimension.

Alternative energy paths are urgently required since the present energy system threatens local environment and global climate (Zingel 1998). Climate change and global warming are scientifically-acknowledged results of the fossil fuel based economies (Heller & Shukla 2003: 111). However, effects on developing countries are particularly severe given their vulnerability and constrained mitigation and adaptation capacities. Water shortages, agricultural disruption and coastal flooding are felt disproportionately by developing countries. Meanwhile, the continuation of traditional power generation, e.g. traditional biomass, and the large-scale electricity generation in antiquated fossil fuel power plants submits dangerous discharges and pollutants. One impact on sustainable livelihood is a disproportionate suffrage from health problems as regulations protecting against polluters are either not in place or not enforced (Ailawadi 2006). Other indirect and longer term impacts are land use restrictions due to localised ecosystems change, overexploitation and disrupted weather patterns. Ultimately, this might be one factor to trigger migration as the only option available to the local population (Barnes 1996).

In today's climate-constrained world developing countries face the cruel choice between developmental and environmental objectives (Heller & Shukla 2003: 112). The need for sustained economic growth in developing countries more or less requires the exponential rise in energy-related greenhouse gas emissions (Ebbinghausen 2005). Consequently, increasing energy demand to meet development objectives vis-à-vis the negative impacts of fossil fuel economies on the environment emphasise the requirements of energy from alternative sources (Varun & Singal 2005).

Economic Sustainability Dimension – Safeguarding Energy Security

Finally, the last linkage between renewable energy deployment and the quest for sustainable development regards the economic dimension. Since renewable energy resources are abundant and have increasingly become more cost competitive (Elliott 2007) they hold the potential to meet challenges of energy security and energy supply.

Particularly developing countries that embark on a development path to catch up with industrialised countries need to secure energy since modern economies depend on reliable and adequate energy supply (Jänicke 1991: 503). This is a precondition to foster socio-economic development at the local level by raising productivity and income generation. Energy security turns into an important objective on the development agendas of countries like India (Divan 2001) as interruptions of energy supply are frequent and cause detrimental financial and economic losses. The finiteness of fossil fuels, and dangers and uncertainties of interrupted supply chains due to geopolitical turmoil (Asif & Muneer 2007) increases the external but hidden costs of conventional resources. As fossil fuels become scarcer developing countries like India will face energy shortages, energy price hikes and augmenting energy insecurity (Dubash 2002). Since renewable energy resources exist in abundance, respective technologies will become vital in order to generate grid electricity and to ensure sustainable economic development (Varun & Singal 2005). As national property and public goods, renewable energy sources have a clear geostrategic advantage and hence decrease energy dependencies. The abundance and the locality of natural resources create options for technological leapfrogging towards more sustainable energy systems.

All in all, this section argued for the relevance of renewable energy in the context of multilevel climate change and development challenges. Sustainable development within a society requires not only the secure but also the sustainable supply of energy services (Dincer 2000: 159). Renewable energies thus play an important role in the quest for achieving sustainable development (Barnes & Floor 1996: 497) particularly in developing countries.

I.2 Realities of Renewable Energy and Governance

Global energy and climate change scenarios trigger the demand for renewable energy, i.e. the need for alternative development options. In the following, the realities of renewable energy are explored globally and with focus on India. This reveals the research puzzle that renewable energy deployment spatially varies. This dissertation will explore the conditions for effective renewable energy governance that is required to embark upon sustainable development paths.

I.2.1 Spatial Variations – The Research Puzzle

One essential advantage of renewable energy sources such as wind, water, biomass and sun compared to conventional fuels is their ubiquity. Nearly all parts of the world are endowed with some natural resource potential that could be harnessed for various applications (Richards 2006). The main puzzle is that despite this resource availability a great variation exists as to whether and how renewable energies are utilised for sustainable energy production.

From a multi-level systems perspective, taking the case of wind power, this is firstly reflected at the global level (WWEA 2006). Regions with a great wind power potential like the Middle East and Northern Africa do not tap this potential, while regions with similar endowments like Europe harness these resources quite effectively. Secondly, at the national level, states like Spain are global wind players whereas Portugal with a similar resource potential lags far behind.

Lastly, narrowing the focus, the local level presents a similar scenario. Taking India as an example, some federal states harness their wind potential while other states barely touch upon this resource (Map A).

Map A: Wind Power Deployment in India

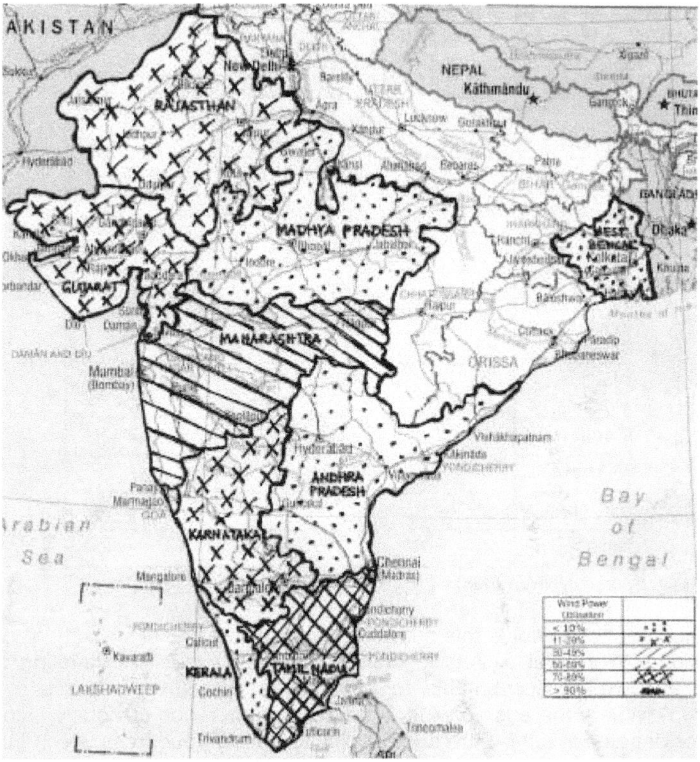

Source: Data from MNRE (2008)

This is puzzling because, the Indian Government has put considerable effort into promoting renewable energy development since the 1970s. In other words, despite the historic importance of renewable energies in India great variances exist with regard to the extent this potential is harnessed by the federal states. Wind energy as the most advanced renewable energy technology contributes roughly 70 % to the entire renewable energy capacity installed in India. The Indian wind sector is described as nearly commercialised (INT 1) and plays an important role in the global wind energy scenario as well as nationally. Hence one might expect similar deployment rates in the states that avail of roughly similar wind power potential. Taking this case, contrasting data of renewable energy resource potential with de facto installed renewable energy capacities gives rise to a remarkable observation. Despite similar natural resource potential, some Indian states harness these resources while others hardly touch upon their renewable energy potential. This is particularly puzzling comparing the state of Tamil Nadu that fully exploited its wind potential with the state of Kerala that barely tapped into its wind power potential (Table 1). Both states are located on the Southern coast of India and are endowed with relatively similar wind resource potential in relation to terrain and population. Since de facto deployment rates differ so starkly compared to the overall Indian scenario and other wind power states, Tamil Nadu and Kerala will be selected as specific case studies[4].

Table 1: Wind Energy Deployment in India

State	Installed Wind Capacity (MW)	Wind Power Technical Potential[5] (MW)	Utilisation Rate (%)
Tamil Nadu	**3873**	**4750**	**81.5**
Karnataka	1011	6620	15.3
Maharashtra	1756	3650	48.1
Rajasthan	539	5400	10
Andhra Pradesh	123	8275	1.5
Madhya Pradesh	188	5500	3.4
Kerala	**2**	**875**	**0.2**
Gujarat	1253	9675	13
West Bengal	1	450	0.2
India	8746	45195	19.4

Source: MNRE (2008)

Several approaches to solve this puzzle discuss technological maturity or contextual preconditions (Worrell et al 2001). Another argument holds the variation of natural resource endowment accountable for differences in natural resource exploitation (Auer 2008). Where there is no wind, wind power generation obviously faces difficulties. Other debates focus on the accessibility to natural resources which in the case of wind concerns the wind regime and the availability of adequate sites (Golait 2007). As a matter of fact, where the terrain does not allow for the construction of wind turbines, wind power generation is severely impeded. However, these intuitive approaches do not deliver satisfactory answers to solving this puzzle. In the context of this dissertation, this will be illustrated alongside the example of the wind sector and specifically the case of the two Indian states Kerala and Tamil Nadu.

All in all, in the light of the renewable energy debate in developing countries, this dissertation sets out to examine the underlying reasons for spatial variations in natural resource deployment for power generation. Taking spatial variations in the case of the Indian wind sector as research puzzle, this dissertation aims to explore the conditions under which wind energy projects are successfully deployed in Indian states. The objective is to map and to examine contextual requirements for effective renewable energy governance in a developing country setting. In this sense, the stakeholder network concept will constitute the analytical and conceptual approach to explore mechanisms and dynamics of multilevel state-market relations. The concrete

[4] Chapter III will go deeper into the justification of this case study selection as concrete research puzzle by empirically excluding other potential explanatory factors such as context-related characteristics.
[5] The assessment of wind resources and the respective technical potential are complex and despite recent methodological improvements still remain subject to uncertainties (Pillai 2006; INT 2).

focus of this dissertation is on the characteristics and dynamics of emerging local governance arrangements and is expressed in the precise research question:

To what extent and how do local stakeholder networks impact on effective renewable energy governance taking the case of wind power deployment in Tamil Nadu and Kerala?

I.2.2 Governance of Renewable Energy

This section will conclude on contextualising research on renewable energy and argue that the wind energy puzzle of spatial variations needs to be explored in the light of a governance framework.

The quest for alternative development options opens a window of opportunity for renewable energy to contribute to socially, economically and environmentally sustainable transformations particularly in developing countries like India. Renewable energies hold many promises as regards increased energy security, reduced environmental damage and enhanced energy access and supply. Scepticism prevails first of all due to the disappointing progress with regard to renewable energy deployment in developing countries (Grubb 1993: 239). Reasons relate to comparatively high costs that reflect the skewed structure of the present energy system and insufficient investment, technological lags due to insufficient innovation and uncertainties about long-term developments (Kjellstroem 2005). Developing countries lack not only resources, knowledge but also the willingness of public actors to support energy alternatives. Development assistance and the international climate change community have attempted to mitigate government failure and to provide adequate instruments to promote renewable energy (Hoekman 2004). However, local barriers and international developments[6] culminated to the dissatisfactory resonance of renewable energy in developing countries (Grubb 1993: 240). The spatial variations in natural resource deployment point towards obstacles inherent in the governance of natural resources.

This dissertation argues that renewable energy deployment is a matter of governance since the delivery of energy services to secure societal well-being depends on how respective governance structures and agency are organised. Energy security is acknowledged a public good (Abbot 2001) and hence the provision of adequate energy services is a government responsibility[7]. Yet, particularly developing countries suffer from a government failure to ensure the effective delivery and to guarantee their citizens' welfare. The emergence of renewable energy solutions on multilevel policy agendas and the promises of sustainable development hence call for revisiting energy governance.

Governance is understood as *"intentional provision of public gods and services to a defined collective"* (Risse et al 2006). Renewable energy governance comprises the provision of institutional structures and regulatory frameworks (output), the deployment of renewable energy projects and the implementation of framework conditions (outcome) as well as the electricity supply to a defined constituency (impact). In the context of this dissertation, renewable energy governance is considered "effective" if

[6] Examples are the lack of primary data, conservatism in adequate technology, failures of transferring existing knowledge to the policy community, lacking vision and an overall low world economic growth.
[7] Even though delivery processes of energy services might often be in the hand of private parties.

respective projects are de facto deployed. This means that successful renewable energy deployment constitutes part of and is outcome of a governance process in which structures and regulations create un-/intended effects on wind power deployment. This dissertation assumes that wind power deployment contributes to the effective delivery of energy services. It explicitly does not consider if addressees, i.e. local people, in fact receive governance services as this goes beyond the research scope[8]. Explorations of the Indian wind power case depart from the assumption that governance works differently than in the Western world where this concept originates from. This means that responsibilities and implementation of energy service delivery do not occur along the same lines and logics as in OECD countries but might involve functionally equivalent structures and processes. The main purpose of this dissertation is to capture these functional equivalents to how renewable energy services are provided, i.e. how renewable energy governance takes place. Consequently, stakeholder network analysis is applied as analytical and conceptual tool to explore how renewable energy is governed in/with functionally equivalent governance arrangements and interactions. Given the lack of research on the governance of renewable energy in developing countries, the governance framework[9] guides the examination of new empirical realities like the Indian wind puzzle alongside the following questions:

- First and foremost, who governs and who is governed? What do governance arrangements involved in the process of renewable energy deployment look like?
  The principal focus is on the stakeholders involved in renewable energy governance and their characteristics.
- Secondly, how and with what means does the governance of renewable energy take place?
  Attention is turned towards the relations and interactions between stakeholders governing renewable energy.
- Thirdly, under what conditions is renewable energy governance effective and efficient, i.e. produces outcomes in terms of successful deployment?
  The exploration of stakeholder networks in renewable energy governance gives rise to conclusions on the conditions under which stakeholder networks as governance coalitions exert certain governance effects.
- Finally, how and to what extent do new modes of governance such as the CDM interact with these renewable energy governance arrangements?
  Local renewable energy stakeholder networks will be discussed as necessary condition for tapping new opportunities like the CDM.

This dissertation will integrate governance perspectives in the exploration of renewable energy deployment and hence contribute to mapping governance arrangements in the case of the Indian wind sector. This allows for implications on the functional equivalents to energy service provision, the conditions for effective and efficient functioning of renewable energy governance as well as the pre-requisites for new modes of governance to mitigate state failures.

---

[8] This potential gap is due to inherent and systemic difficulties of the Indian energy sector and governance reflected e.g. in T&D losses.

[9] This refers to debates in the context of the collective research centre SFB 700 (Risse & Lehmkuhl 2006) that will be elaborated later.

In summary, the first two sections discussed the context and realities of renewable energy governance in developing countries. The next section will elaborate on the research procedures to explore the research puzzle of spatially varying wind power deployment in India.

I.3 Researching Renewable Energy Governance in Developing Countries

The remainder of this chapter will sketch out methodology and methods drawn upon to explore the research puzzle of spatially varying wind power deployment in India. This dissertation adopts a two-stage research approach in order to give justice to the empirical novelty and to ensure the honest representation of concrete empirical realities (Flick 2006: 74). This means that in the first research stage the Indian wind puzzle is examined inductively while the second research stage based on previous empirical results follows a deductive procedure. The inductive and deductive logic of conducting research is also reflected in the structure of this thesis. The reasons for selecting this format firstly arise out of the need for background research on renewable energy realities in the concrete case study context as necessary requirement for an in-depth exploration. Secondly, inductive and deductive approaches imply distinct data needs and research methodologies that suggest a differentiated presentation in the context of this thesis. Lastly, this format of research presentation gives justice to demands on transparency and replicability of qualitative research particularly in non-OECD contexts.

I.3.1 Methods and Methodology

The purpose of this section is to specify how and with what underlying logic the empirical puzzle of spatially varying wind power deployment in India is approached[10]. In the following, the research puzzle is operationalised through a case study design, and the choice of methods is elaborated.

(i) Stating the Case – Case Study Research Design

Point of departure of this dissertation is the empirical observation of spatial variations in wind power deployment in India that do not match the natural resource potential. This constitutes the research puzzle since preliminary intuitive answers do not satisfactorily explain this empirical reality. The main assumption of this dissertation is that wind energy deployment ultimately constitutes the outcome of effective renewable energy governance. The stakeholder network concept is elaborated as analytical and conceptual approach to explore in particular local governance arrangements in the multilevel context of renewable energy deployment in the Indian wind sector.

The precise research question asks:

To what extent and how do local stakeholder networks impact on effective renewable energy governance taking the case of wind power deployment in Tamil Nadu and Kerala?

[10] Due to the uniqueness of the empirical field and data, the kind of qualitative social science research this dissertation engages in is based on a procedural understanding (Flick 2006: 69). This implies that data generation, analysis and presentation take place in a close and circular manner involving methods and theories (Flick 2006: 71).

This dissertation aims at a theory-guided exploration of multi-level determinants of effective renewable energy governance in developing countries taking the Indian wind sector as example with focus on the explanatory power of stakeholder networks. In this sense, contributions to interdisciplinary theory development on governance networks as well as practice-relevant implications are made.

Case study strategy is considered the most suitable approach (Yin 2003: 22) to tackle this research puzzle due to the descriptive but also exploratory nature (Gerring 2004: 347).
The selection of the research topic and the case study is justified given the relevance of renewable energy for sustainable development in the light of the research puzzle. Wind energy is chosen since it constitutes a relatively mature and straightforward technology that has experienced similar development trajectories at global and national levels. In terms of geographical focus, India constitutes a relevant developing country case due to the urgent quest for sustainable energy options and its prominent position, i.e. potential, on the global wind energy market. Since this dissertation aims at an understanding of multilevel and local dynamics, two cases of wind power deployment in Indian states are selected according to the variance on the dependent variable, i.e. different outcomes in wind energy deployment. Tamil Nadu and Kerala are considered most appropriate as they are both located at the Southern coast of India and have substantial wind resource potential[11]. Yet, while Tamil Nadu realised most of its estimated potential, Kerala hardly tapped into wind energy generation at all.
In addition, an embedded multiple case design is chosen to increase the empirical observations and to avoid selection biases inherent in small studies (Collier & Mahony 1996). In this sense, three phases of wind power deployment in India are identified (Figure I):

Figure I: Indian Wind Sector History

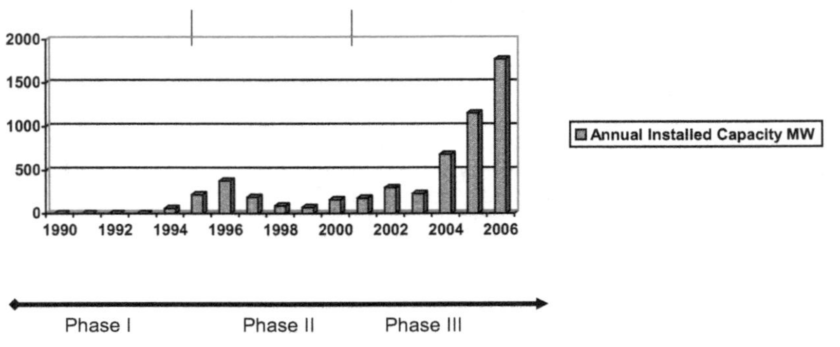

Source: Adapted from Pillai (2006)

[11] In the context of first inductive empirical exploration that encompasses and comparatively discusses all Indian wind power states; this selection is justified as alternative explanations for the puzzle do not hold (cf. Chapter III).

The first phase, initiation stage, starts with demonstration projects in the early 1980s and ends in the mid-1990s. The second phase, development stage, continues from the mid-1990s and ends 2001 with the beginning of the new Millennium. Finally, the third phase, commercialisation stage, extends from 2001 until 2007 – the time of present analysis. The rationales for selecting these periods of wind sector development relate to several factors such as national policy changes, international events and the development of the Indian wind sector (elaborated in more detail in Chapter III). The inductive as well as deductive analyses of wind power deployment will focus on the third phase and the year 2007.
After delineating these case study boundaries, the case study research is operationalised in the following:

(a) Operationalising the Puzzle Step 1: The Dependent Variable
The dependent variable considers the effectiveness of renewable energy governance; precisely, the successful deployment of wind energy projects in terms of installed Megawatt (MW) capacity. This implies that firstly, only large-scale wind power installations are considered that either generate power for the grid and/or produce electricity for captive consumption, i.e. for industries or commercial business. This puts limits to assessing the extent to which local people as governance addresses actually receive electricity services through wind power generation. Yet, this restriction was required out of feasibility considerations, e.g. complexities and novelty of alternatives such as decentralised wind power applications. The second implication is that only the installed wind power capacity is considered and not the actual amount of electricity generated by wind power installations. This constitutes an important and hence mentionable issue as the gap between installed and generated wind power electricity in India is astoundingly high (Narain et al 2008). However, this is a debate on its own ground that goes beyond the scope of this dissertation[12]. The installed capacity is assumed to be the de facto output of wind power generation. The case studies are selected on the variance of the dependent variable (Ganghoff 2005: 2), i.e. regarding the difference in installed wind power capacity. This is based on the assumptions that technological maturity and resource ubiquity are constant, and that internal project characteristics are similar in the selected local settings.

(b) Operationalising the Puzzle Step 2: The Explanatory Variables
During the first, inductive research phase an analytical framework will be derived on the basis of empirical research on renewable energy and respective theory debates. All in all, the four categories of developmental conditions, policy conditions, economic conditions and external conditions are identified as explanatory dimensions guiding the first inductive research stage. Criteria and indicators characterising these explanatory dimensions are specified alongside respective literature debates in Chapter II. The purpose of not engaging in a (theory-guided) pre-selection of just a few independent variables (i.e. deductive procedure) emerges out of two considerations. First of all, this research puzzle constitutes a so-called new empirical reality that due to the lack of pre-existing studies from a similar angle does not suggest a more concrete explanatory focus. The second point is the ambition to mitigate the selection bias of an OECD-researcher by remaining open to alternative and unexpected outcomes.

---

[12] An additional serious constraint is access and availability of consistent data on the actual amount of electricity generated from wind power. Further research could try and obtain statistics on the de facto wind power electricity generated that serves as more appropriate proxy of "successful deployment".

Based on the results of the first inductive exploratory approach, the second deductive research stage will develop and test an explanatory model of wind power deployment in India. This will consider the empirical reality in two concrete Indian states and extend the temporal boundaries. This research stage will draw on previous inductive research that will point towards two main explanatory factors: In the light of the governance framework, these refer to the characteristics of governance actors and their relational interactions. The first independent variable hence considers "actors' attributes" in terms of capacities, interests and capabilities characterising stakeholders and is discussed by organisational sociology. The second independent variable considers "actors' relations" in governance arrangements of different interaction intensity, quality and formality discussed by new economic sociology literature. The main reasons for theory selection are the insights into how economic activities such as wind power deployment embedded in the governance framework are influenced by stakeholder relations and interactions. Literature debates (Chapter IV) will specify and operationally spell out the independent variables. In addition, stakeholder network analysis is elaborated as analytical tool as well as explanatory concept to guide the exploration of these independent variables in the governance framework. As a result, an explanatory model of wind power deployment in India is elaborated that specifies explanatory factors and mechanisms.

(ii) Examining the Case

This dissertation is based on a multidisciplinary approach and integrates backward as well as forward looking research designs (Figure II). In the following, the methodological approaches employed to gather evidence on the empirical puzzle are considered.

Figure II: Research Format

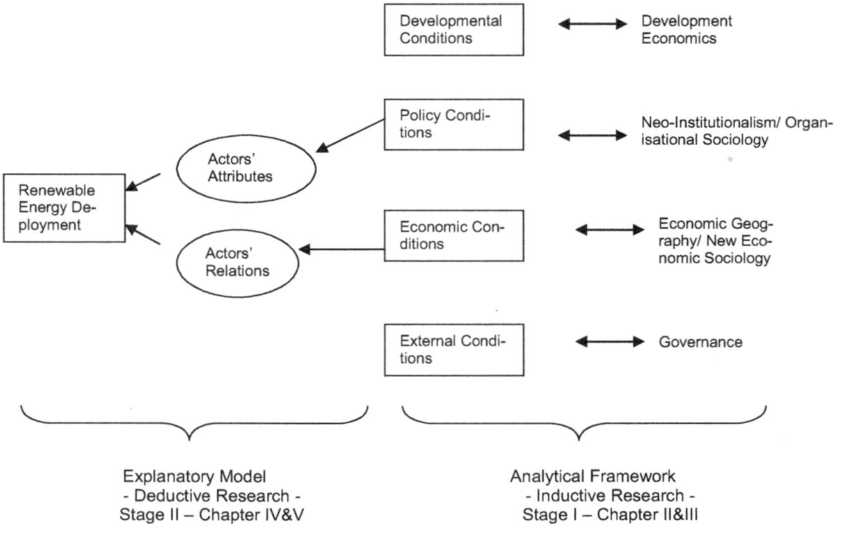

a) Data Requirements – General questions and Issues

Data selection is driven by the research question, strategy and the theoretical framework (Bennett & George 2004: 86) that implies the demand for an overwhelmingly qualitative account[13]. Due to the two-tier research strategy, the data collection process is also differentiated into two stages: During the first inductive exploratory stage, indicators and criteria are derived from literature and theory reviews (operationalised in Chapter II). In the second deductive analytical stage, an in-depth analysis of the same empirical reality requires mostly qualitative and perception-based data as well as ethnographic methods (cf. Chapter IV).

b) Data Collection

Multiple sources of evidence are drawn upon to meet the challenges of data requirement and availability:

Documentary Research

The sources of evidence are based on publicly-available written reports, minutes of meetings, agendas, letters of communication, administrative documents such as proposals, progress reports, formal studies and evaluations as well as newspaper and other media articles.

With regard to sampling and utilising these sources, several weaknesses arise out of retrievability problems and reporting bias (Yin 2003: 86). A selection bias could at times not be avoided when access to de jure publicly available government documents was de facto inhibited which rendered the collection of material inconsistent over time. Apart from the incompleteness of documents, language at times also constituted a problem as particularly local public authorities did not provide for translated records. Since the documentary baseline and reference years differed between the cases, process tracing as well as comparison between the cases was impeded.

Interviews

The main source of evidence relating particularly to the perception-based explanatory factors is derived from interviews and mapping exercises with stakeholders. Stakeholder selection and interviews followed a similar procedure in each case study. The stakeholder group comprises of public sector officials pertaining to the public organisations involved in wind power development and renewable energy governance, private sector business and civil society representatives comprising both unorganised local people as well as NGOs and media officials (cf. Annex I).

The interviews were conducted in a semi-structured format (cf. Annex II) allowing for open-endedness and interpretative freedom of the interviewed. All of the interviews were documented by hand-written notes obtaining prior permission and the majority were taped. However, in some instances taping interviews was consciously considered not an option when e.g. wind energy was perceived to be a contagious issue, i.e. interviewed appeared very hesitant to

[13] One difficulty arises out of the difference in the depth and richness of data required for discussing the explanatory factors. Data sources range from readily-available quantitative statistics to complex qualitative perception-based indices. As the empirical data baseline varies, the comparison and contrast of different explanatory accounts is rendered more difficult. Furthermore, the danger of a cognitive bias is implied since evaluative judgements on the strength and validity of an explanation are subject to the reasoning of the investigator.

talk about their perceptions. Since interviews were a main contributor to exploring and measuring the independent variables, records were verified through other expert interview. For the purpose of confidentiality, interview partners preferred to remain anonymous.
Generally speaking, interview procedures were impeded by at times language barriers, intentional mal-understanding and -interpretation of the research question, accusations of selection biases or irrelevance of questions. In the majority of cases, contacts to interview partners were established through a bottom-up procedure. The main obstacles pertained to language, administrative structures, low status and hence credibility as well as gender of the researcher, time and resource constraints. However, particularly with regard to the private sector, first-time contacts in some instances evolved into further recommendations and introduction into the net of relations.[14]

Observations
The possibility to participate in site visits enabled direct participant observations as well as enhanced the informal interviews with respective stakeholders. Two field visits were undertaken to one wind farm installation under construction in Kerala as well as to one operating wind farm in Tamil Nadu. The problematic issues emerging in context with observations obviously refer to the selectivity of events (Yin 2003: 86) as opportunities and access to field visits were determined by the private companies that hosted these projects. However, these field visits enabled a deeper understanding of explanatory factors and engaging in informal conversation with otherwise inaccessible stakeholders, i.e. on-the-ground staff and local people, that contributed to corroborating evidence.

Stakeholder Mapping – Ethnographic research
This dissertation applied ethnographic research methods in order to derive at appropriate inside accounts of local realities. Behavioural approaches in social sciences introduce perception-based research tools such as mental maps. Previous studies mentioned the lack of local understanding as reason why many assistance efforts or renewable energy projects failed. Hence mental mapping was applied as ethnographic method in order to obtain an inside account of renewable energy stakeholder network structures and processes. Based on this acknowledgement that coming from the outside with certain Western concepts and ideas local conditions are often difficult to assess, the researcher intended not to impose herself as external and objective evaluator.
Concretely, alongside the tool of a puzzle-like mental map (cf. Annex III), each interviewed person was given the task and question of selecting and locating the stakeholders in concentric circles according to their characteristics and importance to renewable energy governance in the wind sector. This process was accompanied by open-ended interviews. The result of this exercise was a mental map or a renewable energy stakeholder network puzzle that allocated the stakeholders according to the perceived importance and relations from the perspective of the interviewed person. The next step in this mental mapping exercise was to a combine the different individual mental maps by corroborat-

---

[14] This was noticed as interesting evidence of stakeholder network dynamics but not adopted as reliable and conscious contact strategy as it implies obvious selection biases.

ing the statements and local mapping of the respective stakeholders. The result of triangulating interview and mapping data was one renewable energy stakeholder network map for the respective local context.
While the strength of this method is the participant-based picture of empirical realities, limitations are the necessity to integrate and corroborate potentially contradictory information. Data had to be reduced for the sake of deriving at a lowest common denominator but comprehensive renewable energy stakeholder network.

c) Data Evaluation

The major strength of case study research is the opportunity to draw on multiple sources of evidence (Yin 2003: 97) required for tackling the broad range of explanatory factors. Consequently, several evaluative procedures were drawn upon in order to make sense of the data and information gathered. Firstly, interview and documentary data were thematically codified into distinct categories in order to structure the data complexity and density (Flick 2006: 310). Secondly, the process of triangulation was drawn upon which combines the different methods, data sources and settings to support arguments and interpretations from various perspectives (Flick 2006: 331). Additionally, validation and cross-checking of data from different sources ensured as correct and objective an account as possible. The last step of the empirical research procedure was matching empirical evidence with the hypotheses suggested (Gerring 2004: 349). Process-tracing was applied in order to give justice to history, path dependency and to increase the number of observations in the case studies (Vennesson 2008: 233). All in all, the overall outcome of this research process and procedure resulted in the elaboration of an explanatory model of renewable energy governance in developing countries that remains to be tested as regards its validity and plausibility.

I.3.2 Looking Behind the Veil – Roadmap to the Thesis

This dissertation embarks upon an exploratory venture of renewable energy governance in developing countries to tackle the puzzle of spatially varying wind energy deployment in the cases of two Indian federal states Tamil Nadu and Kerala. The main objective is to examine determinants of wind energy deployment to derive at specific conclusions on contextual conditions for effective renewable energy governance in developing countries. For this purpose, the research question focuses on the role of stakeholder networks in wind power deployment and their effects on renewable energy governance in the selected Indian states.

The first inductive empirical stage of this dissertation pursues the objective to exclude a multitude of the explanatory factors discussed as barriers to renewable energy deployment. The *second chapter* will hence provide an overview of and contextualise the discussion about barriers and potentials of renewable energy governance in developing countries in respective theory debates. The main issues for renewable energy deployment in developing countries are identified as the four dimensions of developmental conditions, policy conditions, economic conditions and external conditions. These are discussed respectively alongside theory debates of development economics, new institutional approaches of political science, economic geography as well as global governance debates[15]. The outcome of the first literature perusal will be an analytical framework that summarises these potential explanatory factors.

In the *third chapter,* this analytical framework will guide the first exploration of wind power deployment in India. Starting with a broad approach, all Indian states with wind power potential are considered with specific focus on the most puzzling instances of Tamil Nadu and Kerala. The overall outcome of this inductive analysis supports the argument that few of the explanatory factors help to understand why wind power deployment takes place in the one local context but not in the other. Two issues emerge as relevant and suggest further examination. Firstly, stakeholder attributes and secondly, relational interactions between stakeholders seem crucial to get wind power deployment going. In order to examine these aspects, a different perspective is required looking at the network of stakeholders in the multi-level system.

The second deductive research stage departs from the identified shortcomings and lack of explanatory insights provided by the first analytical approach.
Hence the *fourth chapter* retreats to further theory and literature debates with the purpose to understand the two issues revealed as potentially relevant for explaining the Indian wind puzzle. The first theory strand draws on organisational sociology and provides input to the discussion of actors and their attributes. The second theory strand is based in new economic sociology and elaborates on the relations and interactions between stakeholders. Finally, stakeholder network literature is discussed in the light of governance perspectives in order to derive at an appropriate analytical approach. The result of this chapter is the development of an integrative explanatory model that formulates hypotheses on determinants and causal mechanisms to explain effective renewable energy governance in Tamil Nadu and Kerala.

Following from that, the *fifth chapter* takes a deep dive into the empirical realities of spatially varying wind power deployment in Tamil Nadu and Kerala by respectively applying the explanatory model. Stakeholder network analysis serves as participatory ethnographic method as well as analytical tool that integrates the explanatory factors and derives at three main outcomes: Firstly, actors' attributes matter since the characteristics particularly of public stakeholders crucially determine the kind of engagement in renewable energy governance. Secondly, relations between stakeholders matter and influence agency and access opportunities in the process of renewable energy governance. Consequently, local renewable energy stakeholder networks matter and affect renewable energy governance.

Drawing upon these findings, the *sixth chapter* is dedicated to the in-depth discussion of two main aspects. On the one hand, implications with regard to the theory debates drawn upon in this dissertation are considered. These relate to explaining causal mechanisms as well as to understanding why in certain contexts and circumstances local stakeholder networks evolve the way they have. On the other hand, practical implications with regard to the impacts of stakeholder networks on renewable energy governance are examined. This refers to the empirical reality of spatial variations in CDM deployment and draws on the new modes of governance debate. The analysis supports the argument that local stakeholder networks matter as they influence the adoption of emerging opportunities.

---

[15] Theory selection is explained in Chapter II.

The *last chapter* summarises the main research findings. The key point of this dissertation is that without effective renewable energy governance characterised by specific stakeholder networks at the local level, international and national stimuli would not find hold and private sector investment in renewable energies would not take place. Research findings give rise to an explanatory model of renewable energy governance to be tested in other scenarios and settings. In conclusion, contributions as well as limitations of this dissertation are discussed with respect to suggestions for future research.

In summary, this first chapter illustrated the contextual background of this dissertation and elaborated on research methodology and methods to explore the Indian wind power puzzle. The following second chapter will commence the first inductive research stage with an interdisciplinary literature debate on potential determinants of renewable energy deployment in developing countries.

CHAPTER II

Interdisciplinary Perspectives on Renewable Energy Deployment in Developing Countries

CHAPTER II
Interdisciplinary Perspectives on Renewable Energy Deployment in Developing Countries

The spatial variation of wind power deployment in India is an empirical puzzle and raises questions about the prerequisites for successful deployment of renewable energies. As clean energy alternatives are paramount to sustainable development in developing countries, this dissertation aims to explore the conditions under which renewable energy options are effectively governed. The objective of this chapter is to embark upon an inductive first approach to systematically analyse prerequisites for effective renewable energy governance. This is required in order to give justice to the multilevel complexities in which clean energy options emerge, the empirical novelty, and the lack of cohesive and comprehensive previous research. Empirical studies on renewable energy issues in developing countries and India suggest potentially relevant conditions for effective renewable energy governance. In addition, theory debates that are selected according to the extent to which they respond to the respective explanatory levels will discuss how and why these aspects might constitute prerequisites.

Four main dimensions identified on the basis of empirical studies provide entry points to explaining wind power deployment in India: The first set of challenges relates to developmental conditions that characterise the local setting according to physical and socio-economic aspects. These are discussed by development studies/economics literature. The second set of barriers relates to regulatory and institutional aspects, i.e. refers to policy conditions. Institutional approaches of political sciences are drawn upon to provide explanations. The third approach tackles obstacles related to markets and technologies, i.e. economic conditions, and borrows from economic geography literature. Finally, macro-level aspects of finance, research and development (R&D) and global market access are considered in relation to renewable energy deployment. International relations and global environmental governance debates discuss how international mechanisms might stimulate renewable energy deployment. In summary, the dialogue between empirical studies and insights from the theory debates will feed into an analytical framework for renewable energy deployment that will be applied to the case of wind energy in the selected Indian states.

II.1 Renewable Energy Deployment – The Developmental Dimension

In the following section so-called developmental conditions, i.e. natural resource endowment, population characteristics and structures, are suggested by empirical studies as a first approach to explore variations in the effectiveness of renewable energy governance. Development economics will discuss explanations for how developmental conditions might impact on renewable energy deployment in developing countries.

II.1.1 Developmental Conditions – Generic Barriers to Renewable Energy Deployment

Empirical studies on renewable energy deployment which focus on the developmental context differentiate between physical characteristics and socio-economic issues. As will be elaborated in the following, developmental conditions, specifically natural resources and population characteristics, are regarded as decisive for renewable energy deployment.

The first empirical approach looks at physical characteristics and derives at the hypothesis that geography matters. Martinot et al (2002) argue that geographical and meteorological conditions determine the regional renewable energy potential. In the specific case of wind energy, the importance of resource endowment is not to be underestimated. Basu (2007) and Pillai (2006) claim that renewable energy deployment in India faces obstacles due to the different wind regimes and monsoon patters. These vary not only across India but also within federal states dependent on several factors such as differences in terrain, landscape vegetation, wind speed and atmospheric composition. Geography matters for renewable energy deployment since natural resources need to be available and accessible to be harnessed in a sustainable manner.

The second empirical approach looks at the anthropogenic constitution and characteristics of local settings (Martinot 2004). The local context determines the background of renewable energy technology users and consumers. Income levels, education and literacy influence not only the awareness and acceptance of new technologies. Population characteristics also determine the skills and capacities of the local human resources base available to apply and to utilise renewable energies. In addition, population structures play a role in renewable energy deployment. This relates to issues of competitive land uses and property rights. High population density in terms of high urbanisation rates creates a greater demand for power and hence the need for centralised grid-related options. Yet, dense settlement structures with many urban areas put constraints on the availability of land for installing wind projects (Martinot et al 2002).

In conclusion, physical and socio-economic characteristics matter for economic activities. Renewable energy deployment is an issue of economic development and growth in local settings. Development economics and respective studies on economic growth, poverty and (foreign) investment thus provide the appropriate theory approach to suggest explanatory mechanisms and effects.

II.1.2 Developmental Conditions Matter? – Development Economics

Renewable energy deployment as part of providing energy governance services to secure human livelihoods is vital for the social and economic development of countries like India. In the following, development economics will explain how and why developmental conditions identified as relevant determine economic growth and development.

Development economists in a first approach recognise the role of geography and natural resources as important factors driving economic development (Barbier 2003; Rodrik et al 2004). Geography characterises the physical location of specific settings in terms of agro-climatic zones, soil types and natural resources. Proponents of this approach (Sachs 2003, Gallup et al 1998) argue that these features shape economic development not only directly through the exploitation of natural resources for power generation but also indirectly. This means that geography in terms of physical distance bears implications on transport costs and the integration in the world economy (Auer 2008). This school of thought also elaborates on the argument that natural resource abundance correlates with the costs of development, e.g. infrastructure building. Well-endowed regions face less transaction costs and competitive uses in spending on essential development infrastructure due to greater resource availability[16] (Auer 2008). In summary, this strand of development economics refers to an older, neoclassical approach (Dunning 2006: 177) that regards deficiencies of indigenous resources as main impediment to economic growth. Applied to the case of renewable energy deployment, the natural resource potential is considered important not only as input factor for power generation but also given wider ramifications on development and growth.

As second approach, the role of the socio-economic setting as determinant of economic growth and investment in development efforts receives attention by economists. Empirical research that focuses on comparative economic growth exploded during the last two centuries (Krishna 2004) and relates to literature on foreign direct investment (FDI). This section ventures deeper into these debates since spatially varying renewable energy deployment is a matter of comparative economic growth.
Empirically-related theory explanations of economic growth are often based on the experiences of OECD economies during the 1990s and provide evidence on determinants and convergence of economic growth across regions (Barro & Sala-I-Martin 1995). Basically, health and living conditions emerge as decisive growth factors that directly influence the availability and use of labour for productive activities. Higher life expectancy implies the existence of basic health care and infrastructure and indicates the extent to which social and economic development has taken place. Furthermore, literacy definitely contributes to higher growth as it is linked to the creation of employment activities in the productive sector[17]. Urban growth implies an important stimulant of economic development due to implicitly higher agricultural productivity.
Going beyond the OECD context, quantitative assessments on the determinants of economic growth in Indian states (Ahluwalia 2000) find that social factors such as literacy rate, communication, electricity infrastructure and per capita energy consumption correlate with development. Rao's study on determinants of growth (1999) regards per capita GDP, literacy rate and government spending as the central explanatory variables. He argues that higher income levels and lower indices of poverty indicate a larger market that allows government spending to focus on development infrastructure instead of basic amenities for poverty alleviation.

Development economics goes beyond the examination of general determinants of economic growth and focuses on specific sectors. These debates are important since the Indian wind power puzzle demonstrates the specific case of public sector failure to provide governance services in the energy sector. Particularly since the 1990s, investment activities by domestic or foreign private parties have been encouraged (Kirkpatrick et al 2006) in order to mitigate these shortcomings. Consequently, literature on FDI in India adds to the discussion about determinants of renewable energy deployment[18] (Bajpai & Sachs 1999).
The examination of economic development across Indian states (Bajpai & Sachs 1999: 4) derives at the conclusion that the characteristics of social indicators place

---

[16] This dissertation consciously does not discuss resource curse literature or the Dutch disease issue as this would go beyond the scope of analytical and explanatory depth required.
[17] This effect becomes less relevant when urbanisation is taken into consideration.
[18] India receives comparatively little FDI which is a debate on its own and not in the scope of this dissertation. The literature debate is nevertheless useful suggesting a range of generic explanatory factors and is hence drawn upon in this dissertation.

some states in a better position to attract domestic and foreign investment. Literacy, mortality rates and poverty indices range among the factors that account for economic development (Rao & Murthy 2006). The ability of states to attract FDI and to experience economic development is determined by the development of physical and particularly of human infrastructure. Since the private sector is generally not forthcoming to provide the very basic requirements, these factors essentially constitute the overall investment climate. States with less developed social infrastructure are hardly in the position to offer incentives to investors or to reap investment benefits for their local economies (Rao & Murthy 2006: 26). In the more recent literature, labour force characteristics are emphasised as important (Altomonte 2000) since skilled and educated labour is crucial in technologically advanced and flexible production processes. Abundance of labour might be relevant in labour-intensive industries and at initial development stages dominated by industries with high labour demand. However, skilled labour becomes more important when industrial innovation is concerned. As a word of caution, a major problem studying determinants of FDI and economic growth is that despite the abundance of empirical studies and theory debates, little consistency exists as regards the explanatory variables (Globerman & Shapiro 2002). Even though determinants such as population characteristics and socio-economic setting (Nunnenkamp 2002) provide important explanations, other non-traditional determinants like institutional and regulatory frameworks have become relevant (Kirkpatrick et al 2006).

In conclusion, renewable energy deployment takes place given the existence of appropriate developmental conditions. Natural resources need to be available and accessible in order to be harnessed for power generation. In addition, investment decisions in renewable energy activities are taken according to the sufficiency of income growth, high literacy rates, life expectancy and considering population density and urban settlement. Favourable geographical and socio-economic conditions constitute an investment climate attractive for renewable energy deployment. The first explanatory approach to examine developmental conditions for renewable energy deployment in developing countries hence suggests the relevance of geography and socio-economic factors for effective governance.

II.2 Renewable Energy Deployment – The Policy Dimension

The majority of empirical studies on renewable energy deployment in developing countries hold policy and regulatory conditions in one way or another accountable for impeding the implementation process. The aim of this section is to understand how and why policy conditions as regards political, institutional and regulatory issues impact on renewable energy deployment. In the following, empirical explorations of policy conditions for renewable energy deployment will tackle three dimensions. These comprise firstly, institutions involved in renewable energy promotion, i.e. policy outputs; secondly, regulations and policies, i.e. policy outcomes and thirdly, implementation and attitudinal issues as policy impacts. Political science-based institutional perspectives[19] refer to these policy-related dimensions and will discuss the questions

[19] Differences in underlying assumptions and causalities are consciously taken into consideration

how and why institutions[20] and regulations emerge as barriers to renewable energy deployment.

II.2.1 Policy Conditions – Policy and Regulatory Barriers to Renewable Energy Deployment

(i) Political and Regulatory Institutions – Policy Outputs

Taking a bottom-up perspective on policy conditions for renewable energy deployment directs the first point of attention to the institutions, i.e. the governance structures, for renewable energy. Since this dissertation aims to explain the puzzle of spatial variations in a specific sector, i.e. wind energy, the empirical focus is on the corresponding governance actor and structures. To put it concretely, this refers to the power sector in terms of institutional set-up, i.e. institutions involved, functions, roles and responsibilities, financial and human resource endowment.

In the Indian context, energy is a concurrent subject manifested in the Indian Constitution that reflects on energy governance. Power sector governance is in the responsibilities of both the nation state and the federal states. However, the institutional set-up and structure involved in energy governance is similar nation-wide. This allows for the preliminary assumption that institutional structures as such might not account for spatial variations in wind power deployment.

(ii) Regulation and Policy Making – Policy Outcomes

Policies and regulations on the one hand hold the potential to provide supportive frameworks but on the other hand might also impose obstacles to renewable energy deployment. Renewable energy entrepreneurs compete with well-established and entrenched conventional fossil fuel-based energy companies. Painuly (2001) argues that the absence of appropriate policy and regulatory framework confronts renewable energy entrepreneurs with market, financial and economic barriers. Since costs related to renewable energy generation and transmission are higher compared to conventional power, competition takes place on an unequal playing field. One problematic issue is that governments often lack financial resources to support renewable energies in order to mitigate intrinsic cost disadvantages like higher upfront costs. Acknowledging the potentials of renewable energy for power generation and environmental protection, policies have been devised to encourage renewable energy generation, e.g. fiscal and financial incentives; purchase and certificate obligations. Even if support programmes for renewable energy exist, an inappropriate design lacking flexibility render the fiscal and financial incentives insufficient (Ghosh et al 2002). As one example, in many cases legal guidelines for energy production from renewable resources are unclear (Martinot et al 2002).

With regard to federal states like India, the specific challenges to renewable energy dissemination are the non-uniform and unstable policies that vary across federal states. This increases insecurities to investors not only regarding future expansion into other geographical niches but also related to the stability of investment conditions.

Regulatory barriers refer to the fact that tariff, power purchase agreement, transmission and distribution of renewable energy products are either not rigorously specified or not credibly implemented. Even if stipulated in regulatory orders, tariffs might al-

[20] Institutions are broadly conceptualised as formal and informal structural features of society/polity that affect individual behaviour (Peters 1999: 18).

ways be subject to change or contestation by respective public authorities. This might create problems of cost recovery for investors in renewable energy installations (Ghosh et al 2001). A great risk is the breach of contract by public authorities purchasing and/or guaranteeing the utilisation of grid networks. This may result in the reduction or shut-down of renewable energy generation which incurs high losses to investors. The non-payment or delayed payment for electricity services is often due to the poor financial status of utilities particularly in developing countries but translates into severe financial risks for renewable energy investors (Basu 2007). The lack of oversight of or the length of redress procedures against public authorities leaves renewable energy entrepreneurs at the whim of public policy making and regulation. Lengthy and bureaucratic procedures for licenses or purchase agreements constitute high transaction barriers that impede renewable energy investment. Corruption transcending utilities as well as public authorities has been acknowledged a major hurdle (Basu 2007) to effective regulation and policy making. In addition, power sector policies and governance at times promote unsustainable revenue models that favour specific consumer groups or resources. One example is the governments' electoral promise of free power to farmers in some Indian states. This leaves utilities obliged to pay the costs of this scheme that need to be recovered in one way or another – often discouraging and disadvantages investments in alternative energy sources.
The high dependence of renewable energy activities on policies and regulations renders investments extremely vulnerable to changes in existing policy and regulatory settings (Basu 2007). This refers on the one hand to power sector reforms that in the course of liberalisation and privatisation have commenced in developing countries. On the other hand, constant amendments and changes to meet efficiency criteria translate into regulatory uncertainty. Policy uncertainty also arises from changes in the broader polity or the entire policy regime (Singh 2006). Government changes might lead to modifications of existing contracts or debt conditionalities with regard to renewable energy investment. Additionally, political violence and local protests in the context of alternating policy regimes pose unforeseen risks that in the case of Maharashtra stopped wind farm activities in 2007 (INT 1).

(iii) Implementation and Attitudinal Matters – Policy Impacts

Empirical studies agree that the implementation of regulations and policies constitutes the most intangible yet significant barrier to renewable energy deployment. In many cases, the public utilities are the so-called black sheep to blame for impeding effective implementation of policies. For example, utilities persistently object to granting third-party sale of renewable energy power as they fear to loose high-end tariff paying industrial consumers to the more reliable renewable project investors (Rajsekhar et al 1999). This makes it difficult for new actors such as independent power producers (IPP) to enter the renewable energy market, which would scale up renewable energy dissemination. The integration of renewable energy into energy policies is aggravated by the lack of rules regarding the connection of renewable power installations into the grid. This is due to the fact or rather the fiction upheld by many utilities that renewable power particularly from intermittent resources such as wind neither fits into day-to-day running nor into long-term generation planning (Rajsekhar et al 1999). In sum, even if policy stipulation and legal obligation provide support for renewable energy, the de facto implementation impedes the deployment of alternative energy sources.
One explanation for these implementation deficits refers to inherent difficulties to attain acceptance (Kacker 2007). Past experiences of renewable energy deployment

also in the context of international assistance are dominated by stories of project failure due to improper technologies, inappropriate installations or lacking maintenance practices. These contributed to developing and sustaining mental barriers and objections among stakeholders involved in respective energy governance (Rajsekhar et al 1999). Prejudices against renewable energy exist on the levels of households, private sector, power utilities and government (Martinot 2004). As regards the private sector, lack of information and capacities to install and service renewable energy projects impede the willingness to invest. Governments prefer to stick to large-scale conventional energy projects they are accustomed to and shy away from new, risky and uncertain ventures. The lack of societal acceptance of renewable energy (Painuly 2001) negatively affects the market size due to limited demand for renewable energy and renders investment activities less viable.
One example of how attitudinal obstacles challenge renewable energy dissemination in the case of wind power in India is provided by Jagadeesh (2000). While technological hardware is largely available, an appropriate institutional framework for planning and coordinating integrated programmes and mobilising support is lacking. This is due to the multiplicity of institutions involved in renewable energy undertakings, the unclear division of roles and responsibilities and the lacking political will of public stakeholders. The bureaucratic structure and the target-oriented approach of many institutions have created rigidities, inflexibilities and delays. Even though wind energy was pushed as potential source for the augmentation of power on the high policy agenda, little effort was extended to strengthen the institutions involved at the local level to provide an enabling investment environment (Jagadeesh 2000).

In summary, this review of empirical studies on policy-related determinants of renewable energy deployment in developing countries gives rise to several concluding observations. Policy conditions matter not necessarily as regards political and regulatory institutions but as regards the design of regulations and policy outcomes. Implementation practices by public authorities such as governments, regulators and utilities are crucial to render policies and regulations effective. Attitudes and perceptions particularly of public authorities determine policy and regulatory outcomes and practices. This analysis of policy conditions identified three key features namely policy/regulatory outcomes, institutional performance and effectiveness, organisational reputation and political will. In the following, institutional theory debates will provide insights into explanatory factors and causalities relating policy conditions to effective renewable energy governance.

II.2.2 Politics Matters? – New Institutionalist Approaches

Policy conditions for renewable energy deployment consider institutions per se, institutional outcomes such as regulations and policies, and institutional impacts relating to functional organisational behaviour. The study of institutions is central not only to political sciences but also to other social science disciplines (Koelble 1995) and is hence relevant to examine policy conditions. New institutionalist perspectives (Picot et al 1999) agree on the fact that institutions matter since they determine social and political outcomes (Hall & Taylor 1996) as well as constrain individual behaviour (Peters 1999: 18). Institutions affect preferences and goals in decision-making about renewable energy issues and determine outcomes of policy processes (Koelble 1995: 236). Institutionalists take two distinct approaches to respond to the question how institutions matter. On the one hand, the so-called calculus approach argues that insti-

tutions affect behaviour by providing certainty and altering actors' expectations. On the other hand, the so-called cultural approach holds that institutions provide moral and cognitive scripts for action (Hall & Taylor 1995: 939).
The following section will start exploring public policy literature with focus on policy regimes and political parties as a macro-perspective on policy conditions. In addition, the three main new institutionalist debates will provide different accounts on how and why empirically-described policy conditions determine renewable energy deployment in India.

(i) Policy Regimes – The Role of Polity and Parties

The first approach to explore linkages between effective renewable energy governance and policy conditions departs from the assumption that the external institutional environment defines institutional efficiency and performance. Formal organisations produce certain rationality and encourage certain institutional behaviour in order to increase their survival prospects (Meyer & Rowan 1977). Consequently, the external institutional environment shapes organisational structures and functioning.
Empirical studies on renewable energy deployment in developing countries point towards the importance of policy regimes for deployment-related activities. Policy regimes shape general attitudes, policy-related agency and organisational behaviour to support respective activities.

Public policy literature gives justice to this argumentation and looks specifically at two aspects that explore the impact of policy regimes on renewable energy deployment.
The first discussion refers to policy regimes in general (Przeworski & Limongi 1993). The main point is that political institutions matter for growth and hence polity or policy regimes affect developmental activities such as renewable energy deployment. This argumentation goes back to debates on the relations between institutional performance, economic development and democracy, i.e. polity. The not uncontroversial fact that democratic regimes tend to create a more favourable framework for private-sector activities supports the positive impacts of democratic regimes on development (Przeworski & Limongi 1993). Arguments against the impact of democratic policy regimes on economic growth state that democracy does not guarantee property rights protection. Public investment in these supportive policy conditions is undermined by the need to respond to increasing pressure for immediate consumption that goes hand in hand with democratisation. All in all, Przeworski & Limongi (1993) acknowledge that thinking solely in terms of policy regimes does not capture the full picture. This is supported by the inconclusive statistical evidence on the precise relation between the level of development and the nature of policy regimes.
The second debate focuses on the role of political parties as constitutive elements of the policy regime (Schmidt 1996) and their impact on policy conditions. This is contingent on the policy regime, i.e. the type of democracy as well as the power of institutions. The main point is that the government's party composition explains variations in policy choices and outputs. This argument elaborates that parties provide choices to voters that determine policy making and implementation through voicing their preferences in elections. The different roles parties play in a policy regime are reflected in the policy instruments applied and the nature of the policy outputs. However, the role of political parties is confined and circumscribed by constitutional rules and by the autonomy of social and economic life (Schmidt 1996). General political and economic constraints to public policy cast doubt on the impact of parties on policy outcomes. Following this line of argumentation, constraints to party influence depend on the

power resources held by social classes. Not the political parties but the regime type influences how organisations are set up with regard to resources and according to what logic they function, i.e. what interests and attitudes prevail. This ultimately determines the effectiveness of organisations in terms of designing and implementing favourable policy conditions for renewable energy deployment.

In summary, public policy debates suggest the potential relevance of polity, i.e. policy regimes and political parties, for shaping policy conditions for renewable energy governance.

(ii) Rational Choice Institutionalism - Strategic Interactions in Politics

Rational choice institutionalism as second approach looks at the role of strategic interaction to explain policy conditions for renewable energy deployment. Rational choice institutionalism[21] considers policy conditions for renewable energy deployment as political outcomes of government and market institutions (Koelble 1995: 232). This means that not institutions per se determine policy conditions but institutions affect individual choice and action that shape respective policy conditions. Individual actors are assumed to have a fixed set of preferences and to behave strategically as to maximise their interests (Hall & Thomas 1996: 944). Certain institutional arrangements evolve due to the perceived gains of collective action (Koelble 1995). The central role of institutions in determining policy outcomes lies in their capability to produce some predictability and regularity that structures actors' behaviour and decisions. This does not only benefit the members of an institution but also clarifies the range of decisions available to all societal actors within this institutional framework (Peters 1999). Institutions are hence conceptualised as sets of positive inducements and negative rules.

As point of departure, institutions are defined as formal rules, informal constraints and the respective enforcement mechanisms (North 1990). Institutional approaches call upon two main mechanisms through which institutions affect policy or economic conditions, performance or compliance (Williamson 1985; North & Thomas 1973):

The first mechanism is based on transaction cost theory that looks at exchanges and transactions between actors (Picot et al 1999). The key argument is that transaction costs arise in interactions and in corresponding actor constellations. This is due to the bounded rationality and opportunistic behaviour of actors as well as the uncertainty and complexity of the environment (Williamson 1975: 40). According to North (1990), institutions provide the rules of the game that shape human interaction in society. They consist of formal constraints such as economic, political rules and contracts as well as informal constraints like codes of conduct, norms of behaviour and conventions. Institutions shape policy conditions through enabling the supply of information, decrease transaction costs and increase trust between organisations and individuals which reduces uncertainties.

The second mechanism draws on principal-agent theory and considers exchange relations between principals and agents within institutions or institutional settings, e.g. employer and employee. Protagonists of principal-agent theory such as Douglas North depart from the assumption that institutions are created by utility-maximising individuals with clear intentions. Institutions affect individuals' choices and actions (Peters 1999) by shaping individual decisions through rules, constituting contracts, or pay-offs. Relations between individuals and institutions are characterised by chal-

---

[21] Rational institutionalism of political sciences is most intimately connected with new institutionalist economics. Differences and commonalities are not explored in this context as this would not further the main argument that rationally acting institutions shape policy conditions.

lenges of coordination and motivation that are reflected in certain agency costs (Picot et al 1999). Decision-making and exchange is looked at from the principal's perspective who is interested to get agents act on her behalf. The problematic issues arising for principals are the hidden characteristics of agents in the situational context, the agent's hidden intention and the hidden action. Since these shadows of agency are outside the scope of the principal's surveillance, agents have the possibility to act not necessarily or not fully in the principal's interest but to pursue own agendas, i.e. creating moral hazard (Andriesse 2008: 39). The problem underlying these skewed principal-agent relations is the disjuncture between individual and collective rationality. This refers to situations in which agents have an individual interest to shirk, i.e. act in disfavour of the principal even though this agency might not be rational regarding collective objectives. The shape of policy conditions through institutional performance hence depends on the capacity of respective principals to map a common set of preferences that does not threaten the overall legitimacy of the political system (Peters 1999). Consequently, for public bureaucracies, the main challenge is to find ways to ensure actors comply and adopt the institution's view of politics (Peters 1999).

In summary, the rational choice perspective explains policy conditions for renewable energy deployment in terms of the structure of relations and characteristics of interactions in as well as between involved institutions. Firstly, principal-agent relations emerge among public authorities that dependent on information and knowledge asymmetries are favourable or detrimental to renewable energy governance. While policy makers as principals aim to provide investor-friendly conditions setting the right tariff and facilitating bureaucratic procedures, implementing agencies, e.g. regulators or utilities, as agents might not follow suit, i.e. disagree to policy suggestions or impede the implementation. Due to vested interests in maintaining the status-quo, these agents hence have an incentive to shirk. Because accountability mechanisms are weak or not timed according to electoral cycles, principals have insufficient information or lack appropriate enforcement mechanisms. Secondly, transaction costs within and among public authorities might lead to failures in effective policy making. Dependent on how well-established, efficient and transparent the information flow and exchange mechanisms are, transactions are impeded or facilitated. Due to the lack of knowledge, expertise and resources to assess the costs of support for renewable energy, public actors rather stick with business-as-usual activities instead of promoting supportive policy conditions for renewable energy deployment.

(iii) Historical Institutionalism - Organisational Capacities and Legacies

Historical institutionalism as third approach to explain policy conditions departs from similar arguments as rational choice approaches. Institutions impact on individual agency, on the one hand, through interest calculation and, on the other hand, through preference formation on the basis of bounded rationality. Policy outcomes such as renewable energy policies and regulations are considered the product of interactions among various groups, interests, ideas and institutional structures. Historical institutionalism amounts a significant role to institutions in shaping policy actions which are also affected by previous policy choices. Even though institutions shape individual action, institutions are also affected by collective and individual choices (Steinmo, Thelen & Longstreth 1992). Scaling up the perspective, the two different logics with which institutions affect individual behaviour apply to the ways in which social and political institutions structure interaction and generate distinctive national trajectories.

On the one hand, institutions provide certainty of behaviour; establish enforcement mechanisms and influence information and expectations. On the other hand, institutions constitute a moral template that allows for the interpretation of situations and affects identities. Historical institutionalists regard institutions as one element in the causal chain of socio-economic developments[22] (Hall & Taylor 1996: 942). Policy choices are hence not only determined by existing capacities but also by legacies (Thelen & Steinmo 1992).

Historical institutionalism provides three entry points to explain how and why policy conditions determine the effectiveness of renewable energy governance:
First of all, in line with rationalist approaches, historical institutionalists presume that organisationally embodied routines play a crucial role in allocating resources, restructuring incentives, options and constraints (Pierson & Skocpol 2002). In order to explain (economic) policy choices in a specific setting, political histories are prerequisite for understanding policy conditions (Hall 1986). Douglas North for example examined enduring effects of economic institutions shaping economic outcomes after initial decisions were taken to create these institutions. Secondly, in line with institutional economics the concept of path dependency explains how institutional change takes place (North 1990). Once political and economic organisations have come into existence, it is in their interest to perpetuate the existing framework. This produces a perception bias and certain mental model in favour of policies that serve the interest of existing organisations (North 1990). Thirdly, historical institutionalism in line with sociological approaches considers the independent role of norms and ideas in shaping policy conditions (Peters 1999). In this regard, historical institutionalists look beyond formal structures, e.g. bureaucracies, to consider concepts like ideas that define the existence and agency of institutions.

The two characteristic features of research on policy conditions in the tradition of historical institutionalism are that they take time seriously and that they analyse macro contexts (Pierson & Skocpol 2002). The basic idea is that policy choices made when institutions are formed or when policies are initiated will have a long-lasting and determining influence that create national trajectories (Skocpol 1992). Institutions produce certain paths in structuring responses to changes on the basis of existing capacities and policy legacies (Weir & Skocpol 1985). The notion of path dependency implies that once persistent patterns are created organisations routinise their activities. Unless sufficient pressure is exerted, institutionalised routines will be difficult to overcome (Peters 1999). Despite this institutional inertia, policy conditions might change via two mechanisms. Firstly, critical junctures are one instance during which a variety of internal political forces trigger change (Collier 1991). Conjunctures are created through the interaction effects of distinct causal sequences that join up at some point in time. The sequence of interactions matters for changing policy conditions due to the creation of self-reinforcing processes. These transform the consequences of later developments and hence generate irreversibility by removing some options (Pierson & Skocpol 2002). Secondly, policy conditions might also change through learning processes as institutions respond to new ideas and information that leads to reframing policy issues (Peters 1999). In all manners, change requires the

---

[22] Historical institutionalists often set limits to the applicability of their causal arguments (Pierson & Skocpol 2002) as institutions are only one part of the causal chain together with other features, e.g. socio-economic development or polity.

presence of threshold effects or a critical mass (Pierson & Skocpol 2002). Key causal factors that trigger change in the first place are inherent in the structural preconditions (Pierson & Skocpol 2002), i.e. certain configurations that arise out of specific interaction effects between types of regimes, eras, regions and cultures.

In summary, historical institutionalism adds value to the explanation of institutional development and policy-making in renewable energy governance. The characterisation of organisational paths determines not only the design but also the effects of policy conditions on renewable energy deployment. Support for renewable energy deployment within public organisations ranging from favourable policy making and tariff stipulation to facilitated clearance procedures is entrenched in so-called organisational routines. This means that staff and resources are allocated and educated towards the end of promoting renewable energy deployment. According to historical institutionalists, policy conditions are hence shaped by organisational routines as well as previous paths.

(iv) Sociological Institutionalism - Roles and Norms in Politics

Sociological institutionalism as fourth approach considers policy outcomes such as renewable energy policies and regulations as products of individual decisions embedded in a larger frame of reference. Institutions constitute and legitimise actors and provide them with consistent behavioural rules, conceptions of reality, standards of assessment, ties and endowments (March & Olson 1995). Policy conditions are shaped by the effects of institutions on individual and collective behaviour through mechanisms of socialisation and internalisation of respective norms. This means that policy-relevant behaviour is influenced by the meanings, cognitive scripts and models through which actors are socialised into institutional roles according to which they behave (Hall & Thomas 1996: 948). Scaling up the perspective, institutional behaviour and change are determined by the ways in which organisations and participants obtain social legitimacy. Since only those institutional practices emerge that are acknowledged socially legitimate, at times, practices are dysfunctional for the achievements of formal goals (Campell 2002). Sociological institutionalism describes three main causal mechanisms of how and why policy conditions relate to renewable energy governance:

The first causal mechanism holds that institutions define certain actions by the rules of appropriateness (Powell & DiMaggio 1991) that put behavioural expectations on individuals within institutions (Peters 1999: 30). This means that institutions function to reinforce behaviour considered appropriate for the respective role and punish or sanction deviations (March & Olson 1989). Even if there are no formalised means for sanctions, some enforcement mechanisms exist for this logic to be effective (Peters 1999). Often, informal means prevail through which individuals are pressured to conform. These are built into the structures through socialisation rather than through external enforcement mechanisms. Institutions work to produce certain outcomes mainly through a mix of compliance and conformity. Institutional change is inherent in the non-conformity of organisational behaviour through conscious resistance or through leadership and individual efforts.

The second mechanism explaining institutional performance and behaviour is entrenched in routines that arise naturally once individuals start interacting within the context of the institution (Peters 1999). Routines define the nature of the organisation since they constitute the channels through which members minimise transaction and decision-making costs participating in institutional practices. Institutions require routi-

nised responses to problems in order to cope with the demands and ultimately enhance and justify their own efficiency (Peters 1999: 37). Changes to organisational routines or practices will only occur given stimuli or pressure for change.
The third causal mechanism relates to the way in which institutional capabilities and structures affect the flow of information, the kinds of search undertaken and the interpretations made of the results (March & Olson 1999). Institutions do not determine political behaviour straight away but work through a process in which rules are translated into behaviour. This depends on the constructive interpretation of policy actors as well as the on available resources. The latter point is important to keep in mind as resources determine an actor's capability to act appropriately to a certain situation. Even though actors may know what appropriately to do, they are limited by the complexities of demands and by the distribution and regulation of resources, competencies and organising capacities (March & Olson 1989: 252). Sociological institutionalists argue that it is difficult to determine institutional effectiveness or to assess institutional performance since institutional capabilities are defined on the basis of internal norms. Institutional functioning can only be assessed according to the extent to which internal standards are met. One way of judging the effectiveness of institutions is to look at the normative integration and creation of collective values within an institution. Going one step further, one might ask whether the internal culture indeed appropriately responds to the challenges facing an institution and the tasks it performs (Peters 1999). Failures to adapt to challenges result in organisational dysfunctions and pathologies.

In summary, sociological institutionalists provide insights into how organisations in the institutional setting work towards either supporting or impeding the framing of policy conditions for effective renewable energy governance. Enabling policy conditions for renewable energy deployment are the result of the institutionalisation of norms and values that recognise these undertakings as legitimate and socially beneficial. This translates into certain routines, working culture and also attitudes supportive of renewable energy deployment.

In a nutshell, empirical studies on renewable energy deployment in developing countries conclude that policy conditions constitute an important explanatory dimension that is differentiated into four aspects. Firstly, the policy regime or polity appears of overall relevance shaping policy conditions with regard to institutional structures and agency. Secondly, policy outputs, i.e. the kind of institutional arrangements are regarded as essential since they reflect the policy regime and dynamics which impact on policy making and implementation. Thirdly, policy and regulatory frameworks as formalised rules or policy outcomes are considered important for shaping the policy context for renewable energy deployment. Fourthly, with respect to the policy impacts, policy implementation related to the behaviour of respective policy actors affects renewable energy deployment. Theory debates based on new institutional perspectives and public policy suggest explanations and suggest factors and mechanisms that shape opportunity structures for renewable energy deployment (Table 2).

Table 2: Policy Conditions – Explanatory Mechanisms

What? Empirical Dimension	How & Why? Institutional Explanations	Theory Approach
Policies and Regulations	Strategic Interaction: Principal-agent relations and transaction costs	Rational Choice
Policies and Regulations	Legacies and Capacities: Path dependencies and organisational routines	Historical Institutionalism
Implementation and Attitudinal Issues	Norms and Values: Appropriateness of action	Sociological Institutionalism
Design & Structure of Institutions	Policy Regime and Parties	Public Policy

II.3 Renewable Energy Deployment – The Economic Dimension

Renewable energy deployment is based on investment as well as location decisions by private and at times also public actors. The World Development Report 2009 (World Bank 2009) highlights the importance of space in terms of distance and density for shaping economic opportunities and growth. Consequently, opportunity structures for renewable energy deployment matter in relation to a specific local context. Spatial conditions concern on the one hand, market characteristics that create the demand for renewable energy services and on the other hand, supply factors such as the availability and accessibility of basic inputs. In the following, empirical studies on renewable energy deployment in developing countries will emphasise the importance of market characteristics and highlight the problem of supplying adequate input factors. Economic geography will respond to the questions how and why these features determine location decisions and how demand/supply conditions interplay to shape spatial opportunity structures for renewable energy investment.

II.3.1 Economic Conditions – Techno-Economic Obstacles to Renewable Energy Deployment

Empirical studies on renewable energy deployment in developing countries focus attention on so-called economic conditions that might pose obstacles to deployment. These include a range of aspects from economic and technological issues as regards input factors to market-related distortions characterising demand conditions. The problem of finance is not considered in the following analysis of concrete location decision that assumes the availability and access to finance as precondition to investment. This is justified by the focus on macro-level determinants of renewable energy deployment. Additionally, in the Indian case financial institutions operate at the national level which is why less variation in terms of access and availability exists at the local levels of individual states[23]. The following exploration of empirical studies as regards economic conditions for renewable energy deployment will explicitly focus on the supply of location factors, market-related or demand barriers and general opportunity structures[24].

[23] Since access to and availability of credit nevertheless is an essential pre-requisite to economic activities this will be discussed in more depth in the following section.
[24] Investment and location decisions are of course subject to cost benefits analysis. Since technology costs are assumed constant (cf. Chapter I), the exploration of economic conditions explicitly focuses

(i) Supply Conditions - Location Factors

Empirical studies on renewable energy deployment argue that location factors shape spatial opportunity structures that provide potentials but might also constitute barriers to specific investment activities. Any investment decision is based on an assessment of the extent to which location factors are available and accessible. Renewable energy deployment has specific requirements on inputs that relate to infrastructure, technology and the industrial structure.

Starting with a macro-economic investment perspective, as first aspect, the industrial structure bears two important implications for investment and location decisions. On the one hand, the degree of industrialisation indicates to what extent basic location factors are available to enable economic activities. High degrees of industrialisation generally imply that generic inputs such as labour, institutions and basic amenities are available. Furthermore, doing business is assumed not only feasible but also profitable due to the agglomeration of economic activities. This implies that procedures necessary to forego before new investment activities such as renewable energy deployment are taken up are likely to meet less hurdles than in industrially undeveloped regions. Even though high industrialisation implies more competition for perhaps scarce input factors such as skilled labour, it will also drive innovation and expansion of activities. On the other hand, thriving industrial activities imply a high demand for power that indirectly creates a niche for renewable energy activities. With regard to renewable energy deployment, activities are constrained by the lack of appropriate input factors such as labour, technical equipment and components and the regular supply and maintenance of technology. In this sense, the characterisation and density of industrial structures matter for location decisions and influence chances for availing and accessing these inputs.

The second aspect concerns the role of infrastructure in creating opportunities for renewable energy deployment. Empirical studies discuss land use restrictions due to environmental regulations (Ghosh et al 2002) and the effects on the availability and access to the essential input factor land. Unavailability of land or saturation of identified sites (Rajsekhar 1999) leads to scarcity situations with regard to potential wind sites available for exploitation. The general competition for land uses due to the designation of areas or property issues poses barriers to renewable energy deployment. Land issues also matter in terms of the available transport infrastructure (Martinot 2004). Renewable energy activities are confined to and only take place where resources, e.g. wind, are located. This is often the case in remote or far-off areas which may not be accessible in a manner that does allow for the transport of heavy construction material. Furthermore, many components necessary for the installation of renewable energy projects are not produced in developing countries. The development of port and rail networks required as well as the proximity to these facilities constitutes an important cost factor that determines location decisions.

The third aspect concerns technology which of course is the most critical input factor for renewable energy projects. Obstacles to renewable energies in developing countries arise out of the dissatisfactory performance of projects due to the poor quality of technologies (Martinot et al 2002). Even if parts are not imported but produced and sold in developing countries they might not suit local conditions as the production is geared towards export purposes.

---

on concrete supply and demand factors as constitutive investment cost components. Since this dissertation adopts an explicitly qualitative methodology, no quantitative cost analysis is undertaken.

In summary, empirical debates on renewable energy deployment derive at the conclusion that location factors such as the degree of industrial activity, infrastructure and sector-specific input factors matter for shaping supply conditions.

(ii) Demand Conditions – Energy Market Characteristics

In the case of renewable energy deployment, the characteristics of energy markets play a crucial role for investment decisions. Energy markets determine the existing and future demand for capacity addition from alternative energy sources.

Particularly in developing countries, energy markets and the energy sector face the endemic problem of power shortages (Golait et al 2007). Increasing economic growth prospects create pressure for future supply security and stimulate the demand for harnessing all available power resources. This is particularly pronounced for highly energy-intensive industries that suffer from high industrial power tariffs and the low quality and reliability of power supply.

The way in which energy markets are structured with regard to power mix, policies and institutions indicates the extent to which energy market distortions or opportunities exist (Martinot 2004). Opportunities for renewable energies arise out of the increasing demand for power generation. Obstacles are inherent in pricing mechanisms and intrinsic systemic issues that favour conventional energy sources and disadvantage renewable energies. Market distortions that impede renewable energy deployment arise out of a combination of mainly two factors. Firstly, these concern the subsidies for conventional fuels to protect industries and secure power supply. Secondly, this reflects the inability or unwillingness to internalise external costs of conventional power into the pricing mechanism (Ghosh et al 2002). This often leads to unfavourable pricing rules disadvantaging alternative energy generation. Furthermore, the revenue model of the power sector has a great influence on energy market structures and the risks to renewable energy deployment (Basu 2007). The financial status of the power sector is determined by the availability of state government subsidies which are often tied to political obligations of utilities such as free-power provision to certain consumer groups. In addition, the historical performance of the power sector also bears implications as chronically underfinanced power sector governance systems are not likely to become more viable in the near future (Basu 2007). This implies risks to renewable energy investment such as non-payment or breach of existing contracts. Market distortions through unfair pricing mechanisms and high control increase transaction costs that impede investment and location decisions for renewable energy. This culminates into a vicious circle in which renewable energy markets remain small and market-related barriers are retained.

Another important energy market-related feature shaping supply but also demand for renewable energies refers to the electricity infrastructure (Martinot 2004). Renewable energy projects with grid-related applications often face problems with regard to interconnection, access, transmission and distribution of the electricity generated into the electricity system. Transmission and distribution losses due to theft or technical problems (Rao 2005) characterise the existing electricity infrastructure in many developing countries. This does not only pose barriers to the provision of key input factors to investment but also affects demand for renewable energy services by public sector market participants.

In the specific case of wind energy, the greatest market-related barrier relates to the unstable financial situation characterising the energy governance system. In addition, infrastructural issues increase investment risks as particularly for wind energy, poor grid quality and lacking evacuation facilities constitute a severe constraint to deployment (Rajsekhar et al 1999).

(iii) Opportunity Structures and Support Networks

Empirical studies on renewable energy deployment agree that failures of renewable energy installations in terms of technical insufficiencies, shut-downs, operational delays, interconnection problems are often due to the lack of technical and/or commercial skills.

In specific local markets in which renewable energy is to be deployed, skilled personnel to install, operate and maintain renewable energy technologies might simply not exist (Martinot 2004). Entrepreneurial project developers lack essential technical, financial and business development skills due to absent expert and support networks (Ghosh et al 2002). Alongside the supply chain of the renewable energy business, important stakeholders such as engineers, lenders, investors are either not aware of arising opportunities or hesitant to engage in new uncertain and risky undertakings (Painuly 2001). This reflects the lack of information about renewable energy technology, economic and financial costs and benefits, operation and maintenance (O&M) experience and financial sources. Martinot et al (2002) elaborate on the fact that support networks for renewable energy undertakings are missing in many developing countries. Only very few organisations support renewable energy through information dissemination and lobbying, and are confronted by strong interests vested in the conventional energy sector. Mechanisms and institutions dedicated to information dissemination and capacity building for renewable energy are lacking (Painuly 2001). This leaves producers as well as consumers in the vague about the potential advantages of renewable energy.

In summary, empirical studies on renewable energy deployment in developing countries argue that economic conditions characterise the opportunity structure for investment activities and drive location decisions. How and why certain investment and location decisions are taken will be subject to economic geography debates in the following. Since the initial research puzzle concerns the varying geographical distribution of wind energy activities, spatial economics is explored as explanatory approach to renewable energy deployment.

II.3.2 Economics Matter? – Economic Geography and Spatial Economics

Economic geography perspectives are drawn upon to provide causal explanations and suggest explanatory models since space matters for location decisions on renewable energy deployment. Economic geography has experienced a recent revival as the relevance of location and spatial transformation for economic growth and development is central tenor of the World Development Report 2009 (World Bank 2009).

Economic geography explores economic activities in space, their political, institutional, social and cultural causes and their implications in and on space (Bartels 1968). Location patterns of e.g. renewable energy projects are explained by spatial parameters, and their economic equivalents. Issues like distance serve to explain the spatial distribution of economic activities on the grounds of transportation costs related to transport infrastructure. Regional sciences also focus on explaining spatial distribution and location decisions (Schätzl 1998). In this tradition, the locational analysis of economic activities systematically compares locational requirements of firms with location factors available in or particular characteristic features of a region (Bathelt & Glückler 2003: 122).

The various strands of location theory that are based in economic geography share a common interest in economic activities and economic loci of decision-making such as firms and space (Bathelt & Glückler 2003). Among manifold spatial economic perspectives, industrial location theory focuses on specific industries of different sectors such as power or renewable energy. Going beyond ambitions to explain and understand grand complexities of industrial activity in the spatial dimension, the elaboration of spatial models aims to predict present and future economic decisions and spatial developments. Industrial location theory provides two broad sets of approaches (Bathelt & Glückler 2003) to explain how and why location decisions are taken: On the one hand, location choice is examined according to the overall rationale of cost minimisation. On the other hand, behavioural models are drawn upon in order to explain this location choice.

Considering the first approach, the best location for industrial activities is chosen according to the primary objective of minimising the costs of doing business. These costs are most and foremost related to transportation of components for assembly and to specific input factors required for production. As regards renewable energy, these refer to the costs of transporting raw material as well as the assembly of plant components and construction materials. In the specific case of wind energy deployment in India, wind as natural resource is free and does not incur transportation costs. The location of wind power installations close to most suitable wind regimes that maximise wind farm outputs in terms of generated electricity emphasises questions about the availability of infrastructure and transport of equipment. Transport costs of technical parts depend on the access and proximity to appropriate transport infrastructure and respective production centres. A second cost component that matters during construction and operation phases is that of labour. Labour quality and skills become of greater importance than labour quantity during the operation and maintenance of installations. The last issue that reflects on the costs of wind power relates to agglomeration effects (Bathelt & Glückler 2003). Location theory argues that the location of industrial activities close to demand hubs reduces overall costs as it decreases the length of transport for the required products or services. Due to the urbanisation effect, the presence of similar activities increases the potential to share costs of components or other input factors. In the case of wind energy, proximity to agglomerations implies the availability of infrastructure facilities. The presence of competitors might facilitate bargaining with public stakeholders, allow for the shared utilisation and investment in labour quality and the establishment of supply networks.

The second approach to explain location decisions recurs to a behavioural model of economic decision making (Bathelt & Glückler 2003). This argues that location decisions are takes on the basis of available information and the number of participants involved in specific economic activities. Behavioural approaches depart from the overall assumption of rational decision making in a certain context on the basis of fully available information (Bathelt & Glückler 2003). Taking the Indian wind case, the economic market environment for wind energy is characterised as relatively certain (cf. Chapter I). Despite the competition with the established conventional power sector, wind power constitutes by far the most mature and commercialised renewable energy technology. The relatively high number of stakeholders engaged in wind power deployment influences the information flow and contributes to the perceived certainty of the economic environment.

In summary, spatial economists argue that industrial location decisions are based on factors such as the cost-minimising accessibility of raw material, energy, transport, labour, markets and agglomeration given the specific industry requirements. Location theories differentiate hard and soft factors as decisive for decision making. While hard location factors describe quantitative input factors for industrial production, soft factors refer to qualitative aspects such as environmental and living conditions, business climate, image, regional milieu and political climate (Bathelt & Glückler 2003). With regard to soft factors, business economics in particular touches upon the question of why despite the availability of similar hard location factors, industrial structures and activities develop distinct spatial specialisations.

One empirical application of economic geography theories is the examination of industrial location and spatial inequality in India by Lall & Chakravorty (2005). They argue that history, natural resources, human capital, political economy and culture are the main factors contributing to the spatial inequality of industrial location. The most significant implication is that the spatial inequality of industry location is a primary cause of income inequality in developing countries. These empirical studies (Lall & Chakravorty 2005) also show that of all economic geography variables, proximity and local economic diversity have the greatest cost-reducing effect on firms. Proximity to industrial or other location implies better opportunities to access a pool of subcontractors, skilled labour and business services. This refers not only to the presence of similar activities but to benefits from knowledge transfer across industries. Important location criteria are identified as market access, infrastructure availability, agglomeration economies, progressive and pro private sector state regulations and the general level of political support. Particularly in India the latter point is significant as not the localised incentive system but the sense of political support for private investment in the region is crucial for location decisions (Lall & Chakravorty 2005: 59).

In summary, renewable energy deployment is explained by location decisions taken by rational economic actors on the basis of the availability of and access to location factors such as infrastructure, industrial activity and opportunity structures prevalent in the power sector. These determine the investment costs in supply factors as well as the demand for investment in renewable energy and hence create the economic opportunity conditions for deployment activities.

II.4 Renewable Energy Deployment – The External Dimension

In the following section, behind the background of previous empirical studies, finance, technology and R&D are acknowledged as barriers to renewable energy deployment. This dissertation assumes that finances and technology constitute similar factors considering the national level (as will be shown in the following and Ch. III). Yet, this might vary looking at the international level that provides windows of opportunity for external stimuli such as financial incentives or technological innovations to assist developments in the renewable energy sector. These empirical realities are hence discussed as external dimension in the light of international relations and global governance debates. Arguments about new modes of governance and ecological modernisation will make the case for external stimuli to meet financial and technical barriers.

II.4.1 External Conditions – Stimuli to Overcome Internal Barriers

Empirical studies agree that particularly the three issues of finances, technology and R&D imply both, barriers due to the absence, and a window of opportunity for external support to renewable energy deployment in developing countries.

(i) Finances and Investment

The first factor refers to the issue of finance (Rajsekhar et al 1999). In contrast to conventional energy sources, markets for renewable energy are relatively underdeveloped (Martinot 2002) despite the increasing demand for power generation. The number of renewable energy market actors is limited due to the comparatively early stage of renewable energy technologies and due to uncertainties about future developments, technology issues and the political context faced by investors. Investors suffer from a lack of capital to overcome the high transaction costs in developing, installing and running renewable energy systems. Often national finance institutions have reservation about extending credit schemes to renewable energy activities. Unfair competition against conventional energy sources also aggravated the financial position of renewable energy investors (Martinot 2002). Due to higher upfront costs and greater project risks, mobilisation of private capital from domestic institutions remains a difficult challenge particularly in developing countries considering the negative past experience of project failure (Martinot 2002). Lacking financial institutions and instruments tailor-cut to renewable energy investment limits the supply of renewable energy technologies (Painuly 2001). This constitutes a vicious circle since the persistently small market size impedes the development of healthy competition to increase pressures for project effectiveness which prevents realising economies of scale[25].

Jagadeesh (2000) argues that in the case of wind energy in India, the general lack of adequate finance has been a chronic problem for the installation of projects mainly in the past. Yet, relatively high upfront costs still constitute a serious obstacle to wind energy deployment. Despite falling wind turbine prices world wide, capital costs for wind installations in India have been increasing. This is due to a phenomenon called 'gold plating' (Pillai 2006) which describes the promotion of export and sale of equipment globally at the costs of encouraging the adaptation and installations of wind turbines in local settings. Unanticipated operation costs (Prengel 2004) arise out of technological failure and lack of operation and maintenance and constitute commercial risks. These lead to difficulties to obtain credit for wind power installations. Finance institutions are still hesitant to lend to wind entrepreneurs due to the uncertainty that the investor's cash flow will be adequate to meet its financial obligation (Basu 2007). Consequently, companies derive financial resources out of own budgets or are able to obtain credits due to known balance sheets. Particularly for newcomers and small players, the hesitant attitudes and lending practices of financial institutions impede potential investment activities. This makes a strong case for external interventions to support renewable energy deployment activities.

(ii) Technology

The second factor pertains to the technologies applied in renewable energy installations which due to the inappropriateness to local conditions lead to the poor perform-

[25] Even though the Indian wind sector is a relatively mature market dominated by turbine manufacturers with independent in-house financing models, additional financial incentives through international finance options presumably stimulate wind sector activities.

ance (Rajsekhar et al 1999). In the case of wind energy, the utilisation of adequate technologies is partly due to the lack of soft and hard technologies. The greater blame is put on existing incentive structures and performance benchmarks (Prengel 2004) that discourage efforts to deploy adequate installations and employ proper practices. As a result, low technical efficiency and poor implementation practices create performance risks that impede wind power deployment.

(iii) Research and Development

Regarding R&D as third factor, particularly during the initial stages of renewable energy development, appropriate skills and training for operating and maintaining projects are limited and often not locally available (Amin 2007). While technologies that more optimally used renewable energy often exist in industrialised countries, in developing countries the local penetration and interconnection of renewable energy hardware and software is often missing. Country risks or risk perceptions as well as high transaction barriers impede constructive exchange and transfer on the parts of both developing and industrialised countries. The oligopolistic nature of international renewable energy markets renders the establishment of local learning and manufacturing hubs redundant. Local incentive structures in developing countries lack the supportive institutional conditions, and professional and experienced staff (Jagadeesh 2000).

In summary, specifically wind energy has pretty soon become an international matter after the initiation phase of technological development. On the one hand, this relates to the limit of national markets, the benefits of pooling expertise due to the high costs of initial R&D and issues of scale and scope benefits. On the other hand, this is pushed by global changes of greater dimensions related to the increasing importance of debates about climate change and energy geopolitics. This contributed to three important developments that stimulated the external support for renewable energy development in industrialising economies. Firstly, development assistance and global finance have shown an early and continued interest that goes beyond altruism to motivations of pushing national champions. Secondly, global market developments and multilevel network formation constitute opportunity conditions for pushing the renewable energy industry. Thirdly, persistent obstacles to effective renewable energy penetration in developing countries call for new modes of governance to tackle global challenges such as climate change and energy poverty. Impediments to renewable energy deployment in developing countries relate to finances, technology and R&D and are still difficult to come by in the local contexts. This creates the demand and necessity for external stimuli and assistance.

Acknowledging the challenges and potentials of external conditions, the following section will discuss theory approaches to examine respective factors and mechanisms that might impact on renewable energy deployment.

II.4.2 External Stimuli Matter? – International Policy Diffusion and New Modes of Governance

Empirical studies on renewable energy deployment in developing countries touch upon issues of finance, R&D and technology that affect individual investment decisions. Not only official development assistance (ODA) but also new international instruments and actors play an increasing role for stimulating renewable energy deployment in developing countries. The overall discussion of external stimuli for local developments takes place in the context of international relations and global governance debates. To be more specific, three main avenues through which external stimuli relate to renewable energy deployment are explored in the following. These concern international development assistance, diffusion of technologies and policies, and new modes of governance such as the CDM. The channels for external stimuli are discussed in the light of global environmental governance debates and underlying concepts like ecological modernisation, policy diffusion, technology transfer and new modes of governance.

Channel I – International Development Assistance

Traditional development assistance looks back at the longest history of support to developing countries embracing the overall objective to stimulate economic and social development. According to governance debates, the state failure in many developing countries to provide public goods explains the emergence of functional equivalents such as ODA or private actors. Stimulating policy or technology transfer were sought the most effective mechanisms to meet governance challenges.

Development assistance integrates the notion that state failure in developing countries is responsible for the lack of development in areas such as the power sector, e.g. renewable energy deployment. Efforts thus aimed to improve the framework conditions for the effective working of the market (Wolff 2001). World Bank and the donor community adopted a technocratic approach towards strengthening governance in developing countries. The major objective was to improve the management of a country's economic and social resources for development and to strengthen state capacities (Wolff 2001: 5). In the course of many years of ODA, various strategies from import substitution to state-led development have been applied. Without doubt, ODA was crucial sewing the seeds for institutions promoting renewable energy in developing countries. However, the outcomes in terms of stimulating local development such as renewable energy deployment remain ambiguous often due to locally inappropriate strategies and resources deployed.

In addition, not only ODA but also industrialised countries have played important roles as financiers as well as technology development partners through private sector as well as bi- and multilateral governmental institutions (Juma et al 2001: 642). Most of these efforts concentrated on providing resources and financial incentives in order to promote specific policy developments (Biersteker 1992). Science and technology were widely recognised as crucial for the economic development and transformation of developing countries (Juma et al 2001: 630). The accumulation of technology was considered important to close the gap between developing and industrialising countries in terms of real income and productivity. Despite continued efforts to provide international assistance in the increasingly globalised world no significant progress to stimulate local development has taken place (Juma et al 2001: 632).

Channel II – Diffusion of Policies and Technologies

Processes of technology transfer and policy diffusion constitute a second channel through which to respond to governance challenges in the field of renewable energy service provision in developing countries.

The concept of ecological modernisation re-emerged in the context of debates about technology and policy diffusion based on the experiences of industrialised countries in the 1980s. The diffusion of discourses and ideas on ecological modernisation go alongside the recognition of governance failures in environmental politics, emerging environmental movements (Dryzek 1997) and an industry awareness on the business potential of cleaner production methods (Young 1999). In a nutshell, ecological modernisation promises that business can make money and produce in an environmentally sustainable manner if environmental care is internalised by social, political and economic institutions (Dryzek 1997). The expansion of market economies is regarded sustainable only if integrated in the development of environmentally friendly technologies (Jänicke 2000). In this case, meeting prosperity and reconciling environmental concerns are regarded win-win sets. The modernisation of industry along ecological lines (Young 1999) through for example renewable energy deployment is not only desirable but also profitable in order to achieve a different, more sustainable economic growth. The conscious use of market instruments to develop and adapt green technologies constitutes a more viable and appropriate answer to environmental challenges as well as economic development (Jänicke 2000).

Regarding development processes outside the OECD, world ecological modernisationists argue that countries do not necessarily have to pass through dirty stages of industrial growth (Perkins 2003). Instead of adopting technologies characterising the growth trajectories of most OECD countries, developing countries can leapfrog to modern clean technologies. Pressures of population growth as well as rapid economic development (Tukker 2005) create the potential and the demand for leapfrogging options in developing countries. Rapid industrialisation has so far posed a dual challenge of coping with developmental requirements while mitigating mounting environmental pressure (Ho 2005). Leapfrogging in terms of developing technological and institutional capacities to encourage more competitive and sustainable economic development hence provides an alternative development path. The realisation of leapfrogging potentials is conditional upon the transfer of respective environmentally-sustainable technologies as well as the diffusion of associated policies and ideas.

The ecological modernisation discourse dominates not only technological leapfrogging debates in the developing country context but also emerged with renewed strength in the context of climate change negotiations (Langhelle 2000). The transfer of renewable energy technologies is highlighted as important route towards ecological modernisation in developing countries like India that face a dual climate change and energy crisis. Renewable energy is acknowledged as highly relevant due to its potential to improve energy security as well as to stimulate sustainable development by providing energy services and achieving national and international development targets (Wilkins 2002). Since the technical operating environment in countries like India is totally different from industrialised countries, technological adaptation is crucial in order for transferred technologies to reach operational efficiencies (Worrell et al 2001: 39). However, numerous empirical studies provide evidence on the fact that environmentally sound technologies do not transfer rapidly (Worrell et al 2001: 30). This relates to several barriers such as lack of information, skills and knowledge; limited capital availability (Ghosh 2002); inadequate national environmental, trade, investment and intellectual property policies; uncertainties about the regulatory environment and future decision-making. The low diffusion of innovative technology into/within developing countries markets (Levine et al 1994; Worrell et al 2001: 36) constitutes a central challenge. This threatens the potential of external stimuli to encounter technological, financial and knowledge barriers.

In addition to the importance of technology transfer, supportive policy and governance processes are acknowledged relevant for stimulating ecological modernisation. Policy diffusion literature (Tews 2003) argues that both promotional agency and the economic, political and social structure determine how and why intra/inter-state diffu-

sion of technological innovations and respective policies takes place. In the empirical context of renewable energy deployment in developing countries, particularly later stages of technology development are crucial for moving from demonstration towards full utilisation. During these stages, the lack of capacities often constitutes a major issue. Empirical studies in developing countries (Mabuza, Brent & Mapako 2007) show that renewable energy technology transfer is unsuccessful as long as it concerns the mere delivery of high-tech equipment. Consequently, not only hardware components of plants, machinery and equipment but also the soft technologies of know-how and organisation are considered important. Even though transferring renewable energy technologies hold the potential of leapfrogging to alternative development paths, the success does not only depend on the appropriate hardware but also on the availability of technical and managerial skills as well as institutions and networks (Wilkins 2002). Implementing technological leapfrogging as alternative development strategy hence calls for policy leapfrogging towards new modes of governance such as self-/co-regulation and market-based mechanisms in order to stimulate entrepreneurial and innovative capacities (Perkins 2003).
In summary, global environmental governance debates inherent in ecological modernisation discourses suggest the diffusion of policies and technology as options to assist renewable energy deployment in developing countries.

Channel III – New Modes of Governance
Ambivalent experiences with international development assistance, international private initiatives as well as technology transfer give rise to the consideration of new approaches to assist sustainable development in developing countries. These call for new modes of governance to respond to global challenges in the field of environmental and climate policy faced by nation states. Within the multilevel environmental governance context, non-state actors and non-hierarchical modes of steering play an increasing role both in market regulation as well as quasi-policy vehicles. Consequently, new modes of governance such as market-related international policy instruments like the CDM are considered a third channel to trigger processes such as renewable energy deployment in developing countries.
Point of departure for the design of such policy devices is the objective to internalise environmental damage in market instruments as an alternative to public command and control or private self-regulation, i.e. creating new mode of governance (Fuhr et al 2007). Conscious approaches to integrate private actors in the provision of public goods or services depart from the assumption that due to privatisation, liberalisation and deregulation processes worldwide, private actors are capable to exert policy impacts (Brühl et al 2004).This implies that private capacities are consciously used for the stimulation of renewable energy deployment in developing countries. One example of new modes of governance in the realm of international climate politics is provided by the Clean Development Mechanism (CDM). Under the framework of the UNFCCC, the CDM is institutionalised as a market-based policy mechanism that ultimately tackles the issue of climate change. The CDM constitutes a project-based mechanism between industrialised and developing countries. It facilitates investments in climate change-mitigating projects in developing countries that are rewarded with certificates to be accounted to specific greenhouse gas reduction obligations (Figure III).

Figure III: The Clean Development Mechanism

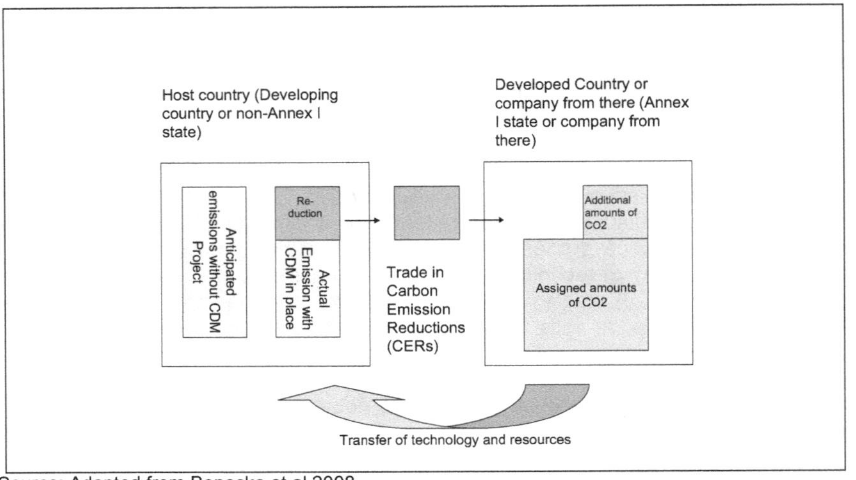

Source: Adapted from Benecke et al 2008

The CDM as climate change mitigation instrument integrates two main objectives: On the one hand, the CDM provides cost-effective means for greenhouse gas reductions. On the other hand, the CDM facilitates sustainable development in developing countries, e.g. through contributing to technology transfer. In this sense, the CDM integrates the ideas of ecological modernisation and promises to stimulate green development such as renewable energy deployment through technology transfer and policy diffusion.

Recent empirical studies examine the extent to which the international climate change regime actually contributes to the transfer of greenhouse gas (GHG-) mitigating technologies (Dechezlepetre et al 2008). Using patent data as a proxy the authors derive at the conclusion that transfer flows of climate change innovations have been increasing and a strong influence of the Kyoto Protocol is felt. With focus on the CDM, studies (Glachant & Ménière 2008) generally conclude that the CDM contributes to technology transfer by financing projects using technologies not available in the host country. This kind of technology transfer refers to the vertical component or the first-time adoption of technology. Uncertainties, however, still exist with regard to the horizontal component as regards the diffusion of climate mitigating technologies within the country. Per definition, technology transfer does not only entail equipment transfer from suppliers located abroad but also the transfer of know-how, information or technical assistance (Dechezlepetre et al 2007). With respect to horizontal technology diffusion this software aspect increases in importance since it creates spillover effects on learning and the adaptation of appropriate technologies and practices. Horizontal diffusion also triggers the diffusion of soft technologies, i.e. norms and practices which might lead to long-term organisational and political transitions towards greener development paths. Given the Indian context and specifically the issue of wind energy, this aspect is important as the main challenges are the effective utilisation and dissemination of respective technologies. The CDM as currently designed prefers vertical technology transfer due to the additionality requirement[26]. Considering the advantages of horizontal technology transfer in contrast to challenges of disseminating wind energy in India raises the question about the extent to which the CDM contributes to horizontal technology diffusion and hence sustainable development paths.

With regard to the Indian context (Banik 2000), conventional technology transfer through FDI has so far been rather limited due to restrictive technology policies with regard to technological licensing as well as policy objectives to promote indigenous R&D. The CDM opens a new avenue for technology transfer particularly in the renewable energy sector in which private participation is encouraged by the Indian government. As external stimulus the CDM might tackle financial, technological and knowledge barriers to renewable energy deployment.

In summary, renewable energy deployment in developing countries is constrained by the lack of sufficient finance, technology and knowledge as well as access to epistemic communities. This creates the demand for external stimuli that through various forms of financing, development assistance, or new kinds of cooperation might facilitate respective activities. Particularly new modes of governance respond to this demand involving multiple actors across the multi-level system in non-hierarchical processes such as technology transfer and policy diffusion.

II.5 Analytical Framework for Renewable Energy Deployment

The research puzzle of this dissertation concerns the spatial variation of wind power deployment in India that does not seem to follow intuitive logics of natural resources distribution. As the first step of an inductive research stage, this chapter discussed this puzzle in the light of empirical and theoretical literature. In the following, the empirical insights and respective theory debates are drawn together into an analytical framework for renewable energy deployment to facilitate and guide the exploration of the Indian wind puzzle.

In a nutshell, empirical studies on renewable energy deployment in developing countries in the context of respective theory debates identified four explanatory dimensions that constitute the skeleton of the analytical framework for renewable energy deployment (Figure IV):

Figure IV: Analytical Framework for Renewable Energy Deployment

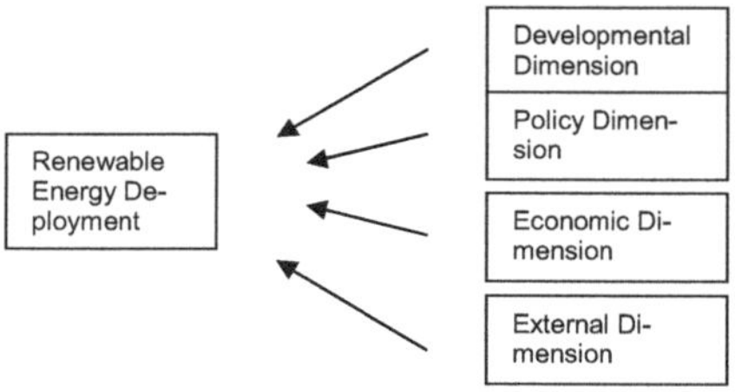

[26] This criterion has come under vociferous debate and criticism particularly with regard to wind energy projects in India.

➢ Impact Dimension I: Developmental Context Conditions

The first dimension considers renewable energy deployment as a development issue. This is based on empirical studies that identify geography and socio-economic structures as well as population characteristics as main barriers to renewable energy deployment. In the context of corresponding theory debates in development economics, a first proposition holds developmental conditions accountable for the success of renewable energy deployment due to related impacts on investment climate and opportunities.

> **Proposition I**
> *Renewable energy deployment takes place given favourable geographical and socio-economic conditions, which constitute an attractive investment climate.*

➢ Impact Dimension II: Policy Conditions

The second dimension regards renewable energy deployment as influenced by policy conditions. Empirical studies suggest that institutional structures, policy and regulatory outcomes, public institutional performance and effectiveness, organisational behaviour and political will determine renewable energy deployment. They refer to the wider policy regime, to organisations involved in wind power deployment as policy output, to policy outcomes in terms of policies and regulations as well as to organisational agency and behaviour as policy impacts. During the first inductive stage of the dissertation, the focus is on public authorities assumed as central locus of policy-relevant decision-making. This leaves out forms of organisations such as firms or civil society[27]. Neo-institutionalist approaches and public policy literature explain why and how renewable energy deployment takes certain directions. Potential causal mechanisms suggest how transaction costs, path dependencies, organisational routines and external institutional environment determine successful renewable energy deployment.

> **Proposition II**
> *Successful renewable energy deployment requires policy conditions to be characterised by polity-shaped institutional characteristics, routines, norms and values that exert effects on supportive policy making, implementation and organisational behaviour.*

➢ Impact Dimension III: Economic Conditions

The third dimension focuses on the understanding of renewable energy deployment as economic activity driven by rational location decisions. Empirical studies identify location factors such as infrastructure and industrialisation, energy market characteristics and support networks as economic conditions. These reflect on techno-economic barriers to wind power deployment in relation to the specific demand and supply conditions that characterise the opportunity structures for investment. Spatial economics based in economic geography theory suggests traditional location factors as determinant of renewable energy deployment.

> **Proposition III**
> *Renewable energy deployment is explained by the rational decision-making according to assessments of industry-specific location factors.*

[27] In the second deductive research stage, however, the research angle is shifted towards acknowledging the broad range of stakeholders involved in wind energy deployment.

➢ Impact Dimension IV: External Conditions

The forth and final dimension takes a global perspective on renewable energy deployment and is based on empirical studies that tackle issues of finance, technology, R&D as impediments to renewable energy deployment. Global environmental governance integrates relevant theory debates like ecological modernisation and concepts such as technology transfer, policy diffusion and new modes of governance. This gives rise to presuming an effect of external assistance and finance as well as new modes of governance on renewable energy deployment.

> **Proposition VI**
> *External finances and resources, and new modes of governance trigger renewable energy deployment in developing countries.*

The following table (Table 4) specifies the propositions that on the basis of the four explanatory dimensions that are operationalised according to specific criteria and indicators. This table is the essence of the preceding empirical literature and theory debates and provides an overview of the causal relations and factors proposed. As such, this summary table renders the analytical framework applicable to the empirical analysis of the Indian wind puzzle.

Table 4: Analytical Framework for Renewable Energy Deployment

Renewable Energy Potentials & Barriers	Indicators	Criteria	Theory Discourse	Causality & Explanatory Factor
Developmental Context Conditions	Urbanisation -Poverty -Education -Income	-Urbanisation rate % -HDI Index -Illiteracy rate % -Rs per capita	Development Studies & Economics	Resource availability & favourable local socio-economic context conditions → Good investment climate → Wind Deployment
Policy Conditions	-Policy regime -Corruption -Policy environment -Power sector – effectiveness -Power sector – efficiency -Power sector – budget -Power policies	-Government party & reputation -CPI -Government/party changes -Rating -SEB deficits in Rs crore -Rs crore for REN -Wind & REN	New Institutional Perspectives & Public Policy	Low transaction costs in principal-agent relations → Favourable wind policy conditions → Wind deployment Organisational routines & path dependencies oriented towards renewable energy → Favourable wind policy conditions → Wind deployment Proactive working culture & attitudes & supportive regime type → Favourable wind policy conditions → Wind deployment Favourable leadership & organisational characteristics/ attitudes→ Favourable wind policy conditions → Wind deployment
Economic Conditions	-Industrial structure - Infrastructure -Investment climate -Power in-	-Industry contribution to SDP -Availability & characteristics - PSR Index -T&D losses -Energy supply gap -Fuel mix – REN	Spatial Economics and Economic Geography	Existence of decisive location factors for renewable investment → Favourable economic conditions → Wind deployment Stakeholder networks structures/ processes& environment & action orientation → Favourable

	frastructure -Power demand -Power competition	share		economic conditions → Wind deployment
External Stimuli	-FDI -CDM	-Rs 1994-2004 -No. Of projects	Governance Debates	Resources & knowledge → Multilevel push & pull factors → Wind deployment CDM → Global incentives → Wind deployment

In the next chapter, the empirical reality of the Indian wind puzzle will be confronted with the analytical framework elaborated above in order to identify factors and mechanisms for understanding and explaining spatial variations. Methodologically, this inductive empirical research will follow a quasi checklist procedure with the intention to eliminate factors not applicable for explaining the Indian case. The main outcome of the first inductive research stage will be to identify relevant explanatory factors and dynamics that shed light on the conditions for effective renewable energy governance in developing countries.

CHAPTER III

Drivers of Renewable Energy Deployment in India – Approaching the Empirics Stage I

CHAPTER III
Drivers of Renewable Energy Deployment in India – Approaching the Empirics Stage I

Spatial variations in wind power deployment in India raise concerns about the potential of alternative energy options to meet both global climate change mitigation and national development objectives. Despite similar sustainable development challenges and natural resource potential different governance responses emerge particularly in the states of Kerala and Tamil Nadu. "*Turning wind into power*" in order to realise sustainable development requires the identification of conditions for effective renewable energy governance. This chapter will undertake the first inductive empirical approach to explore these prerequisites in the case of wind energy in India. The analytical framework for renewable energy deployment elaborated in the previous chapter will guide the analysis.
The chapter proceeds as follows: First of all, a short overview of wind energy in the multi-level governance system will set the context for exploring potentials of and challenges for wind power as alternative energy option in India. Secondly, the analytical framework is applied to identify explanatory factors that account for spatial variations in deployment rates. While initially all the Indian states with wind power potential are considered; ultimately, the puzzling cases of Kerala and Tamil Nadu are attributed special attention[28]. Finally, this chapter will conclude with the main finding that stakeholder characteristics and relations matter for explaining effective renewable energy governance.

III.1 Multilevel Perspectives on Wind Energy

This section will provide an overview of wind sector development and governance in the multi-level setting (Figure V).

[28] This procedure serves to give justice to this case study selection.

Figure V: Multilevel Wind Energy Governance

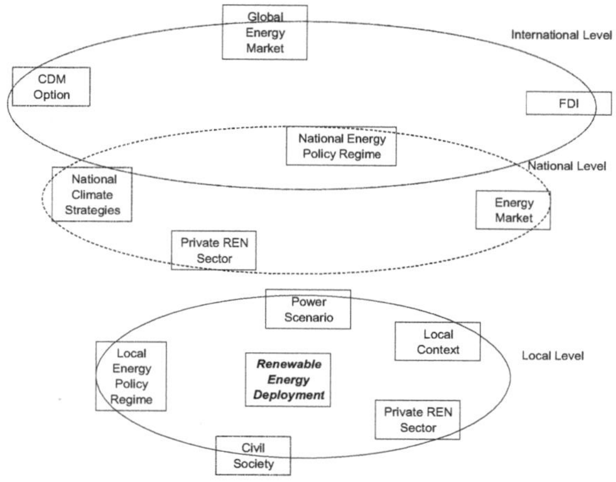

The added value of a multi-level perspective on wind energy deployment is to go beyond an isolated black-box analysis and take into account the impacts of globalisation. Given the interdependencies of mechanisms and developments in energy markets, potentials and challenges for renewable energy deployment are located at international, national and local levels (Bathelt & Glückler 2003).

On the one hand, wind energy deployment is a national and local option to counter sustainable development challenges inherent in specific energy scenarios. The main conditions for successful deployment are hence based in governance structures and processes at the domestic levels. On the other hand, wind energy deployment as an economic activity is subject to global market economics and geopolitics. Drivers for wind power development that shape opportunities in the specific local contexts arise at the global level such as global renewable energy demand or technology supply. Additionally, new modes of governance have emerged at the international level to respond to global and local sustainable development requirements. The Clean Development Mechanism (CDM) constitutes a prominent example of a policy-based market mechanism institutionalised in the context of the international climate regime in order to catalyse green investments. This matters for the wind sector in local contexts such as Tamil Nadu and Kerala since it provides an external impetus for changes in the prevailing opportunity structure for wind power deployment.
In summary, a multilevel approach identifies potential conditional requirements for effective renewable energy governance at multiple levels.

(i) International Level – The Global Context

The first analytical discussion focuses on energy markets and wind sector scenarios at the international level as global market dynamics and the geopolitics of energy impact on national and local developments. The Development Mechanism (CDM) constitutes a prominent example of a policy-based market mechanism institutionalised in the context of the international climate regime in order to catalyse green investments. This matters for the wind sector in local contexts such as Tamil Nadu and Kerala

since it provides an external impetus for changes in the prevailing opportunity structure for wind power deployment.
In summary, a multilevel approach identifies potential conditional requirements for effective renewable energy governance at multiple levels.

Table 5: World Energy Demand

	1980	2000	2005	2015	2030	2005-2030*
Coal	1 786	2 292	2 892	3 988	4 994	2.2%
Oil	3 106	3 647	4 000	4 720	5 585	1.3%
Gas	1 237	2 089	2 354	3 044	3 948	2.1%
Nuclear	186	675	721	804	854	0.7%
Hydro	147	226	251	327	416	2.0%
Biomass and waste	753	1 041	1 149	1 334	1 615	1.4%
Other renewables	12	53	61	145	308	6.7%
Total	**7 228**	**10 023**	**11 429**	**14 361**	**17 721**	**1.8%**

* Average annual rate of growth.

Source: IEA (2007)

Key challenges to sustainable development inherent in the global energy scenario are the increasing global energy demand (IEA 2007, Table 5), aggravating demand and supply gaps and repercussions on environment, climate (Table 6), and energy security.

Table 6: CO-2 Emissions

Figure 5.2: **Incremental Energy-Related CO_2 Emissions by Scenario, 2005-2030**

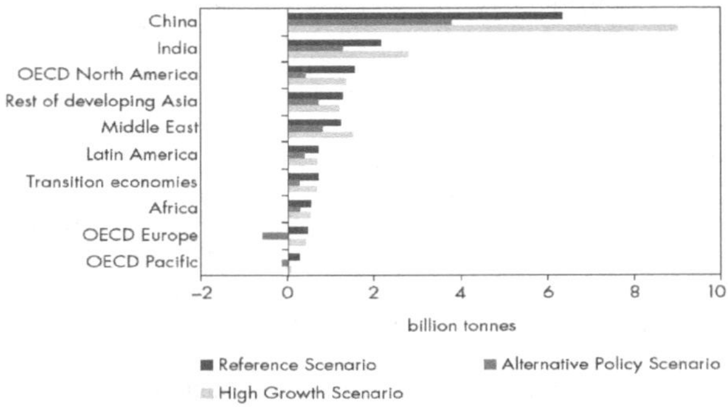

Source: IEA (2007)

The global energy scenario implies a strong demand for indigenous, environmentally less harmful sources of electricity generation. Among the range of alternative resources for power generation, wind energy looks back at the most significant success story (Pillai 2006). While wind energy today contributes only about 1 % to global electricity generation (WWEA 2006), its global growth rate reflects an impressive success story. The momentum was gained after the oil shock of 1973 due to realising geopolitical dangers of energy dependence and limits to fossil fuels. Particularly the last ten years saw an amazing growth of wind power that is attributed to many enabling factors such as the changing awareness about the potentials and promises of wind energy. Technological progress and innovation with regard to wind technologies and resources assessments lastly stimulated the evolution of conducive policy and regulatory frameworks (Pillai 2006). Compared to other renewable energy technologies, wind energy has reached a level of maturity and commercialisation that translated into the evolution of a booming global wind energy market (Table 7).

Table 7: Global Wind Power Development

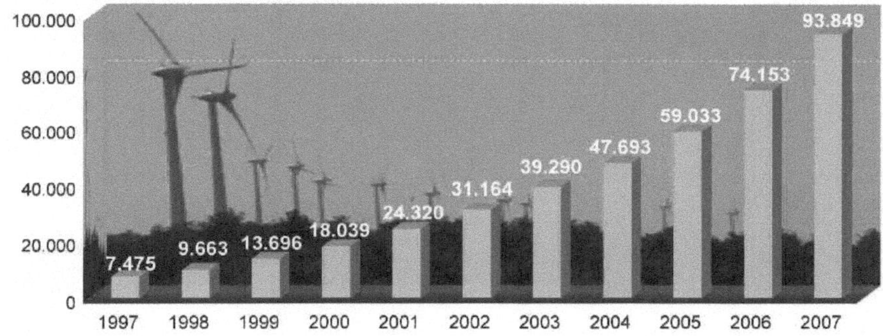

Source WWEA (2008)

Global wind power development clearly sees its origin in Europe (Pillai 2006). Technical innovations were produced in research labs and site facilities in Denmark and Germany and created important stimuli for the adoption and adaptation of wind power technology in other contexts. This has been facilitated by the growth of a specific industry branch based on the manufacturing of wind turbines, the establishment of research networks originating from these innovative milieus (Schamp 2000). The formation of international, regional and national lobby groups shaped the political agenda in favour of wind energy promotion (WWEA 2006). The institutionalisation of the Clean Development Mechanism (CDM) stimulated additional financial incentives for wind farm installation. While at the international level, wind power continued to grow up to 24 % in 2005, trends of geographical diversification have become more pronounced. Even though Europe remains in the leading position in terms of world market share in overall and newly added wind power capacity, Asia is likely to take over with a current growth rate of 48 %. India and China are the major drivers in this region regarding installed capacities and manufacturing facilities while India clearly leads in terms of overall and added capacity.

In summary, global developments and projected scenarios create opportunities for wind energy to meet urgent demands for power generation particularly in emerging economies like India.

(ii) National Level – The Indian Context

At the national level energy markets and governance in countries like India challenge sustainable development in terms of lacking energy access, security and sustainability as well as at the same time provide a niche for wind energy.

First of all, developing economies like India face a massive increase in power demand to fuel economic growth and social development (Table 8). Due to dramatic power shortages India approaches an energy crisis. India's economic growth of roughly 7 % per annum since 2000 is expected to continue (IEA 2006: 425) given a GDP per capita rising by 5.4 % per year in 2000-2005 and projected to grow by 6.4 % in the next decade.

Table 8: Indian Energy Demand

	1990	2000	2005	2015	2030	2005-2030*
Coal	106	164	208	330	620	4.5%
Oil	63	114	129	188	328	3.8%
Gas	10	21	29	48	93	4.8%
Nuclear	2	4	5	16	33	8.3%
Hydro	6	6	9	13	22	3.9%
Biomass	133	149	158	171	194	0.8%
Other renewables	0	0	1	4	9	11.7%
Total	**320**	**459**	**537**	**770**	**1 299**	**3.6%**
Total excluding biomass	*186*	*311*	*379*	*599*	*1 105*	*4.4%*

Average annual rate of growth.

Source: IEA (2007)

Energy demand is likely to rise exponentially particularly in the industry and service sectors but also in domestic consumption (IEA 2007: 467). This increases pressure on power supply and effective energy governance.

Secondly, the expected growth rate of 6.6 % per year is confronted with chronic electricity supply shortages (IEA 2007: 510). As indigenous oil and gas reserves are also short of supply, energy imports and import dependency are increasing. Furthermore, the Indian energy sector is characterised by failures of delivering effective governance services (Reineberg 2006: 79). Current power deficits average 9 % (IEA 2007: 522), and unstable voltages and supply infrequencies are major concerns to industrial and private energy users.

Table 9: Installed Energy Capacity (in MW)

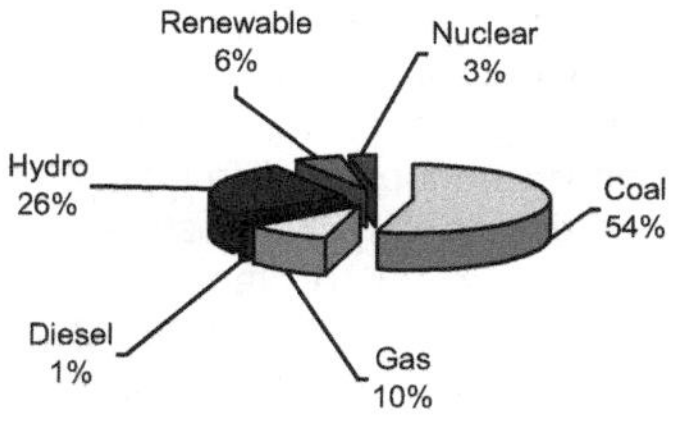

Source: Ministry of Power (2007)

This scenario might constitute the window of opportunity for alternative energy sources such as wind energy to meet sustainable development challenges (Table 9). Already, renewable energies – apart from large hydro and biomass – are expected to see a skyrocketing growth of nearly 12 % per year (IEA 2006: 425). In addition, the Indian government has come forward with the ambitious plan[29] to increase the installed capacity of renewable energy sources to 10.000 MW by 2012 which is roughly 10 % of the aspired capacity addition in the energy sector (bfai 2007: 13).

(iii) Local Level – The Indian Wind Puzzle

In contrast to the conventional power sector, the governance of renewable energies to date is characterised by greater private sector participation in power generation (95.5 % according to Mabel 2008: 1757) while distributions and transmission remains in the public domain.

For the last three decades, the wind energy sector in India experienced variable growth (Pillai 2006: 392) which is differentiated into three phases.
The first phase of wind sector development in India commenced in Gujarat, where RRB Consultants & Engineers, an Indian consultancy company, deployed the first wind turbine installation in 1986. This was a late response or a wake up to the challenges of energy security emerging at the global level with the oil crisis in 1973. Immediately following up on this success story, RRB started commissioning wind farms in Tamil Nadu (Pillai 2006). After the National Petroleum Company (NEPC) entered into a joint venture with Micon in 1994, NEPC received the first private sector order and installed the country-wide single largest wind farm project. This was set up in Tamil Nadu with the purpose of wheeling energy for private self-consumption. This raised not only investors' interest in the potential to procure independent and reliable power. It also mollified utility's reservations concerning the anticipated pressure to initiate investments in this sector out of their own accounts. In the course of global wind market developments, RRB formed a joint venture with Vestas, one of the big international players in the global wind sector, to open turbine manufacturing units in Chennai, Tamil Nadu. This initiated the indigenisation process of wind turbine technology as an alternative to relying solely on imports of components required for wind farm installations. The set-up of a domestic industry branch also provided the condi-

[29] In order to boost renewable energies, comprehensive programmes provide incentives such as credits, low import duties, fiscal triggers and attractive tariffs.

tions for the potential adaptation of wind farm components to Indian circumstances. As a result, this reduced the incidences of operational failures of previous wind projects which were due to the mismatch between imported and local components or due to the unsuitability of imported parts. Most and foremost it provided the preconditions for turbine export and the ensuing global expansion of the Indian wind energy business. These developments slowly encouraged investors' confidence in the longevity of wind farm installations and the viability of investment.

The Indian success story would not have taken place without government support. The Indian wind energy programme was initiated in 1984 (Mabel et al 2008) with the objective to smooth the grounds for private investment through demonstration activities, information and awareness creation and facilitation. In order to open the arena to potential commercial projects, the Ministry of New and Renewable Energy (MNRE) announced capacities at 33 locations in 9 states creating the conditions for the wind energy market to expand. At first, wind farms were installed along the coastal areas of Tamil Nadu, Gujarat, Maharashtra and Orissa (Mabel et al 2008: 1755). Even though the early 1990s saw many new entrants and joint ventures in the Indian wind energy markets (Pillai 2006) by 1992, 20 % of the wind power installed capacities were channelled through demonstration projects and the government. Private sector participation started to set in a big way from 1992 onwards due to the initiation of wind resource assessments, wind farm demonstration programmes and technological development. In 1995 Suzlon as first indigenous entrepreneurial undertaking came into play. Suzlon established a unique business model called "one face to the client" that offered complete solutions to all parties interested in wind energy. This means that the entire up- and downstream processes from wind farm development, installation, implementation and maintenance are taken care of by the turbine manufacturer. Due to this locally-adapted business model Suzlon managed to crowd out foreign competitors and to capture roughly half of the market share in India and about 40% of the wind energy market in Tamil Nadu. This business model is appropriate to Indian conditions as it evolved in response to the demand for all-around services from project design to operation by potential investors in wind energy. This is due to the fact that private parties invest in wind power alongside three main rationales that are not necessarily related to wind power generation (Rajsekhar 1999: 669). Firstly, tax-credit seekers are those companies whose investment decisions are based on tax liability considerations. Even though they constitute 75 % of private sector participants in the wind energy business, they only contribute 31 % to installed capacities. This group of investors gave rise to and explain the success of Suzlon's unique business model. Apart from reaping government incentives tax-credit seekers are not interested in wind power and demand convenient solutions, i.e. encourage providers like Suzlon to manage the entire wind business for them. Secondly, power-seeker companies capture 22 % of the investment community and make investment decisions out of the urgent need for additional power supply. Thirdly, manufacturing companies contribute only 3 % to the de facto investment activities in wind power projects. This is because they service the tax-credit or power-seeking companies by delivering and operating wind farms for them.

The second phase of Indian wind sector development set in after 1997-98 and was characterised by a sharp decline in wind power and is generally associated with the world economic crisis. Moreover, changes in the national incentive system through the introduction of minimum alternate tax (MAT) and the reduction in marginal corporate tax tariff to 35% decreased the attractiveness of investments in wind power plants (Rajsekhar 1999). This scared of tax-seeking investors who were initially allured by promises of high profits for investments in easy undertakings such as wind energy projects that require not too much involvement. The wind industry finally adapted to the changing policy scenario given the surging power demands and technological innovations and business model adjustments.

This led on to the third phase of wind sector development in India. Between 2001 and 2003 the Indian wind industry went through a speedy recovery process. During the last couple of years, the wind energy sector saw constant capacity additions. In 2004, further improvements in wind turbine technologies stimulated the growth of this industry (Mabel et al 2008: 1756). The Indian Government continued to provide a range of promotional measures such as fiscal concessions (80 % accelerated depreciation), tax holidays, concessional customs and excise duties and liberalised foreign investment procedures (PowerLine 2007: 53). Furthermore, at the national level, a range of crucial power sector reforms were implemented that also encouraged new investments in the wind sector[30]. The Electricity Act 2003 as the most important piece of legislation integrated various previous efforts to stimulate competition, restructure the electricity sector and introduce better regulation. As a result, more and more regulators stipulated tariff regimes to provide investment security and to promote the installation of wind farms. Additionally, technological development stimulated the growth of Indian wind power. Wind monitoring stations were commissioned in eleven states to provide more reliable and precise data. Wind turbines and components were now exported and R&D made rapid progress. Finally, a high-level advisory group on wind energy was institutionalised in order to promote wind energy at the highest governmental level. In summary, the development of the wind sector increased installed wind energy capacity during the last few years which renders India the fourth largest wind market globally (Table 10).

Table 10: Global Wind Market – India's Position

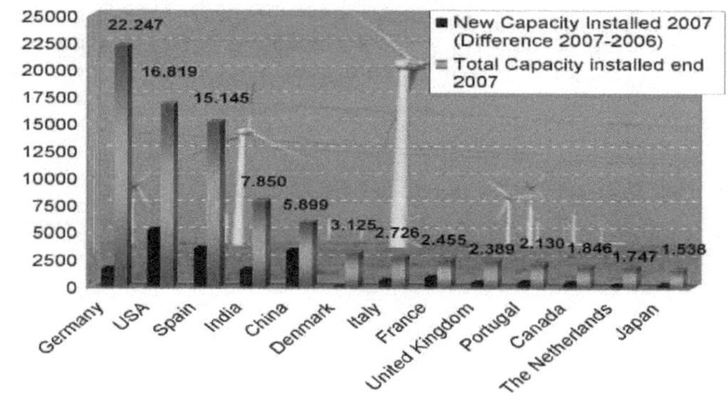

Source: WWEA (2008)

In the multilevel context, international and national opportunity structures like external CDM stimuli or national wind power policies provide similar incentives for respective

[30] Since power is a concurrent subject regulatory and policy matter for renewable energies, e.g. tariffs, are set at the state level. The energy governance system and its impacts on wind power deployment will be elaborated later in th course of this dissertation.

deployment activities all across India. As Table 11 illustrates, the deployment of wind energy varies widely between different Indian federal states. Indeed, spatial variations emerge as the characteristic feature of the Indian wind sector. Highly successful wind power states like Tamil Nadu, Maharashtra and Karnataka contrast states with little deployment activity such as Andhra Pradesh, West Bengal and Kerala. This is puzzling since these low performers are endowed with great technical potential that, if realised, would for example render Andhra Pradesh India's top wind state.

Table 11: Wind sector constitution

State	Installed Wind Capacity (MW)	Wind Power Technical Potential (MW)	Utilisation Rate (%)
Tamil Nadu	**3873**	**4750**	**81.5**
Karnataka	1011	6620	15.3
Maharashtra	1756	3650	48.1
Rajasthan	539	5400	10
Andhra Pradesh	123	8275	1.5
Madhya Pradesh	188	5500	3.4
Kerala	**2**	**875**	**0.2**
Gujarat	1253	9675	13
West Bengal	1	450	0.2
India	8746	45195	19.4

Source: MNRE (2008)

Yet, the starkest difference in the extent to which natural resource potentials are harnessed for power generation emerges between the two neighbouring Southern states of Tamil Nadu and Kerala. This is not simply explained by natural resource conditions as both states are similarly endowed with good potential wind sites (Map B).

Map B: Wind Power Density

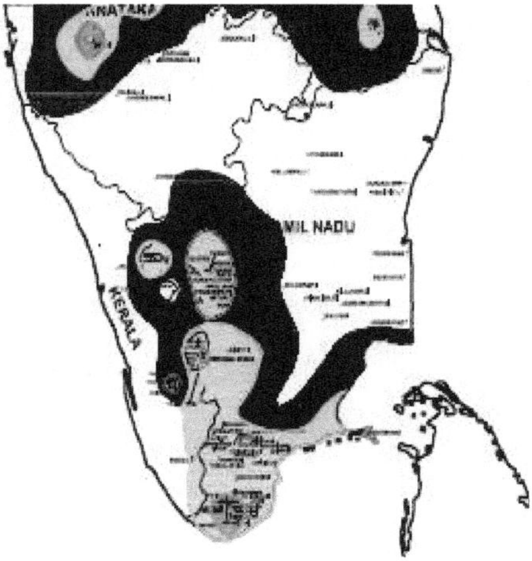

Standard Wind Class definition

CLASS	50 m HEIGHT		INSTALL-ABLE MW*	WIND POWER DENSITY W/sq.m
	WIND SPEED m/s	WIND POWER W/sq.m		
1	< 5.6	< 200		
2	5.6 - 6.4	200 - 300	43106	
3	6.4 - 7.0	300 - 400	4680	
4	7.0 - 7.5	400 - 500	397	
5	7.5 - 8.0	500 - 600	16	
6	8.0 - 8.8	600 - 800	-	
7	8.8 - 11.9	800 - 2000	-	
		TOTAL	48199	

Source: C-WET (2007)

Tamil Nadu as model state for wind sector development in India has experienced a stable and continuous growth in terms of capacity installation. Particularly since 2005 the wind energy sector has taken off rapidly (Table 12). Today, wind energy contributes 13% to the electricity addition in Tamil Nadu out of which 4% are purchased by the state and the rest is used for captive consumption.

Table 12: Wind Energy Development in Tamil Nadu

	Up to 1992	1992-93	1993-94	1994-95	1995-96	1996-97	1997-98	1998-99	1999-2000
Installed Wind Capacity (MW)	63.9	68.6	72.3	151.3	426.1	702.1	779.8	894.9	1155
	2000-01	2001-02	2002-03	2003-04	2004-05	2005-06	2006-07	2007-08	
Installed Wind Capacity (MW)	1095.8	1245.7	1305.5	1361.6	2037	2894.6	3492.7	3873.4	

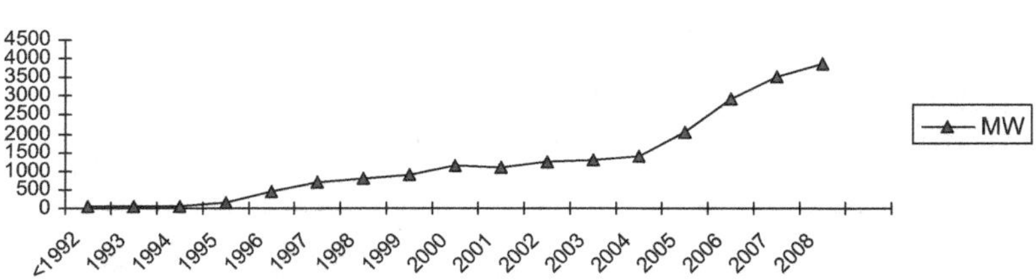

Source: Tamil Nadu State Planning Commission (2002)

Assessments of wind resources potential and technical feasibility are favourable for potential wind sector activities in Kerala. So far, 16 sites are identified for wind farm development. However, until today very little wind power deployment has been seen. Only one 2 MW demonstration project was set up by the state nodal agency ANERT and the state utility KSEB (Figure VI). Wind farm development by private sector stakeholders is taking place just presently at newly explored wind farm sites, which is expected to add to wind energy capacity shortly.

Figure VI: Wind Energy Development in Kerala

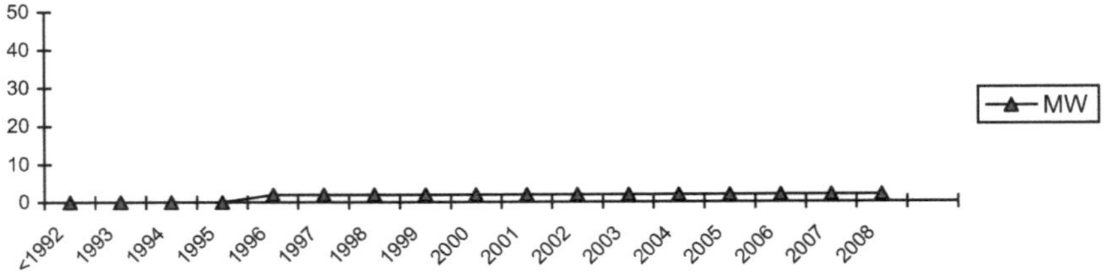

Source: Kerala State Planning Board (2002)

In summary, given the global and national energy demand and supply conditions, opportunity structures for wind energy deployment in India are actually favourable for wind power development. However, in the Indian context, spatial variations in wind power deployment arise despite similar resource potentials. Particularly the cases of Tamil Nadu and Kerala constitute extreme instances of this research puzzle. This empirical observation justifies the general research question of this dissertation that enquires about the governance conditions for successful renewable energy deployment.

The following section embarks upon the inductive empirical exploration of the research puzzle. The purpose is a detailed scrutiny of factors derived from previous literature discussions explaining why and how wind power deployment varies given similar circumstances. In order to understand conditions for successful wind energy deployment, the potentials and barriers specific to the local context of Indian states are discussed. The analytical framework for renewable energy deployment elaborated on the basis of empirical studies and theory debates will guide this exploration.

III.2 Analysing Wind Energy Determinants in India

Nowadays, wind power deployment receives new impetus in the context of multilevel opportunity structures and given the need to respond to sustainable development challenges of power shortages and climate change (Purohit 2007: 2). India constitutes a puzzling case since despite substantial wind energy potential some states do not harness the available natural resources.

The following section will commence the first inductive empirical exploration of the conditions for successful wind power deployment guided by the analytical framework as elaborated in the previous chapter. The examination of requirements for wind sector development will consider developmental conditions, policy conditions, economic conditions and external conditions with regard to the Indian context and focus on Tamil Nadu and Kerala. This first empirical encounter will derive at two explanatory entry points related to the characterisation and dynamics of state-market relations in local stakeholder networks. This gives justice to the argument that functionally equivalent governance arrangements produce different governance responses to similar sustainable development challenges.

III.2.1 Socio-Economic Requirements for Wind Energy Deployment

This first subsection will explore developmental conditions as explanatory factors for wind energy deployment in India, specifically in Tamil Nadu and Kerala.
As elaborated in the analytical framework, empirical studies discussed in the light of development economics suggest that geography and socio-economic factors explain variations in wind energy deployment. Renewable energy deployment takes place given that certain geographical and socio-economic conditions provide an attractive investment climate.

The following inductive analysis of the Indian wind sector will explore natural resource endowment and developmental characteristics in the context of the Indian wind sector in general and with focus on Tamil Nadu and Kerala[31]. The analyses will demonstrate that neither resource endowment nor socio-economic characteristics stringently account for differences in deployment rates and hence do not sufficiently explain the Indian wind power puzzle.

(i) Geographical Factors

Pre-requisites for successful wind energy deployment amount to geography, i.e. the availability and accessibility of natural resources. This relates to issues such as existing infrastructure facilities like road networks, the proximity to populated areas for the transportation of construction parts and labour as well as the possibility to purchase the land for construction (Jagadeesh 2000: 158).

Scientific assessments derive at the conclusion that natural resource conditions qualify all the above-mentioned nine Indian states for wind sector development. Consequently, natural resource endowment per se is not the reason for different utilisation rates of wind energy potential. As this implementation bias is most striking in the cases of the two neighbouring states in Southern India, Tamil Nadu and Kerala, these will be examined in more depth:

Early wind resource assessments established the fact that particularly Tamil Nadu benefits from good windy sites (Jagadeesh 2000: 158), i.e. locations where wind density is very high. Especially during summer monsoon the wind energy season peaks. This meets problems of power shortages and cuts in the power system during this period of the year and makes the case for wind power. The more advanced wind resource assessments; also due to new research facilities locating in Tamil Nadu; the more fine-grained location analyses, i.e. the greater the scope for additional wind energy deployment. Tamil Nadu was the first Indian state to introduce wind monitoring systems that rendered site identification and selection more reliable. The terrain in

---

[31] As elaborated in the methodology section (Chapter I), the empirical analysis focuses on 2007 and does not consider temporal variations for now.

which windy sites are located is flat and not situated in hilly areas (cf. Map B and Map C). This facilitates the installation of wind farms, the connection to the grid for power transmission and distribution as well as site accessibility for day-to-day maintenance and operational works. Availing of land for wind farm construction is not problematic as land ownership is dispersed among local people and areas are not designated to other purposes (INT 2). This renders land acquisition easier as developers can purchase land directly from the owners at mutually agreed rates. However, the increasing competition for land has recently led to price hikes and resulted in the practice of land point purchase for single installations. This incurs the inefficient utilisation of land resources and endangers the future expansion and dissemination of wind power on a large scale.

Map C: Physical Map of Tamil Nadu

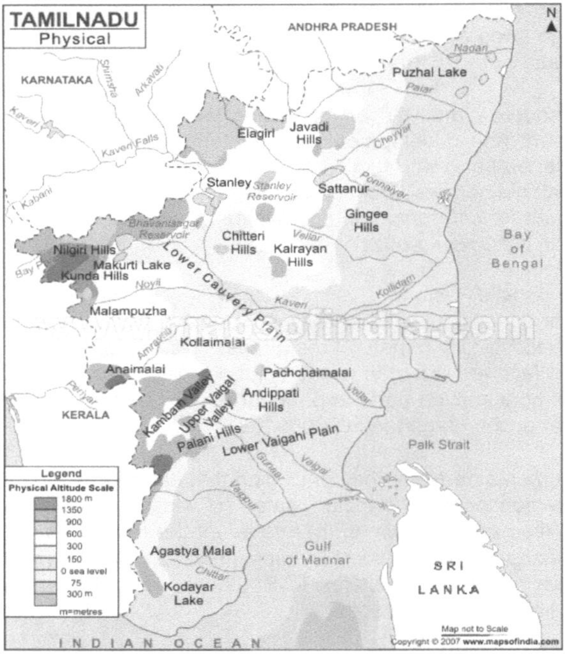

Source: Map of India

As for the Kerala case, natural resource conditions for wind power deployment with regard to wind speed patterns and wind density are assessed as favourable (INT 1). However, the fact that windy sites exist does not say much about the extent to which they can be utilised for wind farm installations. In contrast to Tamil Nadu, the sites identified for wind farm deployment are nearly all located in hilly terrain (Map D) and some are designated as nationally protected natural habitat.

Map D: Physical Map of Kerala

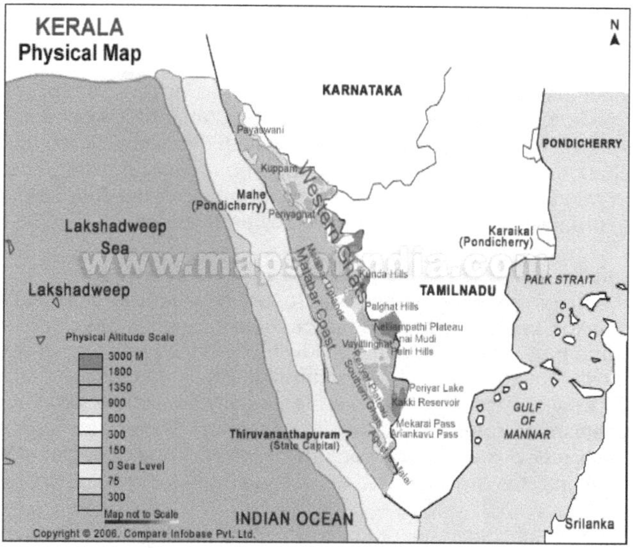

Source: Map of India

This bears several implications: Firstly, species of the surrounding tropical forest areas such as elephants intrude upon construction sites and newly-built roads. This was mentioned and perceived as quite a serious issue by project developers as it disrupts operations and destroys infrastructure (INT 3). Secondly, areas suitable for wind farm installation are often designated as government-protected national habitat. This complicates the application process for wind farm installations as environmental impacts need to be assessed and considered. This requires not only additional bureaucratic procedures but also prolongs the deployment process as more stakeholders have to be acknowledged. The third point relates to the fact that land ownership in these regions is not only left in the private possession of local people. Due to the status of protected natural habitat, certain government restrictions and legislation render land purchase and bargaining for land prices more difficult. Similar to Tamil Nadu, land ownership is highly fragmented. However, local people are much more sensitised due to the influence of environmental lobby groups and higher education levels. They often object to selling parcels of land at throw-away prices and tend to reconsider existing contracts which leads to lengthy and difficult negotiations with project developers.

In summary, this analysis suggests the accessibility and availability to natural resources like wind sites as potential explanation for spatially varying wind power deployment. As a matter of fact, land ownership, area designation and construction procedures do not amount to natural resource endowment per se. Accessibility and availability only become contentious issues dependent on domestic complexities and how for example state officials decide to utilise the existing potential and manage natural resources. Interviews confirm that increasingly access issues are "dealt with" and do not pose significant obstacles to wind power deployment in Kerala any more

(INT 4). Consequently, this dissertation will not consider natural resource endowment a major explanatory factor for wind sector development in India.

(ii) Socio-economic Factors

As elaborated in the previous chapter, factors such as poverty, literacy, urbanisation, religion and income characterise the development status of a local setting. Development economists argue that the more favourable the socio-economic setting, the greater the chances for development activities and growth to take place. The following analysis of wind power deployment will explore the purposefully selected explanatory factors of urbanisation rate, poverty, literacy and income.

a) Urbanisation Rate

Development economists suggest that with regard to urbanisation rates, higher levels of urban population imply a higher status of socio-economic development. In the Indian context, one has to bear in mind that low urbanisation rates have severe implications for the power sector. In all of the Indian states in which the rural sector is important, electricity services for farmers are either heavily subsidised or even supplied for free. High urbanisation rates hence do not only imply higher socio-economic development but also reflect less supply-constraints on the power sector. Consequently, higher urbanisation rates are expected to be favourable for wind power deployment.

Initially, the correlation between urbanisation and wind power deployment is confirmed looking at the states of Maharashtra, Tamil Nadu and Karnataka that have high urbanisation rates and experience wind energy deployment. The reverse is true of West Bengal with low levels of urbanisation and hardly any wind sector activity. Statistics (Table 13) reveal that Tamil Nadu and Kerala are the two states with the highest urbanisation rates in India. Even though Tamil Nadu has a slightly higher urbanisation rate than Kerala, several arguments caution against the conclusion that urbanisation per se is an explanatory factor for successful wind power deployment. This is proven in the case of Gujarat with high urbanisation and insignificant wind power utilisation rates show. Furthermore, Kerala has experienced an immense increase in urban population over the last couple of years and of date is only second to Tamil Nadu. Given Kerala's smaller territorial size, the rural population and associated with that the significance of farming activity need to be considered in relative terms and are less likely to impact on wind sector development. Hence urbanisation rates are not considered relevant explanatory factors for spatial variations in wind power deployment.

Table 13: Urbanisation Rate

	Urbanisation Rate %	Wind Utilisation Rate %
Karnataka	34	15.3
Maharashtra	42.2	48.1
Rajasthan	23.38	10
Andhra Pradesh	27.08	1.5
Madhya Pradesh	26.67	3.4
Gujarat	37.35	13
West Bengal	25.30	0.2
Tamil Nadu	44	81.5
Kerala	60	0.2

Source: Census of India 2001

b) Poverty

High poverty levels are assumed to impede development activities such as wind energy investment due to the absence of demand for or the purchase ability of wind power. As discussed in the previous chapter, high poverty levels have wider implications on education and other investment conditions.

Examining Indian poverty statistics in relation to wind power deployment, the overall picture looking at the % population below the poverty line is ambivalent (Table 14). In states with low figures on the population below poverty lines such as Rajasthan and Gujarat some wind sector activity exists while other low-poverty states such as Kerala or Andhra Pradesh see no wind energy deployed. In contrast, in higher-poverty states like Tamil Nadu and Maharashtra wind deployment takes place. Between Kerala and Tamil Nadu, poverty levels measured in terms of population below poverty line (i.e. HPI or HDI) are relatively similar since both states are among those suffering least from deprivation in India. Counterfactually speaking, if poverty was the explanatory factor for wind sector development, one might expect Kerala to be leading the Indian wind sector due to its outstanding success in poverty alleviation. Consequently, poverty levels do not hold as explanatory factor for wind energy deployment.

Table 14: Poverty

States	Population below Poverty Lines %	Human Development Index (Rank 1-14)	Wind Utilisation Rate %
Karnataka	20.2	7 of 14	15.3
Maharashtra	25.02	4 of 14	48.1
Rajasthan	15.28	9 of 14	10
Andhra Pradesh	15.77	10 of 14	1.5
Madhya Pradesh	37.43	12 of 14	3.4
Gujarat	14.07	6 of 14	13
West Bengal	27.02	8 of 14	0.2
Tamil Nadu	21.2	3 of 14	81.5
Kerala	12.72	1 of 14	0.2

Source: UNDP 2001; Census of India 2001

c) Literacy

One discussion in development economics associates poverty with literacy as any kind of economic activity depends on some degree of basic education of the local population. Specifically the technical and financial intricacies involved in wind project management as well as operation and maintenance of installations demand for skilled labour, i.e. locally-available human resources with some basic level of education. The analytical framework proposes that high levels of illiteracy are likely to occur in states with very little wind energy deployment and vice versa.

Table 15: Illiteracy Rate

	Illiteracy Rate %	Wind Utilisation Rate %
Karnataka	32.96	15.3
Maharashtra	27.73	48.1
Rajasthan	38.97	10
Andhra Pradesh	38.89	1.5
Madhya Pradesh	35.89	3.4
Gujarat	30	13
West Bengal	30.78	0.2
Tamil Nadu	26.53	81.5
Kerala	9.08	0.2

Source: UNDP 2001; Census of India 2001

At first sight, this proposition seems to be confirmed as states with high illiteracy rate such as Andhra Pradesh and Madhya Pradesh experience little wind sector activity (Table 15). Also Karnataka and Maharashtra as successful wind states are not characterised by very low illiteracy rates. The case of Kerala refutes this proposition as despite being the state with highest education level across India, hardly any wind energy deployment takes place. Concerning Tamil Nadu, illiteracy is higher but in comparison to other Indian states still features less significant. Hence education and literacy per se does not explain wind energy deployment.

d) Per Capita Income
Discussions on the impact of income levels in terms of state domestic product per capita on economic development are closely related to debates about poverty and education. Low income inhibits public and private spending on basic goods and services such as education. Furthermore, paying for investments in basic infrastructure such as electricity is aggravated. The analytical framework suggests that low income levels imply little wind energy deployment.
Indeed, the cases of Maharashtra and Tamil Nadu reveal that higher income goes hand in hand with wind sector development. This is contradicted by the cases of Andhra Pradesh and Kerala that see no wind energy deployment despite higher per capita income. In spite of similar income levels, Tamil Nadu and Kerala experience different outcomes in terms of wind sector development. Hence income levels are also not significant for driving wind energy deployment.
In conclusion, developmental characteristics to a great extent shape the general local context for investment activities in the wind sector. Without doubt, features like poverty, education, and income as well as natural resource endowment characterise the developmental conditions of the local setting. The above analysis derives at the conclusion that none of these factors taken on their own accounts satisfactorily explains deployment activities. Developmental conditions are not the prominent drivers for renewable energy deployment and hence do not explain the research puzzle of spatially varying wind power deployment between Kerala and Tamil Nadu. In the Kerala case, potential disadvantages relating to land access and terrain are irrelevant considering the global renewable energy demand and other driving factors discussed in the following. What seems to be important particularly in relation to accessibility issue is what actors make of the local contexts and conditions. The following section will pay attention to the political setting and dynamics in which wind sector activities are embedded.

III.2.2 Policy Determinants for Wind Energy Deployment

As second dimension, the analytical framework for renewable energy deployment elaborates on policy-related aspects of renewable energy deployment in developing countries. Successful renewable energy deployment depends on policy conditions as polity-shaped institutional characteristics, routines, norms and values that exert effects on policy making, implementation and organisational behaviour.
The impact of the policy regime on shaping policy conditions is considered the first, macro-level aspect. Since societal and political orientations determine the overall course of development, policy regimes might have an impact on wind energy deployment. The way in which organisations work efficiently and transparently is influenced by certain policy regimes and determines institutional outcomes. The effec-

tiveness of policy regimes in terms of stimulating developmental activities such as wind power deployment is hence related to the level of corruption.
The second aspect characterising policy conditions regards the institutional set up and structures of respective organisations involved in power sector governance. In this sense, the characteristics, quality and effectiveness of governance institutions constitutive of the power sector are assessed. The higher the perceived quality and the more efficient the design of respective institutions, the greater the chances for new ventures such as wind power to obtain institutional support.
The third aspect recognises that it is not sufficient to look at policy regimes and respective institutional structures but to integrate them into the respective policy areas, e.g. investment and economic policies. In this sense, policy outcomes related to regulations and policies presumably account for variations in wind energy deployment.
The last aspect tackles the behaviour of organisations in terms of constructing and maintaining supportive frameworks and implementing policy decisions for wind power.

In the following, the aspects of policy regime, policy outputs and policy outcomes constitutive of the policy conditions pertaining to the wind sector will be analysed for the different Indian states. In this first inductive encounter with the empirical realities, these aspects characterising policy conditions will emerge as insufficient for explaining the wind puzzle of Tamil Nadu and Kerala.

(i) Policy Regime

The policy regime is considered as potential explanatory aspect for wind sector development at the macro level. Public policy literature argues that policies and to some extent also organisations are shaped by socially-embedded ideologies. Private-sector friendly and stable policy regimes hence support renewable energy deployment through institutional back-up and efficiency. In order to draw conclusions on the impact of policy regimes on wind power deployment, ideological orientation and corruption are analysed.

Contrary to the assumption that wind power deployment goes hand in hand with centre-right policy regimes, Gujarat and Karnataka with relatively well developed wind sectors went through left-party governance in the course of time. The opposite argument that non-leftist policy regimes are accountable for wind sector development does not hold either. Despite the distinctiveness of policy regimes in the cases of Andhra Pradesh and Madhya Pradesh, wind energy deployment is still not taking place. Additionally, the correlation of political stability with wind sector development is also questionable when it comes to states like Maharashtra and Karnataka that saw many changing government coalitions and thriving wind sectors. Comparing policy regimes of Tamil Nadu with Kerala might easily tempt to conclude positively on the supposed correlation to wind power deployment. Since the early 1990s Tamil Nadu has been governed by the regional party DMK with no changes in the policy regime. This party incorporates a social democratic and populist stance which is associated with openness to economic activities, private sector investment associated with contributions to equality and social justice. Kerala is currently governed by a communist party, the CPI-M, which renders Kerala apart from West Bengal the only Indian state with a democratically elected socialist government. Due to the ideological orientation towards the nationalisation of private means of production to guarantee equality to each citizen, this policy regime has the reputation of being averse to private sector participation and investment. Since the 1990s Kerala has experienced frequent policy regime changes between left-wing governments and the national Congress party INC. Due to the instability and ambiguous reputation of the policy regime, this aspect is commonly assumed as relevant for explaining wind sector variations.

To venture deeper into the discussion, the policy regime-related impacts on economic, investment and private sector policies are considered in terms of potential relevance for wind power deployment. It goes without saying that as one of the most prominent investment destinations in India, Tamil Nadu has supportive industrial policies accompanied by streamlined administrative procedures. Contrary to many preconceived expectations, also Kerala has from the early 1990s onwards been very active as regards economic liberalisation and privatisation. Kerala announced a considerable set of proactive IT, Industrial and Labour Policies and actively supports and encourages investment and private sector activities notwithstanding political concerns for social justice and welfare. West Bengal with a Socialist-governed policy regime also belongs to the leading industrialised states. Similar to Tamil Nadu facilitated administrative procedures in the context of economic and investment policies are established; yet still, little wind energy deployment takes place. The same is also true for states like Madhya Pradesh that proactively seeks investment or Andhra Pradesh with the best reputation for reform orientation and good governance. Both states see little wind sector activity. In contrast, Rajasthan and Maharashtra with thriving wind industries have not stipulated specific industrial or investment policies.

In summary, this dissertation argues that policy regimes and economic and investment policies do not sufficiently explain wind power variations since they do not hold when contrasted to the success of the Indian wind states.

The second policy regime characteristic with regard to the quality of governance in specific sectors pertains to the level of corruption. Corruption is associated with inefficiencies of the overall policy regime that affect organisational outcomes such as policy making and administrative procedures in the wind energy sector. High levels of perceived corruption do not only imply unfavourable wind energy policy conditions but due to the anticipated additional transaction costs also deter potential investors.

Table 16: Corruption Perception Rate (CPI)

	CPI	Wind Utilisation Rate %
Karnataka	17 of 20	15.3
Maharashtra	5 of 20	48.1
Rajasthan	16 of 20	10
Andhra Pradesh	4 of 20	1.5
Madhya Pradesh	18 of 20	3.4
Gujarat	3 of 20	13
West Bengal	8 of 20	0.2
Tamil Nadu	12 of 20	81.5
Kerala	1 of 20	0.2

Source: Transparency International 2006

As Table 16 shows, the causalities associated with corruption and wind sector development do not hold for the Indian context. Looking at the CPI that measures the level of perceived corruption, Kerala astonishingly shows the least level of corruption while Tamil Nadu lies in the middle range of corrupt states. Low corruption and low wind sector activity might even suggest a potential causal relation, if it was not for Maharashtra with low levels of corruption and high wind energy deployment. Stating that high corruption leads to high wind power deployment rates appears contradictory; yet, considering the Indian context this argument might not be entirely nonsensical. India is characterised as clientelistic state dominated by a logic of patronism (Randeria 2003). Corruption is more or less part of the institutional and societal system based on the exchange of favours. Low levels of corruption imply that wind sector participants and investors have a much more difficult time to access the institutional and policy setting since they cannot reciprocate the support they demand for facilitated wind energy deployment. However, this logic does not hold persistently as demonstrated in the cases of Madhya Pradesh with high corruption rates and no wind energy deployment or Gujarat with low corruption and emerging wind power activities.

In summary, these arguments call into question the relevance of the policy regime explanation for spatially varying wind power deployment. The impact of policy regime on economic and investment politics and hence private sector activities in the wind sector emerges as insignificant. As comparisons to other states reveal, different wind power deployment rates in Tamil Nadu and Kerala are neither due to the policy regime nor to governance stability and quality.

(ii) Policy Output – Power Sector Constitution

This section focuses on the power sector as the institutional output characterising policy conditions. As wind energy is generated for the purpose of electricity contribution to the grid, wind energy deployment is based in the overall context of the power sector and cannot be decoupled from the energy governance system. With reference to new institutional perspectives, mechanisms like transaction costs, certain path-dependent developments and organisational routines entrenched in institutional structures and agency affect wind power deployment.

As analytical point of departure, the power sector is characterised by its institutional set up and quality, financial situation and institutional agency and development with regard to power sector reforms. The power sector constitution as regards governance structures or institutional set-up is similar in all Indian states and determines both supply as well as demand for alternative electricity generation from wind energy. In order to explore power sector characteristics in the following, studies by the Ministry of Power are drawn upon that assess power sector policies, reforms and the efficiency of respective institutions in the Indian states. On the one hand, one might argue that the better the shape of the power sector in terms of efficiency and financial position, the more scope for wind sector activities as private investors are attracted by a favourable policy setting. On the other hand, advantageous power sector conditions might deflect investments in wind energy as the conventional-power-based economy is well established. Demand for alternative sources of energy is not sufficiently high to stimulate new investments in renewable energy. Given the overall Indian context of chronic power shortages, this dissertation assumes that the better the shape of the power sector and the greater the openness to private investment, the better equipped to accommodate alternative sources of energy capacity addition.

Table 17: Power Sector Rating

	PSR	Wind Utilisation Rate %
Karnataka	4 of 29	15.3
Maharashtra	12 of 29	48.1
Rajasthan	11 of 29	10
Andhra Pradesh	1 of 29	1.5
Madhya Pradesh	19 of 29	3.4
Gujarat	2 of 29	13
West Bengal	8 of 29	0.2
Tamil Nadu	5 of 29	81.5
Kerala	16 of 29	0.2

Source: Ministry of Power 2006

A comparative look at the Indian wind states (Table 17) suggests that power sector characteristics per se are ambiguous and insufficient explanatory factors. While Maharashtra with a favourable power sector and few utility losses experiences great wind sector development, states with high financial losses like Rajasthan also benefit from wind power investment. Additionally, the power sector rating per se appears of little explanatory value since good ratings for West Bengal and particularly for Andhra Pradesh contradict the proposition due to lagging wind sector development. Power sector characteristics in states like Andhra Pradesh that embarked fastest upon power sector reforms and privatisation and experience strong regulation and energy capacity addition, do not satisfactorily explain the state of wind power development.

Power sector characteristics in Tamil Nadu in contrast to Kerala support the assumption that well-established institutional structures promote wind energy deployment (Table 17). In general, both states are characterised by a similar power sector system. Both states have not undertaken significant steps towards power sector reform as regards the unbundling of state electricity boards into the functional units of generation, distribution and transmission. Kerala is somewhat even more advanced in the structural change process than Tamil Nadu. Not only has the electricity utility KSEB ventured towards reform by establishing profit centres along functional divides, but also does the utility incur fewer losses than the TNEB. Other similarities are noticeable such as the generally limited financial support by state government and the lack of reforms. Striking points of difference exist as regards the regulatory process that is lagging behind in Kerala as well as the paltry policies and efforts undertaking in terms of capacity addition from alternative energy sources.

The last feature characterising the power sector system and governance relates to policy issues that put a high demand on the power sector and hence shape respective structures. Studies on the Indian power sector often remark that state power sectors are in dismal shapes due to the constraints to offer subsidised or free services to agricultural or rural consumers. This affects power sector structures as it entirely distorts tariff structures and throws power sector institutions into financially and politically weak positions. The existence of political obligations to provide agricultural subsidies might be associated with the propensity of the power sector not to be open to accommodate or support alternative sources of energy such as wind. This is related to the fact that additional generation from alternative resources demand different grid qualities and management, i.e. additional investment. Among the wind power states in India, only Tamil Nadu, Andhra Pradesh and Maharashtra suffer from the additional policy demands of the agricultural sector. Yet, while Andhra Pradesh reflects shortcomings in terms of wind sector development, the other two states experience no difficulties to accommodate both issues.

In conclusion, the state of the power sector in terms of structure, financial shape and overall efficiency does not provide a satisfactory explanation for spatial variations in wind energy deployment. However, this preliminary inductive analysis indicates some aspects that are worth a closer look. These refer to the functioning of public organisations in the power sector with regard to how regulatory processes are shaped, how support is extended to energy investors and how power sector entities institutionally fit with involved organisations.

(iii) Policy Outcome – Policies and Regulations

Zooming into the processes of power sector governance considers the policies and regulations for the renewable energy and wind sector. The analytical framework proposes that the more favourable the policy outcomes, the more wind energy deployment is expected to take off. New institutionalist perspectives argue that this is due to the creation of path dependencies and organisational routines that are expressed in supportive policy outcomes.

Some of the features repetitively mentioned in empirical Indian wind sector studies remark upon the differences in policies and regulations between the federal states (Table 18). The lack of explanatory significance with regard to the effect on wind power deployment is given justice to by taking a comparative closer look. The regulatory design does not allow for conclusions on wind energy deployment since states with high wind energy deployment like Maharashtra display favourable tariff structure as do states with less wind sector activities such as Gujarat (Table 18). Both states elaborate very detailed policy specifications on renewable and wind energy. Strikingly, particularly Gujarat with the longest path dependent history of wind technology and policy development has not realised its wind energy potential to nearly as satisfactory level as Rajasthan that came out with general renewable energy policies in 2006.

Table 18: Regulatory Policies

	Andhra Pradesh	Karna-taka	West Bengal	Madhya Pradesh	Ma-harashtra	Rajast-han	Tamil Nadu	Gujarat	Kerala
Tariff in Rs	3.5	3.4	Case (Cap 4.0)	4.03-3.36	3.5	3.71 / 3.67	2.9	3.37	3.14
Annual Escala-tion				0.17 Rs four years	0.15				
Period	10 yrs	10 yrs	n.n.	n.n.	13 yrs	n.n.	n.n.	20 yrs	20 yrs
Wheel-ing	5%	5%	0.3 Rs	2%	2%	10%	5%	4%	5%
Banking	No	No	6 months	Yes	Yes	3 months	5% 12 months	Monthly settled	9 months
RPS	N.N.	N.N.	N.N.	N.N.	N.N.	3-6%	N.N.	N.N.	N.N.
Wind Utilisa-tion %	1.5	15.3	0.2	3.4	48.1	10	81.5	13	0.2

Source: MNRE and SERC (November 2008)

Empirical observations contradict the proposition that more favourable policy outcomes translate into higher wind power deployment activities[32]. Policy outcomes particularly in the case of Kerala and Tamil Nadu do not explain why Kerala despite a more favourable tariff regime has seen so little wind energy deployment (Table 19):

Table 19: Regulatory Policies in Tamil Nadu and Kerala

Regulatory Policies	**Kerala (24.06.2006)**	**Tamil Nadu (18.05.2006)**
Tariff (average annual)	Rs 3.14	Rs 2.76 for existing and Rs 2.90 for new units
Annual escalation	3.6% for 25 years	4.5% for 10 years
Tariff Period	20 years	Review after 3 years

[32] As mentioned previously, the empirical analyses focus on the third phase of wind power deployment and specifically the year 2007 in order to eliminate temporal variations as explanation for spatial variations. Chapter V will engage in process tracing and give justice to the variations in policies and regulations with the impacts of instability on deployment activities.

Wheeling	5% charges	5% charges
Banking	12 months	12 months
Renewable Portfolio Standards	5% for all renewable sources; out of which 2% from wind	10% for all renewable energy sources

Sources: Tariff Orders on Renewable Energy (SERCs)

The empirical case of Kerala demonstrates that policy outcomes do not one-directionally relate to path dependent developments or organisational routines that promote wind power. Even though policy conditions in Kerala have until recently not been as favourable for attracting private investment to the wind sector, policy specifications on wind power were amended and a regulatory order was stipulated in 2006 and appear favourable for attracting wind sector investments. The case of Tamil Nadu disproves the proposition of the analytical framework since the design of policy outcomes does not seem to stimulate wind power deployment. However, from the late 1980s onwards, policy conditions in Tamil Nadu were oriented towards promoting private sector investment in wind energy. Despite government changes, the energy ministry ensured the stability and continuity of wind energy policy. Consequently, even though policy and regulatory designs does not feature overt support, policy outcomes have so far been perceived as favourable (INT 5; INT 6).
This analysis implies that policy outcomes per se do not sufficiently explain successful wind power deployment. However, the history of these policy outcomes varies due to the different path dependent developments with regard to the policy outcomes exist. While Tamil Nadu benefits from continuous, stable and supportive policy conditions, Kerala has just recently derived at the favourable renewable energy governance framework through an instable and inconsistent policy process. Additional differences emerge in terms of how the policies stipulated are implemented, i.e. the organisational agency. Not policies per se but what and how actors make of governance framework conditions seem to matter for wind energy deployment. All in all, policy conditions characterising the institutional setting and procedures of power sector governance are not sufficient for explaining wind energy deployment in India. Causal mechanisms suggested by the analytical framework give superficial explanatory insights since e.g. path dependent developments just indirectly relate to the current effects of policy outcomes on wind power deployment. Issues that require more attention are the agency and dynamics of organisations in the renewable energy governance context. This concerns the ways in which policies and regulations are implemented and the procedures are followed to put them into practice.

In summary, the first inductive exploration gives rise to the conclusion that policy conditions in terms of polity, institutional output and outcome do not provide complete accounts for why wind energy deployment spatially varies. This approach to examine policy conditions for renewable energy deployment fails to address the policy impact dimension with regard to organisational behaviour. Preliminary observations pointed towards the importance of organisational behaviour and underlying organisational attributes characterising the stakeholders involved in wind energy deployment which requires further in-depth analysis.

III.2.3 Economic Conditions for Wind Energy Deployment

Since empirical studies emphasise the availability of and accessibility to basic production factors as prerequisites to economic activities such as renewable energy deployment this section will explore economic conditions. Economic geography literature looks specifically at the determinants of investment and location decisions according to the opportunity structures of demand and supply factors. The analytical framework proposes that renewable energy deployment is explained by rational decision-making of economic actors according to the assessment of industry-specific location factors. Location factors relevant for renewable energy investment comprise infrastructural conditions, industry structure, and investment climate. Additionally, wind sector-relevant factors such as electricity infrastructure, energy mix and power shortages will be examined. Comparing the economic conditions in the respective Indian wind states, this dissertation derives at the conclusion that the research puzzle is not satisfactorily explained by the economic opportunity structures.

a) Input Factor – Infrastructural Conditions

First of all, empirical studies on renewable energy deployment in developing countries argue that in many cases respective activities are impeded by the absence of basic input factors such as infrastructure and transport networks. Infrastructure networks are important for the transport of technical equipment as well as the workforce to and from markets. According to economic geography this impacts on the costs of doing business since an extensive and well-maintained infrastructure system is prerequisite to any economic activity.

This dissertation derives at the conclusion that infrastructure-related input factors insufficiently explain wind power deployment. The limited extent to which infrastructure conditions play a role in determining wind sector development is reflected in the comparative analysis of the Indian states. In the case of West Bengal, excellent road, rail and port systems provide national and international connectivity. However, obviously this is not related to wind sector development. In Tamil Nadu, an elaborate infrastructure system with roads for transport of heavy machinery and staff is already available or easy to construct which provides conducive conditions for wind farm installation. Another advantage is the existence of sophisticated maritime transport facilities such as Chennai port for importing heavy components and intra-state transportation (Jagadeesh 2000: 160). Generally speaking, Kerala has the reputation for being ahead in infrastructure development. Yet, for the specific purpose of wind energy installations the general existence of well established networks seems insufficient. In this sense, infrastructure conditions in terms of transport networks are problematic as concerns land access for construction purposes. Since the sites identified for wind farm installation are located distantly in hilly terrain, road networks often either do not exist or are insufficient for the transport of heavy construction material. In contrast to Tamil Nadu that profits from flat terrain and established transport networks, road construction or extension is required in Kerala. This implies another cost factor to be born by wind sector entrepreneurs. Apart from construction difficulties, obstacles arise out of specific legislations and contention with the local population. Since land ownership is dispersed and objections to environmental interventions are high, the acquisition of the required terrain encounters difficulties. However, the overall rail and road networks are well established in Kerala even though the port system is not of as great international importance as in the case of Chennai in Tamil Nadu.

Importing heavy wind farm components either from domestic manufacturing units or from international technology hubs hence constitutes another cost factor.
The Kerala case points out that not necessarily infrastructure density, availability and quality but issues of access matter. This implies the importance of agency with regard to what actors make of a given setting. Consequently, infrastructure conditions per se do not provide a satisfactory explanatory account for wind sector development in India.

b) Input Factor – Industrial Structure

The second aspect reflects on the significance of industrialisation and the structure and composition of the local economy as location factors considered in investment decisions. Firstly, industry composition and the general industrialisation pattern imply the availability of input factors to wind power deployment activities. Higher industrial shares of economic activities mean that certain basic production factors and conditions exist. Secondly, industry structure translates into the potential energy demand and provides opportunities for additional wind power capacity. Higher shares of power-intensive industries indicate existing demand for additional investments in the energy sector. The analytical framework proposes that higher degrees of industrial activity explain wind sector development due to immense power demand as well as due to the supply of basic production factors.

Looking at states like Maharashtra and Gujarat that are highly industrialised confirms the proposition that the industrial structure positively relates to wind deployment. Dominated by power-intensive industries like electronics, automobiles, textile and chemicals, electricity demand is met by alternative sources of energy such as wind. However, in the cases of Rajasthan and Andhra Pradesh, the opposite is true which means that industrial development does not explain wind sector developments. Andhra Pradesh's power-intensive industries such as IT, textiles contribute 22 % to the SDP (Directorate of Economic & Statistics; Andhra Pradesh State Government 2004); still, hardly any wind sector activity is stimulated. Rajasthan with low industrial production and productivity nevertheless ventures into wind energy deployment.
Tamil Nadu was one of the first Indian states to experience industrial development under British Colonialism and even today accounts for the highest degree of industrialisation in India. Textile, steel, foundries, cement, automobiles, electronics and fireworks are the priority sectors in which most economic activity takes place. The industrial sector contributes 34% to economic activity in the state trumped only by the tertiary sector with 45%. As mentioned above, the economic structure of the secondary sector is characterised by highly power-consuming branches (Figure VII).

In addition to this fact that already implies high potential demand for power, industrial electricity tariffs are very high, which renders industrial production very costly. As grid supply is expensive, erratic and of low quality, a strong interest in potential energy sources for captive power consumption such as wind emerged. Huge profits made in the textile and cement industries created to an industry sector demand for lucrative investment opportunities that meanwhile benefit the own industry. In this context, the wind business emerged as profitable investment source since respective activities were rewarded by tax concessions and holidays (Jagadeesh 200: 157). This was initially pushed by government that wished to stimulate wind power development and to combat energy shortages[33].

Figure VII: Sectoral Shares of SDP Growth

Economic Activity	1993-94	2001-02
Agriculture & Allied, Forestry, Fishing	24.82	16.65
Manufacturing and Construction	33.72	34.04
Trade, Transport, etc.	20.40	23.83
Banking Insurance, etc.	11.01	13.63
Public Administration & Services	10.05	11.85

Source: Tamil Nadu State Planning Commission (2006)

This analytical narrative confirms that the industrial structure has an impact on the demand for power, investment potential and capacity addition from alternative sources which pushes wind energy deployment. Industrialisation patterns hence explain wind sector development in Tamil Nadu.

Industrialisation in Kerala is lagging behind states like Tamil Nadu. Kerala has traditionally focused on the primary sector and service-oriented industries with 63 % of economic activities amounting to the tertiary sector and 19.8 % to industrial activities (Figure VIII).

Figure VIII: Sectoral GSDP Distribution Kerala

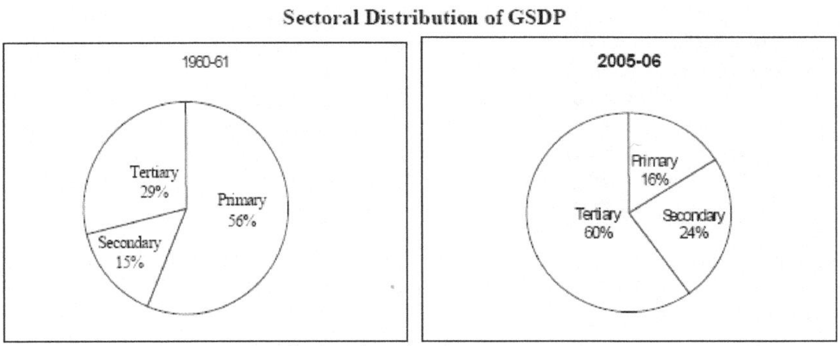

The main industries in Kerala operate in branches such as electronics, textile, agricultural manufacturing, and engineering and information technology. At first sight, this industrial structure features the absence of heavy power consuming industries that hence did not create the crucial demand for power as is the case of Tamil Nadu (Rajsekhar et al 1999). Due to the absence of industrial power consumers such as textile industries or cement, the need for captive power generation utilising wind energy resources did not exist. However, three developments caution against the rushed conclusion that the absence of industrial structure-related demand conditions doomed wind sector development in Kerala. Firstly, high living standards and the rise

[33] At later stages central government rowed back on several support systems, reduced tax concessions, introduced a minimum alternate tax and withdrew capital subsidies. This led to a slump in these industries in the late 1990s and wind energy investments also lost speed (Jagadeesh 2000: 161).

of the service economy in Kerala have and will contribute to an exponential growth in domestic and commercial energy consumption. Kerala has already experienced shortages in power supply and will start looking into alternative means of capacity addition such as from wind energy. Secondly, Kerala's industrial structure also experiences changes. With the full effect of the proactive industrial and investment policies stipulated in the last couple of years, growth of industrial activity is already palpable. With an increase in domestic electricity consumption and the emergence of industrial sectors such as IT, buildings and hotels, energy demand has been outgrowing installed capacities. A third factor is the pronounced interest in lucrative investment opportunities in this local context claimed by the Keralite business community. In conclusion, despite a different industrial structure and set up, the demand for power stemming from economic activities is also prevalent in Kerala.
All in all, the puzzle why states with huge industrial potential and progressive industrial policies such as Andhra Pradesh, Madhya Pradesh and West Bengal have not successfully deployed wind power still remains unsolved. This cautions against the rushed assumption that industrialisation alone accounts for wind energy deployment in Indian states.

Another aspect that speaks against drawing these causal relations between industrial activity and wind energy deployment concerns the issue of agriculture. The primary sector per se is not very energy-intensive. Considering Indian circumstances, agriculture constitutes an invisible yet significant power consumer which creates the demand for investment in alternative energy sources such as wind energy. This is due to the fact that electricity is either heavily subsidised or given for free to farmers and agricultural activities. Taking its origin in the Green Revolution, the system of subsidised agricultural power has translated more or less into perceived common law and constitutes a highly controversial and politicised issue. Consequently, one might assume that counter-intuitively to industry-structure arguments states characterised by high agricultural activities will deploy more wind energy. However, Indian realities do not support this assumption. While some states with high agricultural activities experience high wind power generation such as Karnataka, others such as Madhya Pradesh and Andhra Pradesh see no wind power activity.

In summary, economic conditions inherent in the industrial structure and patterns of economic activity do not satisfactorily account for spatial variations in wind energy deployment in India.

c) Input Factor – Power Demand

The third aspect narrows the focus to energy market-related barriers. Economic geography literature argues that location decisions and respective investment activities are based on assessing cost minimisation potentials in a certain industrial setting. According to the previous analysis, the industrial context of economic activities is not sufficient for fully understanding why wind energy deployment turns out differently in these states. This suggests a closer look at the specificities of the power sector alongside three factors of interests: The first one concerns the concrete demand for power that is reflected in the existing power shortages. The higher the power shortages the greater the demand for additional power generated from wind resources. Secondly, the existence of alternative sources of power generation in the sense of competing fuels is considered. The more limited alternative fuel options, the greater the window of opportunity for wind resources to be harnessed for power generation.

Lastly, the transmission and distribution losses (T&D losses) characterise existing electricity systems. The status of the electricity system implies the capability to integrate wind power into the existing energy mix. This last point is the most ambiguous one as on the one hand, insufficient electricity infrastructure deters investments due to the hidden costs anticipated. On the other hand, this might provide a window of opportunity for investors to set up own, more reliable facilities or to obtain state subsidies for such undertakings. The general intuitive experience suggests that faulty electricity system rather deter wind power investors that combat with many more investment risks. The following section will discuss the extent to which the demand for wind power generation is indeed triggered by energy systems characterised by existing power shortages, the absence of competing fuels and low T&D losses.

Comparing wind energy development in India; at first sight, the local specificities of the respective power sectors seem to explain differences in investment activities particularly in the Tamil Nadu and the Kerala cases. Yet, more detailed observations will call into question rushed conclusions on causal relations between favourable power sector-related economic conditions and wind energy deployment:
The fact that power shortages per se do not account for differences in wind energy deployment rates is true for several local settings (Table 20). Both Maharashtra (-21.4 %) and Madhya Pradesh (-17.8 %) are extremely power hungry. While the Maharahstra is successful, Madhya Pradesh lags behind wind power deployment.
As regards power systems, Maharashtra and Madhya Pradesh also provide the example that despite similar T&D losses (-16.69 %) wind energy deployment rates differ (Table 20). Very little wind sector development takes place regardless the very high T&D losses in Andhra Pradesh (-28 %) or the comparatively low ones in West Bengal (-20.71 %). Power system structures concerning the ability to absorb additional energy generated play a negligible role for wind power investment.
Nearly all of the Indian wind states avail of competing fuels as alternatives to electricity generation in their power mix. While in some states, these may be renewable sources such as hydro power in Karnataka and Andhra Pradesh, conventional power fuels such as coal, gas and oil prevail in Gujarat. The availability of alternative sources for power generation per se does not have much explanatory value for spatially varying wind power deployment. The observation that in all of the states lagging behind wind sector development, the share of renewables in the energy mix is relatively moderate suggests the importance of certain path dependent developments. One might assume that lower levels of experience with alternative energy sources enhances the scepticism towards wind energy and renders this option less likely. However, the cases of West Bengal and Maharashtra caution against hasty causal assumptions. Despite similarly high renewable energy shares in power generation, wind energy deployment rates differ.

With regard to the empirical picture in Tamil Nadu, energy market conditions are shaped in favour of attracting private sector investment in wind power. This is most and foremost due to the high demand for additional electricity generation and the financial potential behind one of the big consumer and investor groups - industry. With regard to its power supply position, Tamil Nadu always experienced a high demand for power and was pressured to increase installed capacities. As depicted in Figure IX, peak demand has been steadily increasing and is met only with increases in newly installed capacities.

Figure IX: Power Sector Development in Tamil Nadu

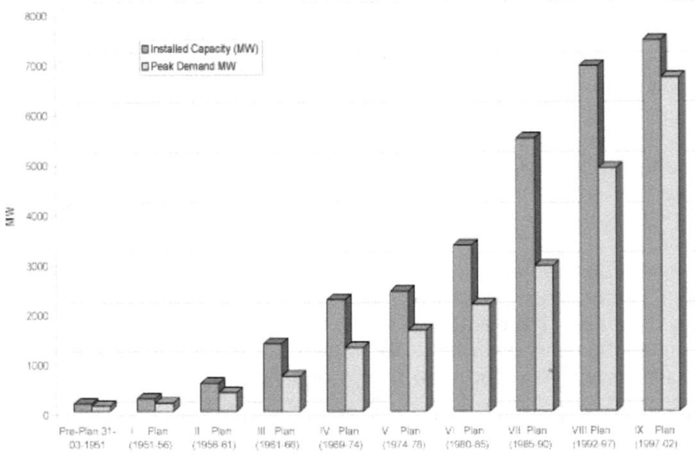

Source: Tamil Nadu State Planning Commission (2006)

Power shortages in Tamil Nadu are about -2.6 % which compared to other states is in the lower range. Alternative fuel sources for power generation such as coal, oil and gas exist but are estimated as insufficient to meet projected power scenarios. The share of renewable energies in the Tamil Nadu energy mix is already impressive (Figure X). Due to the fact that renewable energy already contributes more than 21.9 % to power generation, one might expect that additional wind energy deployment is less likely to be sought. Concerning the status of the energy system; Tamil Nadu experiences relatively modest T&D losses of -16.71 % as compared to overall India statistics. The grid network is characterised as well connected and passes through most wind sites. Since grid conditions are relatively stable, power addition from intermittent resources such as wind causes less problems with regard to grid overload as encountered in other states. Despite available grid infrastructure, issues of evacuation and power planning are occasionally mentioned as obstacles to future wind power expansion (INT 7). This is due to many unplanned and unforeseen wind power installations which contribute to network overload as well as maintenance delays. Market-related demand factors of the Tamil Nadu energy system characterise economic conditions for wind sector investment as advantageous.

Figure X: Tamil Nadu: Power Sector Mix (in M.U.)

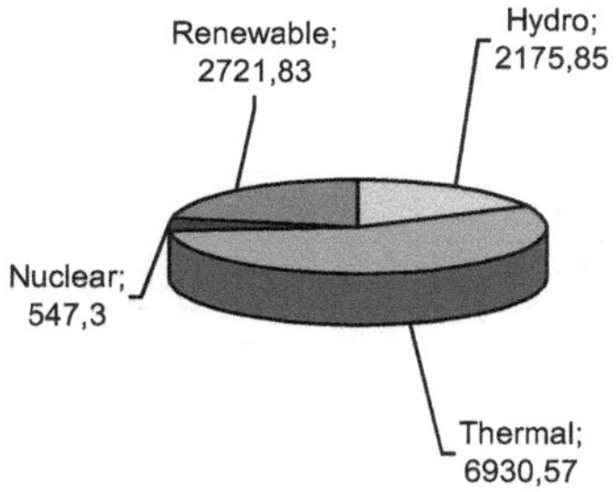

Sources: IndiaStat 2007

With regard to the specific wind-industry context, Tamil Nadu has over the last couple of years become the target state for private actors investing in wind energy. Most of the wind turbine manufacturers and suppliers are located in Tamil Nadu (Jagadeesh 2000: 157) known as the cradle of the Indian wind business model. As a result of these developments, the wind industry today is cartelised and allows for little competition. In conclusion, economic conditions characterising the Tamil Nadu power sector are shaped in favour of attracting wind power investment.

Generally speaking, power sector-related economic conditions in Kerala appear at first sight not very conducive for private sector investment in wind farm installations. This is due to a slightly more relaxed power supply situation and the absence of big industrial consumers that render the state less desperate for additional power. Indeed, until the mid-1990s Kerala did not have many incentives to explore the wind energy potential for electricity capacity addition (Figure XI). Until then, Kerala was not short of power and focussed on harnessing alternative energy resources such as hydro and thermal power. Unlike many other Indian states, Kerala even experienced periods of power surplus during which power exports to other states were feasible. No urgent demand for harnessing all available resources for the sole purpose of additional power generation to meet stark shortages existed that might have focused the attention on wind energy.

Figure XI: Power Sector Development in Kerala

Sl.No.	Year	Available generating Capacity (MW)*	Maximum Demand (MW)		Load Factor %	
			Internal	System	Internal	System
1	81 - 82	1011.50	813.20	921.00	53.60	69.30
2	82 - 83	1011.50	823.20	845.80	54.10	61.70
3	83 - 84	1011.50	812.60	837.40	50.40	52.00
4	84 - 85	1011.50	911.00	939.80	58.30	60.50
5	85 - 86	1271.50	1001.80	1033.80	59.30	61.70
6	86 - 87	1477.00	1004.00	1061.00	53.10	50.00
7	87 - 88	1477.00	1116.80	1075.60	51.72	53.70
8	88 - 89	1477.00	1184.80	1090.20	55.25	60.59
9	89 - 90	1477.00	1270.00	1171.20	55.22	58.22
10	90 - 91	1477.00	1271.60	1147.80	61.06	67.66
11	91 - 92	1477.00	1308.80	1264.60	62.32	64.57
12	92 - 93	1477.00	1403.00	1302.00	60.13	64.75
13	93 - 94	1484.00	1459.60	1235.40	61.46	66.25
14	94 - 95	1503.50	1614.60	1329.80	56.41	62.17
15	95 - 96	1505.50	1651.40	1372.60	55.40	64.20
16	96 - 97	1508.50	1572.00	1235.00	50.86	63.89
17	97 - 98	1775.78	1785.80	1337.00	44.30	60.25
18	98 - 99	2043.10	1612.00	1896.20	53.83	67.40
19	99 - 2000	2350.68	1820.00	2177.00	48.01	62.56
20	2000 - 01	2420.68	1988.00	2316.00	40.01	61.66
21	2001 - 02	2601.62	1940.00	2333.00	42.02	61.54
22	2002-03	2601.62	1868.00	2347.00	33.46	61.69
23	2003-04	2614.22	2088.00	2426.00	24.54	58.83
24	2004-05	2617.22	1860.00	2420.00	38.84	59.99
25	2005-06	2644.22	2124.00	2578.00	40.61	60.30

* Includes capacity of BSES,KPCL,Kuthungal,Maniyar and Kayamkulam also

Source: KSERC (2006)

This picture has changed in the last couple of years and Kerala now experiences power deficits of -2.3 % which are in the same range as in Tamil Nadu. The share of renewable energy in the energy mix is very low taking only 1.2 %. Early decisions to exploit hydrological resources created a certain path dependent bias towards competitive alternative renewable energy resources for power generation. This puts wind power into a less advantageous position than in Tamil Nadu (Figure XII). As monsoon patterns do not coincide with wind power generation, Kerala does not profit from the same seasonability benefits as Tamil Nadu. This means that wind power is not generated at points when demand is high and/or the contribution from other energy sources is low. This is aggravated by the fact that power generation cycles from wind energy fall together with those of hydro energy. Furthermore, the energy system relies upon electricity infrastructure facilities which are less well-established than in Tamil Nadu. On the one hand, T&D losses of 20.3 % are higher which implies lower capacities to absorb additional energy generated from wind power resources. On the other hand, the costs of additional infrastructure development are also supposed to be greater due to the hilly terrain. Evacuation facilities, substations, extension lines for the transmission and distribution of electricity generated from wind energy still need to be constructed in order to connect wind farm installations to the grid. This does not only constitute another cost factor for private developers but is also a point of discontent for public utilities that are in charge of grid management. All in all, demand-related factors inherent in the power sector characteristics and structures do not feature favourably for wind energy deployment in Kerala.

Figure XII: Kerala - Power Sector Mix (in M.U.)

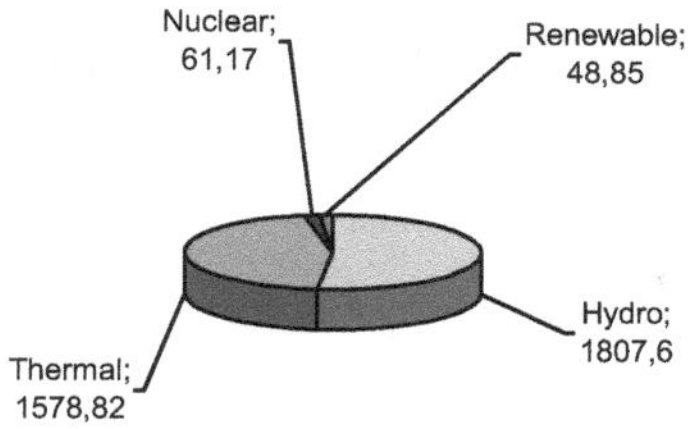

Sources: IndiaStat 2007

Additionally, with regard to the concrete wind industry context, Kerala has a comparatively weak position. Apart from the lack of industries with a high captive demand for power that could be met by wind energy, little private sector interest in wind power investment exists. This is due to the absence of big private investors that would be keen on availing of tax benefits and due to the energy system characterised by less severe power shortages, more fragile electricity system and available competitive fuels. This seems to confirm the proposition that unfavourable demand conditions inherent in the power system setting translates into less wind energy investment. Recent trends caution against overestimating that demand-related economic conditions explain wind power deployment. To date, the first private sector participants in the wind sector start opening offices in Kerala in order to broaden their geographical scope of activities and to venture into a quasi-virgin but potentially profitable area. Furthermore, increasing investment demand and power requirements are noticeable.

Table 20: Power Deficit

	Power Shortages %	Wind Utilisation Rate %
Karnataka	- 2.8	15.3
Maharashtra	- 21.4	48.1
Rajasthan	- 3.5	10
Andhra Pradesh	- 4.5	1.5
Madhya Pradesh	- 17.8	3.4
Gujarat	- 6.9	13
West Bengal	- 1.4	0.2
Tamil Nadu	- 2.6	81.5
Kerala	- 2.3	0.2

Source: India Stat (2007)

In summary, the overall economic conditions characterising the power sector settings of Tamil Nadu and Kerala as regards shortages, energy system and alternative options are roughly similar and per se do not constitute explanatory factors for wind energy deployment. Tamil Nadu as well as Kerala nowadays experience power shortages (Table 20). Both states Tamil Nadu as well as Kerala, suffer from relatively similar T&D losses (Table 21). With regard to alternative sources for power generation crowding out wind power alternative, competing options exist in both states, Tamil Nadu and Kerala.

The analytical framework suggests economic conditions as decisive for investment and location decisions in the wind sector. The first inductive analysis of the Indian case revealed no clear causal relations between economic supply and demand factors with regard to energy markets and wind power deployment. Further refinement is required looking at the wind industry context and the history of alternative energy development. This raises the assumption that path dependencies in the institutional development of the power sector as well as stakeholder interactions matter for constituting favourable economic opportunity structures for wind sector development. This calls for a deeper dive into respective literature and ensuing empirical analysis of relations and networks between stakeholders engaged in wind power deployment.

Table 21: Transmission & Distribution Losses

	T&D loss %	Wind Utilisation Rate %
Karnataka	18.7	15.3
Maharashtra	16.69	48.1
Rajasthan	24.4	10
Andhra Pradesh	28	1.5
Madhya Pradesh	16.69	3.4
Gujarat	20.81	13
West Bengal	20.71	0.2
Tamil Nadu	16.71	81.5
Kerala	20.3	0.2

Source: Central Electricity Authority, Government of India (2002)

III.2.4 External Stimuli for Wind Energy Deployment

This section explores to what extent external conditions stimulate wind power activities by providing finance, technology and R&D. Empirical studies conclude that the lack of technology, R&D and finance constitute barriers to wind energy deployment that go beyond the capacities and capabilities of developing countries. Global environmental governance debates discuss options such as ecological modernisation, technology transfer and new modes of governance as alternatives to meet these existing challenges. External assistance in terms of ODA, traditional external financing devices and new modes of governance are emphasised due to their potential trigger effects on renewable energy investment. The following section will examine the level of FDI and the extent of CDM activities under the aspect of providing capital supply and external support for wind energy deployment.

Taking a comparative look to other Indian wind states speaks against preempt correlations between external stimuli and wind power deployment in Kerala and Tamil Nadu. The renewable energy sector in India is characterised by almost entire private sector participation, which stands in stark contrast to the Indian conventional power sector (Pillai 2006). A demand for foreign capital to be invested in wind power installations exist all across India and general supply conditions are shaped as to welcome foreign private sector participation. Without doubt, foreign investment is

Table 24: Foreign Direct Investment (% share of total approved FDI 1991-2004

	FDI %	Wind Utilisation Rate %
Karnataka	11.16	15.3
Maharashtra	23.63	48.1
Rajasthan	1.4	10
Andhra Pradesh	6.35	1.5
Madhya Pradesh	4.58	3.4
Gujarat	8.71	13
West Bengal	4.31	0.2
Tamil Nadu	11.59	81.5
Kerala	0.72	0.2

Source: Ministry of Commerce and Industry (2002)

attracted to local contexts that are characterised by and considered as favourable investment locations relating to demand and supply constraints. One might hence expect wind energy deployment to take place where FDI is high and where CDM business thrives (Table 24).

At first sight, this proposition is confirmed looking at Maharashtra and Karnataka that as FDI leaders experience high wind energy deployment. Meanwhile, Madhya Pradesh and West Bengal do not see much FDI, and wind sector development does not take place either. However, this picture becomes blurred in the cases of Gujarat and Rajasthan. These states reveal extremely different FDI rates but reflect similar wind energy deployment activities. Gujarat for example is a famous FDI investment hub but plays a small role in the Indian wind energy market. A similar assessment is made with regard to the distribution of CDM activities in India (Figure XIV).

Figure XIV: CDM Projcts and CDM Wind Projects

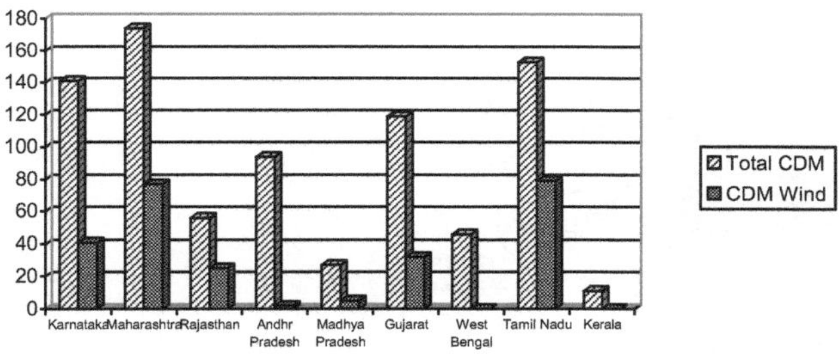

Source: UNEP Risoe (November 2008)

Even though states like Maharashtra see many CDM activities in addition to wind energy deployment, other states like Andhra Pradesh experience CDM investment but hardly any wind sector activity. These empirical observations call into question the extent to which external conditions in terms of FDI flows and CDM activities support and stimulate wind sector activities.

The case of Tamil Nadu confirms the proposition that external stimuli contribute to successful wind power deployment. From a multilevel governance perspective, global wind market developments have right from the beginning stimulated wind sector development in Tamil Nadu. To date, Tamil Nadu draws on close ties to international technology providers and assistance in the area of technical expertise and capacity development. With an Indian wind business model and own expertise and capacities, the Indian wind business plays an active role in regional and also global wind markets. The proneness of Tamil Nadu to become a cradle for wind sector development is not only explained by the excellent wind conditions but also by openness to private and foreign investment reflected in the receipt of FDI (Table 22).

Table 22: FDI in Tamil Nadu

State/FDI	Total approved FDI (Rs mio) 1991-2004	Share FDI 1991-1998	Share FDI 1999-2004
Tamil Nadu	250,720	10.98%	12.41%

Source: Ministry of Commerce & Industry (2002)

Given the fact that Tamil Nadu is the Indian state with the highest level of FDI it does not seem much of a surprise that external financial opportunities also stimulated investment in profitable and innovative niches such as wind power.
As concern the transfer of finance and technology in the context of new modes of governance such as the CDM[34] a similar picture emerges. Tamil Nadu is one of the states most active in tapping the potential provided by the CDM. In comparison to other Indian states, Tamil Nadu has the highest share of CDM projects of which 50% take place in the wind sector (Figure XIV):

Figure XIV: CDM in Tamil Nadu

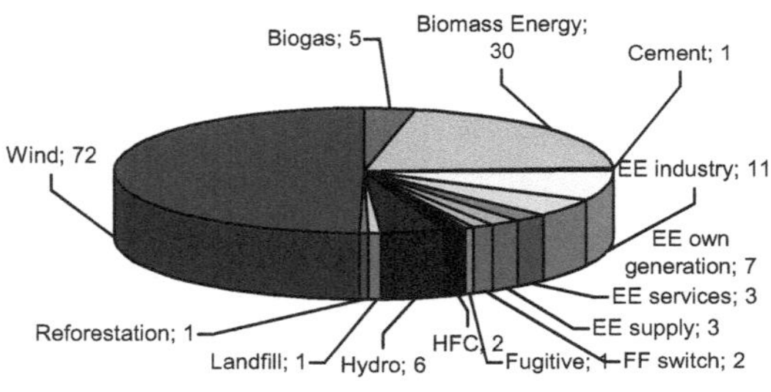

Source: UNEP Risoe (September 2008)

All in all, the observation that Tamil Nadu receives both high levels of FDI and CDM activities gives justice to the proposition that external factors stimulate wind power investment. The availability of financial resources and the opening of channels to international knowledge and technology create multiplicator effects that promote developments in new economic niche sectors such as wind power.

The case of Kerala at first sight seems to tell the opposite story since the absence of external factors potentially stimulating investment activities in wind power confirms the proposed causal relation to unsuccessful wind sector development:
Departing from a multilevel governance perspective, external stimuli for wind energy deployment in Kerala have not taken place and/or found hold. Unlike Tamil Nadu, Kerala has not linked up to international developments in global wind markets. Networks with international knowledge and technology sources are still missing. Early

[34] This dissertation takes CDM as external stimulus and potential explanatory factor even though it consciously acknowledges that most of particularly wind sector projects are unilateral activities. This implies that not foreign investment directly flows into wind power. This dissertation argues that, nevertheless, the CDM provides external stimuli through the creation of new global business contacts, green image considerations, and reaping additional even if negligible financial fruits.

guidelines for public and private sector participation in wind energy development issued from federal agencies such as the MNRE have not been adopted by state government. State government to date has not put additional efforts into implementing a favourable framework for attracting private investors and tapping these opportunities. Even though fiscal and financial incentives for investments in the wind sector issued by the Government of India also accrue to Kerala, neither domestic nor foreign entrepreneurs have ventured into the wind business.
With regard to external finance, factual foreign direct investment (FDI) is still quite low (Table 23). This situation has changed and investment interest from Keralite citizens working abroad exists. Yet, the de facto impact of external finance is negligible and does not translate into multiplication effects of financial stimuli for new and innovative activities such as wind power deployment.

Table 23: FDI in Kerala

State/FDI	**Total approved FDI (Rs mio) 1991-2004**	**Share FDI 1991-1998**	**Share FDI 1999-2004**
Kerala	15,520	0.48%	1.03%

Source: Ministry of Commerce & Industry (2002)

New modes of governance such as the CDM as a matter of fact intend to fill this gap between lacking conventional external stimuli and potential local demand and opportunities. One might assume that given the high potential for wind power deployment states like Kerala constitute excellent opportunities for new modes of governance to mitigate shortcomings of traditional finance instruments such as FDI. However, the Kerala case surprisingly proves true exactly the opposite: Additional potential revenues promised by the CDM still have to materialise and to date have not been able to attract private sector interest in wind energy in Kerala. Even though eleven CDM projects take place in Kerala, none of them is related to investment in the wind sector. Opportunities and developments in the global wind sector and institutionalised in the Kyoto mechanisms did not impact on respective private sector investment. All in all, the analytical assessment of the Kerala case also gives justice to the proposition relating the absence of external conditions to the lack of wind power deployment.

In summary, at first sight one might assume that external conditions such as FDI and CDM explain spatial variations in wind sector development in Tamil Nadu and Kerala. However, four main arguments caution against rushed assumptions:
Firstly, Kerala disposes of alternative sources of foreign investment in terms of remittances. This point is not to be underestimated as remittances constitute roughly 20%, and as confirmed in interviews are available and ready to be invested in the wind sector as profitable capital investment (INT 8). In this sense, also Kerala obtains external financial support that might have stimulated investment in wind energy – even though not officially accounted for.
Secondly, assuming that external finance is attracted by seemingly profitable niches such as wind power is rushed since related government incentives promised to investors are in many cases not available to foreign parties.
Thirdly, the factual level of FDI as determinant for wind sector investment is debatable since firstly, the FDI figure says nothing about the direction of investment. This means that external finance might not actually relate to investment activities in sectors such as wind energy. The only relation between FDI and wind power deployment would emerge taking into consideration implications on general business climate and

the investment context. In addition, India has so far not become a major FDI destination which implies that foreign money is not likely to stimulate or attract wind sector activities.
The fourth and last point relates to the discussion that FDI and CDM levels are positively associated with increased wind activity. This encounters the famous chicken-and-egg question of whether external conditions cause wind energy deployment, or wind sector activities attract external stimuli. This refers to preconceptions that external support only takes place where respective activities happen anyway, and calls into question the direct causal relation between external stimuli and wind energy deployment. In order to give justice to this argument, the following brief analysis will examine the extent to which a favourable investment climate attracts external stimuli.
External conditions are not explanatory factors but factors explained and conditional upon specific local settings. Usually, foreign money and activities take place given a favourable investment and business climate. Advantageous assessments of investment climate conditions are assumed to benefit investments in wind energy. The reason for considering the overall perceived business climate is the fact that renewable energy technologies are a capital-intensive business. Due to the specific risk profile and the incalculable political and regulatory uncertainties, domestic financial institutions have so far been hesitant to lend money to wind sector entrepreneurs. The only option apart from drawing on own accounts is to obtain finances through foreign participation or any other foreign financial source. Considering the general investment climate casts doubt on both the impact of external stimuli on wind energy investment as well as on the existence of wind power deployment as condition for external support[35].
With regard to the international investment climate, Kerala until recently suffered from a bad reputation of being politically averse to private sector investment and economic liberalisation. Due to the changes of the polity regime between Communist and Congress party leadership, the business environment is not perceived as particularly attractive to private investors. However, this constitutes a predominantly attitudinal barrier given the de facto existence of investor-friendly policies (Subrahamanian 2000). This is also reflected in the ranking of Economic Freedom indicators between Indian states according to which Tamil Nadu (rank 1 out of 20) and Kerala (rank 3 out of 20) are characterised by roughly similar and favourable conditions of doing business. Economic freedom is expressed in the openness to foreign investment and progressive industrial and investment policies. Comparisons to other Indian states also confirm that investment climate constitutes only one factor in the overall picture. In states in which wind energy has taken off such as Maharashtra and Karnataka, economic freedom ratings are not as favourable as in states lagging behind wind sector activities such as Andhra Pradesh and Madhya Pradesh. This implies that investment ranking and business climate perceptions in general say little about the specific investment-related structures and dynamics in the wind sector.

In summary, empirical assessments of linkages between external conditions and wind energy deployment reveal an ambiguous picture. While at first sight, causalities supporting the proposition emerge in cases of Tamil Nadu and Kerala, this conclusion is rejected as rushed in comparison to other Indian states. External conditions per se are not considered explanatory factors for spatially varying wind power de-

---

[35] The discussion of mutually constitutive relations between external stimuli and local conditions will be subject of Chapter VI.

ployment despite undoubtedly existing interrelations with the renewable energy governance.

III.3 Synopsis of Main Findings and Implications

This chapter conducted a first inductive analytical examination of wind power deployment in India guided by an analytical framework and looking specifically at Tamil Nadu and Kerala. The four main conclusions are (Figure XV):
Firstly, with regard to developmental context conditions, factors such as geography, population and socio-economic characteristics do not satisfactorily explain differences in wind power deployment. Both Tamil Nadu and Kerala are endowed with abundant natural resources, and population characteristics constituting the socio-economic setting are relatively similar. Hence the impact of developmental conditions proposed as pre-requisite to economic growth by development economics does not sufficiently account for wind energy deployment.
Secondly, policy conditions were examined in relation to wind sector activities. The first preliminary assessment is that neither the policy regime, nor the institutional power sector set-up nor policies and regulations explain differences in wind power deployment. However, not the de facto institutional structure but the way in which organisations function in implementing and managing processes related to wind governance seem to matter. This draws attention to the characteristics of governance actors, i.e. so-called "actors' attributes", which require more detailed analytical exploration and further literature debate.
Thirdly, the majority of empirical studies point towards market-related barriers to renewable energy deployment in developing countries that are reflected in economic geography debates. Economic factors constituting the demand for renewable energy and shaping opportunities for investment activities do not sufficiently explain spatially varying wind power deployment. Despite the similar availability of important input factors such as energy market demand and supply in Tamil Nadu and in Kerala, deployment rates differ. The preliminary inductive analysis pointed to the fact that not necessarily the availability but the accessibility and utilisation of input factors play a role for explaining how and why states harness their wind power potential. This draws attention to questions how governance takes place in the context of relations and interactions between governance actors. "Actors' relations" in stakeholder networks seem to determine how production factors and potential resources can be and are used.
Fourthly, assessments of external conditions for wind energy deployment provide no satisfactory account for wind power deployment. External conditions per se do not constitute explanatory factors for spatially varying deployment rates. However, the empirical and theory-based linkages and effects between external interventions and wind power deployment call for a more detailed examination.

These observations bear several implications that shape the further research procedure of this dissertation in order to gain a more comprehensive understanding of the explanatory factors and dynamics accounting for the spatial variations in wind power deployment:
As a result of this first exploration, two aspects referring to actors' attributes and relations emerge as potentially relevant explanatory factors for spatial variations in wind power deployment in Tamil Nadu and Kerala.

Figure XV: Analytical Framework – Empirical Findings

	Tamil Nadu	Kerala
Developmental Context Conditions	-Geography: favourable – competition for land -Socio-Economy: favourable	-Geography: favourable – hilly terrain Socio-Economy: very favourable
	No explanatory factor – only as concerns access to resources allowed/facilitated by public authorities	
Policy Conditions	-Polity: stable, somewhat corrupt -Output: favourable sector rating -Outcome: less conducive regulations	- Polity: changing, integer -Output: less favourable sector rating -Outcome: conducive regulations
	Emerging explanatory aspects: implementation issues, practices and procedures by public authorities, attitudes and interests, accessibility of institutions and coordination	
Economic Conditions	-Infrastructure: well established -Industrial structure: high industry share -Power demand: existing	-Infrastructure: dense – under construction -Industrial structure: high tertiary share -Power demand: establishing
	Emerging explanatory aspects: relations between interested parties, strategic interactions	
External Conditions	-FDI: high share -CDM: high share -Business climate: favourable	-FDI: remittances -CDM: hardly -Business climate: favourable policies versus ambiguous reputation
	Related explanatory aspects: proactive and active seizing of opportunities available	

As summarised in Figure XV , the essence of economic geography analysis is that firstly, what business actors make of given circumstances and opportunity structures, and secondly the way in which interactions with the political setting happen are worth further exploration. Interview statements confirm the relevance of the nature of relationships between public institutions and private actors, the positioning of public vis-à-vis other public or private institutions and power relations between stakeholders. Issues such as interaction density, trust, mutual understanding and coordination particularly vertically amongst private or public but also horizontally between private and public stakeholders were emphasised as crucial for wind energy deployment. In summary, actors' relations that consider private and public stakeholders in interactions and embedded in the local context emerge as potential explanatory variable for wind energy deployment.

Secondly, with regard to policy conditions, research interviews and publications emphasised the role of public institutions and non-governmental actors that have a stake in the renewable energy sector. While some stress the importance of pro-activeness and leadership in regulatory and government organisations others direct attention to the general active positioning and political clout of respective institutions. Once public institutions have a reputation of being in favour of wind energy, are assigned with staff that is known to work efficiently and utilise existing capacities for renewable energy activities, wind energy has a better standing vis-à-vis conventional fuels. Consequently, these actors' attributes that characterise private and public stakeholders as governance actors emerge as one significant variable for understanding wind energy deployment.

In conclusion, since neither markets nor hierarchies satisfactorily account for wind energy deployment in India a governance perspective is required in order to explore who governs and how governance takes place. In order to apply this governance perspective to the empirical realities of spatially varying wind power deployment, stakeholder network analysis provides an analytical approach. A network perspective acknowledges the importance of both aspects and similarly to the governance approach circumvents a market/hierarchy dichotomy.

The following two chapters will take up these preliminary research findings and embark upon the second deductive research stage that aims to prove the explanatory relevance of characteristics and constitution of governance arrangements, i.e. stakeholder networks.

Concretely, the next chapter will go back to literature and theory in order to examine characteristics and mechanisms underlying the factors that emerged as potential explanatory variables. For this purpose, organisation sociology will be discussed in order to account for issues identified as actors' attributes, while new economic sociology debates will tackle actors' relations. A governance framework will structure the discussion who governs and how governs takes place in the context of wind power deployment. The network concept will be applied to analyse and to conceptually integrate these aspects. As an interdisciplinary approach, this is considered most suitable for understanding the multi-faceted dynamics of renewable energy deployment in a non-OECD context. Consequently, the next procedural stage of this dissertation will embark upon a deductive approach to explore conditions for effective renewable energy governance.

CHAPTER IV

Explaining Renewable Energy Governance in the Framework of Sociological Perspectives

CHAPTER IV
Explaining Renewable Energy Governance in the Framework of Sociological Perspectives

Spatial variations of *"turning wind into power"* are not simply explained by certain prerequisites such as developmental, policy, economic and external context conditions. This chapter will elaborate a deductive approach to explore who governs and how the effective governance of renewable energy takes place with focus on the characteristics and dynamics of local stakeholder networks.
The inductive theory-guided empirical analyses of the previous Chapter III fell short of satisfactorily explaining spatial variations in wind energy deployment but pointed towards the importance of governance arrangements. As additional conclusion, Kerala and Tamil Nadu turned out the most striking instances of the Indian wind puzzle as spatial differences cannot be accounted for by other factors. The second research stage will focus solely on Tamil Nadu and Kerala for the purpose of in-depth explorations of local stakeholder networks.

This chapter embarks upon the second research stage and takes a deductive stance on the basis of previous research findings that highlighted the explanatory potential of stakeholder characteristics and relations in local governance arrangements. Relevant theory and literature will be discussed in the light of the research question that seeks to explore the role of stakeholder networks in renewable energy governance. The purpose is to specify how the potentially relevant explanatory factors identified previously might contribute to effective renewable energy governance, and why stakeholder networks differ in the local contexts of Tamil Nadu and Kerala. Firstly, organisational sociology literature will explore characteristics of actors' attributes in renewable energy governance arrangements. Organisational sociology is selected as relevant theory debate because it looks inside organisations and aims to understand what characteristics and processes render their agency functional. Secondly, new economic sociology provides insights into actors' arrangements and interactions constituting renewable energy stakeholder networks. This theory debate is selected as it adds dynamics and explains outcomes like renewable energy deployment in terms of relational interactions and emerging constellations. Thirdly, a comprehensive approach to explain effective renewable energy governance is developed. Stakeholder network analysis will be drawn upon as analytical and conceptual tool, and governance debates provide the overall framework for exploring who governs renewable energy, how and with what means. This will not only enable the empirical exploration of how renewable energy services are delivered in functionally equivalent manners. It will also integrate the sociological theory debates to conceptually understand renewable energy governance. This will culminate into an explanatory model of wind energy deployment in India that will be tested alongside the empirical cases in Chapter V.

IV.1 Who Governs? – Stakeholder Characteristics in the Light of Organisational Studies

The first section will discuss organisational sociology literature in order to gain insights into the constitution and the characteristics of "who" governs and who is governed in the renewable energy context.
The inductive approach applied new institutionalist perspectives to examine the impact of policy conditions and suggested answers to how institutions determine what actors do (Scharpf 2000: 770). Confronted with the empirical realities, three aspects were identified as important entry points. Firstly, history and the overall institutional environmental context seem important in shaping certain expectations and behaviour, opening windows of opportunity and creating development paths. Secondly, structures and characteristics of organisations appear relevant for organisational agency. Thirdly, attitudes, interests and leadership at the organisational level emerge as influential for driving positions and activities in the wind energy sector. More detailed explorations require looking inside institutional contexts constitutive of stakeholders in renewable energy governance. For this purpose, organisational sociology is drawn upon. This strand of literature promises comprehensive insights into the logics of organisations in relation to how agency or functional behaviour is explained and affects renewable energy governance, i.e. wind energy deployment.

In the light of governance debates, this dissertation aims to explain different governance outcomes, i.e. why wind energy deployment spatially varies despite similar opportunity structures. Previous analyses point towards the importance of stakeholder agency, specifically public bureaucracies, and the characteristics of who governs. This subsequent discussion focuses on the governance actor defined in the context of this dissertation broadly as all stakeholders involved in renewable energy governance[36]. The main interest is to find out how actors derive at certain choices that shape as well as constitute the governance context for wind energy deployment. For this purpose, stakeholders are examined with regard to how, due to certain characteristic attributes, they engage in different kinds of agency and relations that produce different results. The aim of this section is to explore organisational sociology literature in order to identify the main features characterising actors and the ways and underlying logics that explain successful wind power deployment. A brief introduction to organisational theory will lead to the discussion of one specific approach to the study of organisational agency.

Organisational theories based in sociology provide valuable insights into actors' characteristics related to impacts on and the interplay with respective governance agency, i.e. organisational behaviour. Organisational sociology opens a variety of entry points to the study of organisations[37] since most of the major sociological schools have contributed to this literature. Three general approaches are distinguished

[36] Even though organisational sociology focuses mainly on organisations, this dissertation will expand the scope to consider public, private and also un-organised civil society as stakeholders.
[37] Even though the following discussion will remain in the organisational sociology terminology and talk about "organisations", this dissertation considers public authorities, private companies, local people as stakeholders or actors in renewable energy governance. Hence organisational sociology as well as new economic sociology debates that focus on private actors are defined as applicable to all stakeholders identified in this dissertation.

(Clegg & Dunkerley 1980) that look at typologies of organisations, organisations as social systems, or organisations as structures of action (Marshall 1998). The objective of this dissertation is to understand how stakeholder agency shapes governance-related opportunities for wind power deployment. Meta-questions of organisational definitions and different systemic aspects will hence not be considered. Organisational theories examine the puzzle of why institutionally comparable organisations, i.e. stakeholders characterised by similar attributes, exert different impacts on renewable energy governance. The theory strand selected as most relevant to discuss who governs looks at what and how organisational attributes structure agency. The focus of this debate is on organisational performance as well as the behaviour of governance actors in specific organisation and studies the so-called dark side of organisations (Vaughan1999). The main interest of this approach is to understand how and why organisations fail to deliver optimal results and hence behave dysfunctionally. The study of dysfunctional organisational behaviour is relevant to understand individual and/or collective agency in the wind sector in terms of the constitutive features and mechanisms. This theory debate is relevant because inductive research concluded that policy conditions in term of public authorities per se or regulations and policies do not explain the puzzle. Instead, organisational practices or organisational behaviour seem to exert the greatest effects on renewable energy governance related to successful wind power deployment.

The strand of literature on organisational deviance distinguishes two different approaches that examine causes and dynamics inherent in governance agency:
In the first sense, dysfunctional organisational behaviour is understood as rational response to predictable pressure inherent in institutional systems. Similar to the mechanisms described by rational institutionalists, Cooley & Ron (2002) hold certain contractual relations, incomplete information, transaction costs and properties responsible for the constraints and pressures that cause dysfunctional outcomes. On the one hand, sub-optimal results arise out of certain principal-agent constellations mostly within but also between organisations. Renewable energy governance in India requires the participation of several governance actors, specifically public authorities. Even though their roles and functions are institutionally delineated and functionally differentiated, overlaps or competency issues are part of the power struggle about influence and gains in the wider policy realm. Within organisations, principal-agent problems are quasi institutionalised in the complex bureaucratic structure of the Indian administrative system (Mehta 2005). This situation is aggravated when newly appointed principals, i.e. directors or heads, cause changes in procedures or personnel that are negatively received and rejected by the agents, i.e. civil servant officials.
On the other hand, issues impeding effective organisational performance are results of multiple-principals constellations. These emerge out of the inherent and implicit incentive structures that are sometimes designed to undermine competitors by concealing information and by acting unilaterally. With regard to the Indian renewable energy bureaucracy this situation arises among public authorities that compete for tight budgets and policy influence (Mehta 2005).
This dissertation argues that rationalist approaches are not sufficient for understanding the complexities of renewable energy governance in the non-Western world. The general criticism of this strand of management and organisation theory is that they are extrapolated to developing countries and disregard socio-cultural differences and local conditions (Jaeger & Kanungo 1990). Rationalist approaches exacerbate this problematic issue as they exogenise environmental differences from the analysis. In

many cases not necessarily rational calculations or predictable actors' constellation motivate agency of and within organisations. Other issues such as mutual recognition and willingness to engage constructively in renewable energy governance in terms of wind power deployment emerge as important.

The second approach to studying organisational dysfunctions acknowledges that formal organisations produce sub-optimal outcomes (Vaughan 1999). According to Weber's conception, formal organisations are designed to stimulate means-ends oriented social actions since the formal structures and processes intend to assure certainty, conformity and goal attainment. However, empirical observations contradict this statement as formal organisations also generate unintended and unanticipated outcomes. Organisational deviance reflects a basic assumption that refers to Durkheim's thesis, namely, that the pathological is an inextricable part of the social system (Vaughan 1999: 274). This means that the same characteristics that produce the bright side i.e. render organisations effective, will from time to time provoke the dark side, i.e. disastrous organisational performance. Vaughan (1999: 273) defines organisational deviance as *"event, circumstance and activity occurring in and/or produced by a formal organisation that deviates from formal design goals or normative standards or expectations"*. Mistake, misconduct and disaster are three situations in which dysfunctional organisational behaviour is openly revealed. These instances differ in terms of their long-term effects; scope and scale, and implications.

In the context of this dissertation, the case of organisational misconduct is considered. Ineffective governance in terms of unsuccessful wind power deployment reflects medium-term and persistent agency with suboptimal but not disastrous outcomes. This requires explanations for organisational behaviour of individuals as well as collective actors engaged in practices of longer-term duration with more durable effects on renewable energy governance. Organisationally deviant behaviour comprises most and foremost obstructive organisational agency that impede the deployment of renewable energy. These range from malpractices in grid management, bureaucratic delays and lengthy administrative procedures and practices as well as retardation in initiating or implementing policies and regulations. Widening the scope to other stakeholders, deviant behaviour also includes operational delays by private actors or destructive protests by local people.

In order to understand the causalities that lead to these systemic failures in organisational agency, the three distinct aspects of environment, organisational characteristics and individual cognitive practices are analysed in the following. The purpose is to identify certain features explaining organisational agency, i.e. renewable energy governance, in relation to the attributes of governance actors. These stakeholder characteristics will hence define who governs and is governed in the context of wind power deployment.

Organisational Environment: A Macro-Level Approach

Organisational theory regards the environment as first level of organisational analysis that comprises the organisational set as well as the characteristics of the social context.

General explanations of why organisations deviate from formal standards consider the impacts of the overall environment in relation to the following three conceptual issues. Firstly, a novel event might occur due to changes in societal, political or economic overall conditions. This poses new demands on existing organisations or challenges prevailing organisational constellations as the status quo appears inappropri-

ate to respond to challenges. For example, external energy crises might trigger calls and policy responses for alternatives, e.g. renewable energies. Secondly, changing power structures in the institutional environment create different incentive structures for organisational performance and requirements on organisational action. For example, energy sector reforms imply changes to the power structure and the embedded governance actors due to new tasks and actors introduced. Thirdly, the cooptation to certain environmental conditions might require dysfunctional organisational behaviour in order for the entire institutional complex to survive. Applied to this dissertation, this relates to slowly changing global discussions and attention to renewable energies that create pressure on existing energy governance. All in all, these instances characterise the overall environment that might set negative incentives for, i.e. cause, organisational deviance.
Narrowing the focus to stakeholder agency, environmental characteristics have feedback effects on the work culture in organisations (Jaeger & Kanugo 1990). Jaeger & Kanugo consider the economic-technological, the political-legal and the socio-cultural environment as decisive for constituting and shaping work culture. Work culture comprises the shared values, beliefs and norms that are influenced by the environment in which organisations are embedded. The role and impact of norms, ideas and interests is a central theme among social constructivists (Finnemore, 1998; Risse 2002). Norms are defined as *"standards of appropriate behaviour for actors within a given identity"* (Finnemore 1998: 891). Norms transfer what is regarded as adequate or appropriate behaviour and are internalised in the actors' consciousness. They are revealed in the context of action and interaction, and influence interests as well as identities. Shaped and constituted by the social environment, norms translate into a kind of structure, i.e. work culture that impacts on organisational action and behaviour (Wendt 1999). Norms and interests do not arise out of the blue in an undefined response to environmental characteristics. Norms are subject to change and actively constructed by some actors – so-called norm entrepreneurs (Ingebritsen 2002).

In summary, norms, work culture and environment interrelate and are mutually constructed. This means that in certain societal constellations or due to specific environmental triggers, organisations develop certain work cultures and specific norms that produce dysfunctional outcomes. Organisational culture constituted of and by certain norms, interests and identities guides the organisational behaviour. An analysis of organisational dysfunctions will look at the work culture of an organisation taking into consideration the specificities of respective environmental conditions. This is based on acknowledging that organisations are embedded in wider societal systems. Organisations have and establish a certain history, build up and integrate specific experiences and knowledge in interaction with the locally-specific environment. Organisational effectiveness depends on how and to what extent organisations function in harmony with the surrounding environment (Rieger 2003). Organisational dysfunctions might on the one hand arise when organisational characteristics reflected in work culture are not in equilibrium with external environmental conditions. On the other hand, organisational deviance might sometimes be required in order to ensure survival given certain challenges and changes in the organisational environment.
Applied to the case of renewable energy governance in India, interrelations between environment and organisational characteristics are ultimately reflected in the history and experience of stakeholder agency in the wind sector. This means that the history of experience in a local setting shapes the work culture that determines stakeholder agency. In this sense, organisational agency is related to how stakeholder are char-

acterised by actors' attributes defined as capabilities with regard to the history of experience.

Organisational Constitution: A Meso-Level Approach

The second level drawn upon to analyse origins and causes of organisational deviance concerns the characteristics of an organisation. The three constitutive elements that describe organisations consider structure, process and tasks. Organisational sociology argues that organisational dysfunctions result from specific structural constellations in terms of labour and power relations, processes and interactions established, and task specificities. Since these features are interrelated, organisational characteristics might ultimately produce certain results, i.e. in-/effective organisational behaviour. In the following, these three features are analysed in more depth:

Firstly, organisations incorporate *structures* and practices that conform to certain institutionalised cultural beliefs with the overall purpose of attaining legitimacy (Meyer & Rowan 1977). In order to be acknowledged as an organisation and as hence necessary pre-condition for organisational agency, an organisation needs to display certain features that accord to prevailing ideas and conceptions. The central dilemma is that these structures, though fulfilling legitimacy requirements, may be inefficient or inappropriate to the tasks demanded of an organisation. For example, complex and elaborated structures of organisations with various hierarchical levels might not suit the requirements for fast-track approvals of certain clearances, e.g. in the case of wind project application procedures that await processing by other organisations. The ways in which labour division, hierarchy and specialisation are incorporated into the organisational structure might segregate knowledge about tasks and goals (Vaughan 1999). The greater and more artificial these differentiations are, the more difficult for knowledge to be created and shared among organisational members. Yet, incomplete or abundant information and uncertainties might produce unexpected negative outcomes. The separation of tasks and staff also results from communication barriers. This aspect refers to the ties between individuals that generate control and trust but also provide scope for deceit and misconduct (Granovetter 1985). In the Weberian sense, an effective organisation is characterised by a strict hierarchy and clearly assigned responsibilities that promises straightforward task completion. Complex and anonymous structural labour divisions render the generation of mutually trusted relations more difficult. However, strict hierarchical divisions also provide scope and incentives to pursue self-interested action that does not necessarily support organisational aims of effectiveness and performance.

Secondly, as regards *processes* within and among organisations, differences in understanding and struggles about subunit goals can produce unanticipated negative outcomes. Not only the kind and complexities of processes but also the ways in which procedures are executed give rise to dysfunctional outcomes. This is the case when parallel processes increase the work burdens on units or when feedback processes within organisations do not work well. Processes to a large extent reflect the previously-discussed structure of an organisation. On the one hand, complex structures and strictly separated hierarchies tend to impede open and flexible processes by increasing not only transaction costs but also mistrust, suspicion and power struggles. On the other hand, hierarchical structures facilitate processes through clearly delineated responsibilities. This supports Gidden's structuration argument that basically holds agency interrelated with structure accountable for organisational deviances.

Thirdly, unanticipated organisational outcomes also result from the *tasks* organisational stakeholders fulfill. On the one hand, this depends on and varies according to the kind of occupational skills, the frequency or performance of skills, and training. On the other hand, the performance of tasks is also influenced by the distribution of risks among professional occupations and the way the organisation delegates or concentrates risks and losses. Technical and scientific uncertainties also pertain to work-related technologies and contribute to risk perceptions that all judgements are made under conditions of imperfect knowledge (Vaughan 1999: 279). The performance of tasks is subject to labour force characteristics as well as the clear designation of responsibilities and functions. Dysfunctional outcomes are generated when organisational constituents, i.e. labour force, might not be up for the demands of the specific tasks or not fully aware of the concrete task requirements.

Applied to the Indian wind puzzle, these discussions suggest that internal structural and procedural features characterise governance actors. Actors' attributes in terms of capacities emerge as explanatory factors for renewable energy governance in India.

Organisational Logics: A Micro-Level Approach

The third level of organisational analysis narrows the perspective to the impact of individual cognition and choice, i.e. organisational logic on producing organisationally deviant behaviour. The organisational logic arises out of the interplay between institutionalised beliefs and individual attitudes.

Institutionalised cultural beliefs reduce complexities as they determine what individuals will consider rational to do in certain situations (DiMaggio & Powell 1983). This means that inherent and unquestioned beliefs and understandings automatically guide individual and collective behaviour by excluding what is considered inappropriate. Individual perceptions and interpretations drive action and reaction in certain situations and characterise routines. Individual cognition hence constitutes an important determinant of dysfunctional behaviour. Drawing on largely unconscious cultural knowledge, individuals formulate and define situations in order to make sense of them in cultural terms. Individuals, i.e. staff or organisations, per se view their action as non-deviant before they actually act. This implies that even agency with dysfunctional outcomes might be in harmony with social expectations and acquires legitimacy.

Individual cognition also relates to individual attitudes that shape positions and influence action. At higher organisational level, organisational elites take decisions on the premises of concrete action. Organisational elites or leaders set up and institutionalise so-called routines so that internal individual agency in organisations becomes the result of rule-following. On the one hand, this establishes a legitimate system of appropriate action that serves as orientation for processes, tasks and individual behaviour. On the other hand, however, this can create extreme rule-mindedness that deflects individuals from alternative action that in some instances would be more beneficial for the organisation (Barnard 1938). Flexibility and taking responsibility for individual decisions on issues not specified by routines is shied away. This impedes finding optimal solutions that accord to individual circumstances.

Applied to the empirical case of this dissertation, this aspect is controversial as on the one hand, every renewable energy project is unique. On the other hand, overall renewable energy governance effectiveness requires an equal level playing field that cannot and must not consider specificities.

Individual cognition at higher organisational level touches upon debates about leadership. Even though leadership is an under-explored category in many academic disciplines, this dissertation discusses two approaches. Management studies provide the first perspective that deals with intra-/inter-organisational interactions. In this respect, leadership is understood as *"mechanism that makes things happen in collaboration"* (Vangen & Huxham 2003: 62). In what ways and how leadership takes effect depends on structures, processes and participants engaged in collaborative undertakings.
The second approach to leadership based in political science focuses on the role of policy or norm entrepreneurs. These are advocates willing to invest their resources such as time, energy, reputation and money in order to promote a venture since they anticipate future gains in terms of material, purposive or solidarity benefits (Kingdon 1995: 179). An entrepreneur is characterised by the extent to which she has a claim acknowledged by others, is known for her connections or negotiation skills and is willing to invest significantly her own resources. The role of entrepreneurs is to advocate own proposals as well as to act as brokers (Kingdon 1995: 183).
Why is leadership important? Hardy (1995) argues that leadership provides a strong centralising force in relation to hiring staff and creating the social fabric of an organisation. Leadership or entrepreneurs within an organisation can steer actions and organisational behaviour towards effective results. Leaders and entrepreneurs are not necessarily located at top hierarchical levels but also at less exposed positions given the possibility to broker and access strategic nodes of decision making. Leadership is important in as far as it shapes organisational reputation. Furthermore, leadership exerts an influence on internal individual cognition and self-perception establishing an organisational culture.
Organisational theory acknowledges leadership as important explanatory factor that affects the influence of organisations such as bureaucracies (Biermann & Siebenhüner 2007). The institutional design and its stability have direct repercussions on the extent to which leaders are flexible and have the discretion to take action. This means that the position of leaders within organisational hierarchies and the designated authority impact on the clout of a leader. The importance of leadership also depends on the leader's capacity which involves resources and infrastructure as well as the relations to other gatekeepers within and outside the organisation (Shepsle and Bonchek 1997: 381). In this sense, the routinisation of charismatic leadership might fail when leadership is not recognised or does not meet up to predecessors (Biermann et al 2007: 12).

In summary, organisational deviances are systemic outcomes of complex structures, processes and individual cognition inherent in and constitutive of organisations. Under certain conditions, organisations can produce unanticipated negative results that deviate from the formal design and normative expectations. Applied to renewable energy governance in India, the attitudes and interests characterising the attributes of governance actors reflect the importance of individual and collective cognitions as well as leadership.

Who Governs? - Stakeholder Networks Characterised by Actors' Attributes
In the first inductive research stage, preliminary empirical analyses of policy conditions for wind energy deployment in India pointed to the fact that organisations matter. New institutionalist theory propositions turned out insufficient to grasp the full empirical reality. In this second deductive research stage, organisational sociology

literature is explored in order to examine how and why organisational behaviour shapes policy outcomes and organise policy practices. In the light of governance perspectives, the role of actors and their characteristic attributes in governance arrangements, i.e. stakeholder networks, constitutes one central element explaining renewable energy governance. Organisational sociology promises theory-guided insights into organisational agency looking inside organisations, i.e. organisational characteristics and attributes.

The first issue of explanatory relevance refers to the importance of institutional history and the embeddedness of public authorities in the institutional environment. These shape not only experiences and attitudes but also drive agency related to renewable energy. Organisational sociology argues that the organisational environment that influences organisational effectiveness and behaviour through mechanism like work culture and norms.

The second empirical issue clearly pointed to the resources and capacities available within organisations. As a remarkable fact, not so much the quantity but rather the quality of resources is emphasised particularly as regards individual skill and tasks. Organisational sociology takes up this aspect in the discussion of organisational characteristics that comprise not only the organisational structure but also include processes and tasks.

Lastly, attitudinal obstacles and interests were aspects of policy conditions frequently mentioned as insurmountable obstacles to wind power deployment. Organisational sociology draws on issues of individual cognition in order to explain organisational deviance and dysfunctional behaviour.

All in all, organisational sociology is considered relevant for characterising and explaining who governs and is governed [38]. This gives rise to hypotheses on the role of stakeholders and their characteristic attributes as conditions for effective renewable energy governance.

In the following, these aspects will be operationalised as the first independent variable "actors' attributes" that explains effective renewable energy governance. Units of analysis are individual and collective stakeholders involved in renewable energy governance with specific regard to wind power deployment. An organisation is understood in Merton's sense as social system which allows for including private actors, various organisational forms of public entities as well as civil society. This dissertation assumes that rationalists as well as constructivists logics of action are valid for explaining organisational behaviour and causalities relating actors' attributes to wind power deployment. This dissertation focuses on actors' characteristics and hence the effects of actors' attributes on agency which are constituted of three characteristic features. The proposed effects of actors' attributes are summarised in the following hypothesis:

Hypothesis on Actors' Attributes

Effective renewable energy governance in terms of successful wind energy deployment takes place if governance actors constituting local stakeholder networks are characterised by favourable attributes, specifically:

[38] Even though theory perspectives depart from distinctive assumptions, e.g. institutional definitions, logics of action, they give rise to more comprehensive explanatory versions which is required for understanding new empirical realities.

- *Firstly, actors' attributes are favourable of wind energy deployment if integrated in an institutionally embedded history of experience with renewable energy deployment as well as a supportive institutional environment.*
- *Secondly, actors' attributes are favourable of wind energy deployment if characterised by existing capacities in terms of resources including infrastructure, skills and other structural and procedural organisational features.*
- *Thirdly, actors' attributes are favourable of wind energy deployment if displaying cognitive values, interests and attitudes that embrace renewable energy. Organisational staff as well as leadership either establish or enforce routines that promote renewable energy activities.*

In summary, this hypothesis suggests that actors' attributes embody distinct logics of action that characterise who governs and is governed, and contributes to shaping governance conditions for wind energy deployment.

IV.2 How is Governed? – Stakeholder Relations and New Economic Sociology

This second section will explore new economic sociology debates in order to derive at insights into how and why relations and interactions between stakeholders in the context of specific networks impact on renewable energy governance. In light of governance debates, this will respond to questions "how" and with what means governance takes place.

In the course of the first inductive research stage, the analysis of the Indian wind puzzle concluded that location factors as described by economic geography and spatial economics literature appear dissatisfactory for fully understanding investment decisions and agency in the wind sector. In most of the Indian states opportunity structures per se are relatively similar; however, what economic actors make of them differs. Aspects emerging as relevant and requiring further research are identified on the grounds of three main observations. Firstly, not location factors per se but the way stakeholders avail of and utilise certain incentive structures such as industrial base, infrastructure and labour force differs. This highlights the engagement with other stakeholders involved in wind energy deployment, i.e. the private sector as well as public and civil society actors. Secondly, the importance of multi-stakeholder networks in the context of wind power deployment is emphasised. Without the support of facilitators, brokers and specialists with expertise and local knowledge, stakeholders are left on their own which increases the overall costs of wind power engagement. Lastly, even though external factors such as foreign investment or incentive mechanisms relate to wind sector activities, a causal mechanism does not seem to be plausible. Rather than facilitating wind power deployment, external incentives such as FDI or the CDM follow suit previous sectoral activities.

All in all, this calls for an in-depth analysis of relations and interactions between stakeholders with regard to the effects on wind power deployment. In the light of governance perspectives, actors' relations explain how and with what means governance takes place in stakeholder arrangements with implications for effective renewable energy governance. The second deductive research stage will explore actors' relations in stakeholder networks in terms of evolving actors' arrangements and interaction constellations. New economic sociology is drawn upon in order to understand the structure and dynamics of actors' relations in governance arrangements.

In the following, relational economic geography and new economic sociology approaches will discuss relational characteristics, interactions and networks constellations to tackle the question how wind energy is governed. This will culminate into the hypotheses on how and under which conditions governance arrangements in terms of stakeholder relations and interactions impact on wind power deployment.

Relational Economic Geography Background

Despite the importance assigned to revisiting economic geography in the recent Word Development Report (World Bank 2009), empirical observations point to shortcomings in the debates. This warrants a different approach with focus on relational dynamics between stakeholders.

The first inductive empirical analysis of the Indian wind puzzle guided by economic geography theory in Chapter III confirmed that to some extent location factors matter for wind power activities. However, these classical geographical aspects only become important with regard to what actors make of the opportunity structures, i.e. through the relations and interactions with other stakeholders. This observation supports the general call for relational thinking in human geography (Massey et al 1999) that also transits to economic geography.

Relational economic geography emphasises the importance of relations among actors as they influence dynamic changes in the spatial organisation of economic activities (Yeung 2002: 3). The analytical focus in economic geography has shifted to unravelling the myriad ways in which social actors, firms and organisation are embedded in different kinds of relations, i.e. networks. Furthermore, attention is paid to inter-organisational networks at the micro-level of individual firms. The emphasis on networks and power relations in economic geography contributed to a transition away from narrow firm-centric approaches and structural determinism that characterised previous studies on geographical industrialisation (Yeung 2002: 6). New relational economic geography aims to identify causal mechanisms for localised economic actions and interactions. Research rests upon critical realism as epistemological foundation which opposes individualistic spatial laws (Bathelt & Glückler 2003: 128). Economic geography offers an economic perspective on conventional problems of spatial distribution and acknowledges linkages with economic theories (Bathelt & Glückler 2003: 178). Relational economic geography puts economic actors, their action and interactions at the core of the theoretical framework. The analytical focus is on explaining localised economic processes and their consequences (Bathelt & Glückler 2003: 124) taking space as a perspective. This means that firms do not act only in response to spatial attributes but contribute to the social construction of spatial structures, e.g. through recruitment and training procedures or lobby activities. In order to render the relational economic geography perspective applicable to the study of renewable energy governance, two conceptual issues will be considered.

Firstly, relational conceptions of economic action in a spatial perspective depart from the assumption that economic action is strongly shaped by the particular context (Bathelt 006: 225). This means that economic agency is regarded as embedded in the structures of social and economic relations (Granovetter 1985). The analysis of economic relations using a geographical lens rests on three main tenants: Firstly, action is embedded in specific settings, i.e. contextuality. Secondly, contextual decisions, actions and interactions shape the context of future action, i.e. path dependency. Thirdly, economic actions are not fully determined and may deviate from existing development paths, i.e. contingency.

Secondly, analysing economic action also calls for a conceptualisation of actors (Bathelt 2006: 228). Based on institutional perspectives (Amin 1999), agency is embedded in structures of socio-institutional relations and actor networks, and takes place according to context-specific rationalities. Therefore, economic relations comprise social, political, cultural aspects that affect actors' decisions (Bathelt 2006: 229). Relational economic geography includes individuals engaged in purposeful activities as well as collective actors, e.g. firms and organisations. This dissertation goes beyond the focus on private actors and integrates public stakeholders as well as civil society in the analysis of agency, relations and networks. Wind power deployment in India is not a matter of decision making on investment, economic activities and location only. It constitutes a complex process that takes place on multiple levels and involves various kinds of stakeholders. Going beyond confinements to the private realm is required for an exploration of the Indian wind puzzle that gives justice to the empirical realities.

Venturing into New Economic Sociology

Relational economic geography is based in the broader context of new economic sociology which goes into more detail with regard to exploring and conceptualising actors' relations.

Governance perspectives provide an overall conceptual framework and allow for scaling up new economic sociology's focus on economic actors towards relational aspects of agency. This means that ultimately the role of actors' relations, i.e. new actors' arrangements and interactions in networks, is examined in the context of renewable energy governance comprising economic and policy-related agency. In order to understand relational agency in governance arrangements, the following debate elaborates on the basic tenants of socially-embedded economic agency.

New economic sociology aims at a comprehensive understanding of location decisions and hence integrates strands of network theory, cultural and sociological sociology (Swedberg 1997). The problem of industrial location is considered a matter of entrepreneurial decisions that are based on certain motives, preferences and attitudes (Witlox 2000). New economic sociology originated in response to criticism on new institutional economics that tackled location decisions alongside Williamson's transaction cost approach (Richter 2001: 14). Granovetter emerged as harshest critic against Williamson's new institutional economics and the assumed contrast between hierarchies and markets. He argues that Williamson overemphasises the notion of hierarchical market power and disregards the individual embeddedness in social networks. In order to understand economic activities such as location decisions not transaction costs but power, trust, social relations and networks matter (Richter 2001: 24).

New economic sociology is based on two closely-related master concepts (Swedberg 1997: 170). On the one hand, this concerns the embeddedness concept that originates in Polanyi's work (1944). On the other hand, the idea of the "social construction of the economy" is advocated by Granovetter (Swedberg 1997: 171). Both of these concepts depart from three main assumptions (Richter 2001: 8): Firstly, economic action is regarded as social action. Secondly, social actions are embedded in the wider context, i.e. cannot be explained by individual motives only. Thirdly, economic institutions are socially constructed and path dependence matters. New economic sociology proposes that social structures affect economic outcomes for the following reasons (Granovetter 2005). Firstly, social networks influence the flow and quality of information as people belief and rely on people they know. Secondly, in social network systems, reward and punishment function more effectively since people personally know each other. Thirdly, in the context of social networks trust emerges in line with a clear balance of incentives. Hence concrete social relations and structures of embedded actions in networks discourage malfeasance (Granovetter 1985: 490).

In order to venture further into new economic sociology, Mark Granovetter's article 'Economic Action and Social Structure: The Problem of Embeddedness' in 1985 is discussed as founding contribution. The article departs from the criticism of conventional economics to fail incorporating social structure into their analysis due to the inherent behavioural assumption that economic actors make decisions independent from their social connections (Granovetter 1982: 2). Granovetter calls for a new kind of analysis acknowledging that economic actions are embedded in concrete systems of social relations and need to be understood in network terms (Granovetter 1985: 504). Granovetter (2005) makes three important arguments that provide the basis for empirical research on socially-embedded economic action. These are elaborated at this stage since they suggest explanatory variables and causal mechanisms to understand the role of actors' relations and interactions in economic activities such as wind power deployment.

The first argument regards the quality of relations between actors and refers to the so-called 'strength of weak ties'. Weak ties describe relations characterised by few qualitative characteristics such as mutual understanding or trust. Within networks of actors, loose relationships or mere acquaintances encourage the flow of novel information and new ideas. Not considered as strategic information that is passed on to trusted partners only, innovations are hence more likely to efficiently diffuse through weak ties.

The second aspect concerns 'network density'. Social networks are characterised by the extent to which and how ideas about the proper way to behave, i.e. norms, are shared in relations between constitutive actors. Network density refers to the kinds of interactions in certain relational constellations. Greater network density creates more channels for ideas, information and influence to travel. The internalisation and enforcement of shared norms discourages free-riding and will decrease deviant behaviour. Yet, the larger the network in terms of involved stakeholders, the less will be the network density due to the fact that people have cognitive, temporal, emotional and spatial limits to the number of relations.

The third argument looks at the entire network and discusses the importance of 'structural holes'. As point of departure, not the quality of particular ties between individual actors but the way different parts of the network are bridged and actors are positioned (Burt 1992) gives rise to certain strategic advantages affecting economic agency. The causal idea behind Burt's concept is that given certain circumstances, e.g. uncertain environments, novel information and innovation is more easily available for brokers located outside dense or inside less densely coupled networks.

Several empirical studies apply new economic sociology perspectives but derive at mixed results regarding the impact of embedded relations on economic agency: Empirical studies of construction industries in the USA (Eccles 1982) argue that long-term relations between contractors and the embeddedness within the construction business community promotes efficiency due to generated standards of expected behaviour. Embedded relations are hence superior to pure authority relations. This proves that social relations between firms are then more important than authority

within firms (Granovetter 1985: 501). This contradicts proponents of hierarchical power that emphasise the importance of formalised control and command regulation. Additionally, surveys on social structure, productivity and compliance confirm that the positioning in a social group influences productivity as serious cooperation and the sharing of tacit knowledge is facilitated (Granovetter 2005).
In contrast, Uzzi (1996) argues against taking the quality of relations as panacea for the effectiveness of economic action. It would be simplistic to assume that long-term strong ties automatically improve the productivity and profit of individuals and organisations. In the case of the New York apparel industry, embedded ties in which repeated interactions create trust and mutual understanding are of advantage in stable environmental situations. However, in periods of change, they might impede adaptation since they lock firms into relationships. Arms' length ties characterised by less intimate relations lack the advantages of trustful interactions but allow for more flexibility when change is needed (Uzzi 1996: 692).
Lastly, studies on innovation processes support Burt's argument about the importance of structural holes. Innovation processes are characterised by the break from established routines which requires pulling together previously unconnected resources (Granovetter 2005). Actors outside closed networks or located within the structural holes of a network are placed best to engage in innovate activities by facilitating the connection and deployment of resources.

Exploring the Concept of Embedded Relations
This section will go into depth discussing new economic sociology's core concept of embedded agency. The embeddedness concept allows analysing local governance arrangements with regard to constitutive and entrenched relations, arrangements and interactions between stakeholders. This will concretely tackle the governance question how and with what means renewable energy is governed in the context of stakeholder networks.

As point of departure, embeddedness refers to the process by which social relations shape economic action (Uzzi 1996: 674) and ultimately reflects upon the fact that actions and relations take place in and are influenced by a certain context. Embeddedness is classified in various ways. Zukin & DiMaggio (1990) for example differentiate between cognitive, cultural, political and structural embeddedness. The most fundamental distinction in terms of actors' relations and interactions emerges between the aspects of structural and relational embeddedness. While relational embeddedness characterises the quality of relations between two actors, i.e. dyadic ties, structural embeddedness refers to the structure of relations between groups of actors, i.e. network structures (Bathelt & Glückler 2003: 190).
Why is it important to differentiate these notions of relationally and structurally embedded economic agency? Apart from the fact that these aspects are intertwined, both perspectives provide different approaches to understanding wind energy deployment in the context of socially-embedded agency. This dissertation will explore actors' relations with regard to the dyadic ties in arrangements and the interactions within structural stakeholder networks. Consequently, the following discussion examines the two perspectives in turn:

(i) Dyadic Ties in Actors' Arrangements

The relational embeddedness or cohesion perspective stresses the role of dyadic ties between actors (Gulati 1998: 296). The purpose of discussing relations between actors is to identify and characterise new actors' arrangements evolving in renewable energy stakeholder networks.

Different approaches suggest different entry points to characterise relations between actors. Granovetter describes relations or tie strength in terms of the amount of time, the emotional intensity, the intimacy and the reciprocal services invested (Granovetter 1973: 1361). Rowley et al (2000: 371) measure tie strength by the frequency of interactions between partners and their level of resource commitment. Uzzi & Lancaster (2003) venture into a classification of relations and make a distinction between arm's length and embedded exchange relations. They argue that this differentiation is based on variations in the quality of informal ties that constitute exchange relations. Taking the conventional view of interfirm ties, so-called arm's length relations are impersonal and actors are motivated by instrumental profit seeking. In contrast, embedded ties integrate the logic of trustful cooperative behaviour and emphasise social attachments in commercial transactions.
Since all of these studies acknowledge a significant difference between informal versus personal ties, the aspect of formality will be further elaborated. Organisational sociology debates also look at the organisation of social structures and are relevant for studying the relations and patterns of activities between various stakeholders. In the context of this dissertation, social systems are understood as the constellation of private, public and civil society stakeholders engaged in renewable energy governance in the Indian wind sector. Taking formal, rationally organised social structures such as organisations as point of departure, Merton (1986) argues that a certain degree of formality and clearly defined social distance are inherent in relations. Formality is defined as social ritual that symbolises and supports pecking orders within social structures that are integrated with the distribution of authority. Formality as characteristic feature of relations within social systems exerts two different impacts. On the one hand, formality structures interactions according to previously defined rules. These serve to minimise friction and to establish calculable behaviour. In this sense, the creation of a stable set of mutual expectations hence facilitates interactions (Merton 1986: 249) and results in optimal and effective outcomes. On the other hand, formality might translate into formalism that implies the unchallenged adherence and insistence upon formalised procedures (Merton 1986: 253). In this case, the primary concern of interactions becomes to conform to and to uphold certain rules even though this might contradict the overall purpose of exchange or organisational objectives. At the extreme, overtly practiced formality results in technicism or red tape that impedes effective governance. This aspect touches upon discussion on forms of corruption. Not only within the Indian administration and bureaucracy but to some extent also prevalent among all stakeholders, corruption is characterised as *"institutionalised"* (INT 1). This means that exchange of so-called service fees or other kinds of favours such as information or posts are integral part of practices and formal procedures. These are applied more frequently and are more widespread than direct forms of corruption, i.e. pays. Particularly in the case of Kerala, the absence of monetary corruption was mentioned (INT 2) due to the prevalence of softer means to exchange favours and to ensure expectation certainty in decision making processes.

Apart from suggesting these characteristics of actors' relations, new economic sociology examines impacts and causal mechanisms with regard to relationally embedded agency. Network ties influence two strategic outcomes in terms of economic agency. Since ties are characterised by levels of trust and are governed by mecha-

nisms of relational contracting (Hite 2003: 11), actors' relations provide critical resources and opportunities unavailable through traditional market exchange. Dyadic relations influence competitive actions and contribute to facilitating or hindering the successful emergence, survival and growth of firms. Given the fact that ties may also limit a firm's strategic autonomy and thus future performance, the critical issue is under what conditions, why and how relationally embedded ties create differential advantages or disadvantages for stakeholders.
The other impact of embedded ties concerns knowledge transfer which is considered promotional for economic activities. Since expectations of trust and reciprocity constitutive of embedded relations provide the assurance of mutual benefits, assets such as knowledge are shared. In some instances, arm's length ties, i.e. disembedded relations might have a positive impact on economic agency or governance. Given that knowledge is unevenly scattered, the brokering of information access is required which foregoes trustful embedded exchange relations. This means that access to novel information and new actors is only feasible when actors are connected via arm's length ties.
The impacts of relational embeddedness on economic agency and governance depend on three major scope conditions. Advantages of relationally embedded ties are related firstly, to specific requirements of economic activities on the type of knowledge, secondly, to the learning purpose and thirdly, to the environment (Uzzi & Lancaster 2003: 394).

Empirical studies[39] taking the relational embeddedness perspective derive at somewhat ambiguous results about the impact of strong and weak ties on economic agency in terms of firm performance. Strongly embedded ties are beneficial as they create trust and enable information transfer (Uzzi 1997, Rowley et al 2000). Strong ties provide organisations with strategic advantages associated with the exchange of information and tacit knowledge as well as the constitution of social control mechanisms based on trust, reciprocity and long-term commitments. Yet, weak, arm's-length ties also display advantages since the diffusion of novel information is facilitated (Granovetter 1973).
Empirical studies of New York-based apparel firms (Uzzi 1996) derive at the conclusion that the context as well as the structure of ties affects social and economic behaviour. Embedded ties fulfil the functions of generating trust that promotes asset exchange and reciprocity among actors. This contributes to fine-grained information transfer and provides joint problem-solving arrangements. Embedded ties stimulate exchange that is independent of narrow economic interests and equals or surpasses market outcomes which creates competitive advantages (Uzzi 1996: 682).

Even though different characteristic aspects of dyadic network ties have been taken into consideration (Granovetter 1982, Uzzi 1997), the majority of empirical studies reduce the classification of dyadic ties to either embedded/strong or not embedded/weak ties. One empirical study of emerging firms in the computer industry ventures into differentiating relationally embedded ties (Hite 2003). Hite's study goes beyond the prevalent assumption in new economic sociology that dyadic ties are homogenous and develops a classification of relational embeddedness. The starting premise is that as social relationships differ, relationally-embedded network ties also

[39] As an empirical application, the importance of tacit and codified knowledge in global-scale production networks as most salient characteristics of emergent systems of industrial organisation has been examined in the case of Silicon Valley (Sturgeon 2003).

differ along a broad spectrum. The main argument proposes that the unique nature and variation of social relationships/ties leads to the different modes of exchange within relations as well as different governance of ties that result in specific outcomes (Hite 2003: 12). Social relationships of relationally-embedded ties are constituted of three main components which are personal relationships, dyadic economic interaction and social capital further defined by specific attributes. Firstly, personal relations are a critical element of social relationships within embedded network ties (Granovetter 1985) and are characterised by personal knowledge, affect and sociability. On the basis of personal knowledge, firms use pre-existing relations in order to obtain necessary resources and to engage in interactions. Secondly, dyadic economic interactions are based upon the history of exchange relations and are characterised by the extent, effort, easy and quality of interactions. Thirdly, social capital is manifested in the attributes of obligation, resource accessibility and brokering capabilities. On this basis, the different characteristics of social relationships identified among firms in the computer industry give rise to distinct types of relationally embedded ties (Hite 2003: 31). The added value of this particular study is that it illustrates the impact of different typologies of relationally embedded ties, i.e. differentiating specific qualities of ties, on economic agency and governance.

In the context of this dissertation, identifying differences in the quality of ties according to these criteria will allow for establishing typologies. This promises insights into how similarly embedded actors' relations impact on wind power deployment in different stakeholder constellations. The relational embeddedness perspective contributes to the identification and characterisation of emerging actors' arrangements and the effects in/on renewable energy governance. In this sense, it adds to the understanding how governance takes place as condition for successful wind power deployment.

(ii) Interactions in Structured Stakeholder Constellations

The structural embeddedness perspective considers an actor's position and the distribution of embedded exchange relations in the context of the overall network structure. In contrast to the focus on the quality of dyadic relations adopted by the relational embeddedness approach, this perspective looks at emerging actors' interaction constellations in terms of the overall architecture of a relational network. The following discussion will elaborate on this approach for the purpose of identifying interactions, new coalitions and the effects on renewable energy governance in the Indian context.

This approach departs from the assumption that the type of network in terms of constellations and interactions between stakeholders defines the opportunities potentially available for embedded actors. Relational embeddedness is acknowledged important since the types of interfirm ties define access to opportunities inherent in stakeholder networks, i.e. effects on economic performance (Uzzi 1996: 675). The structural embeddedness approach considers embeddedness more holistically as generic opportunity structure that reflects both how firms are tied to the each other constituting a network and how they interact in structured networks. The structural embeddedness perspective classifies network characteristics into dense structures, i.e. tight constellations and interactions remaining within structures, and into sparse structures, i.e. open constellations and interactions beyond network structure.

Empirical debates about the effects of structural embeddedness on economic agency also derive at ambiguous conclusions:

According to Coleman (1988), dense networks promote trust and cooperation between members thus enhancing the efficiency and effectiveness of economic activities. The social capital debate (Rowley et al 2000: 372) similarly argues that better connectivity to the network yields competitive advantages. In contrast, at times dense networks of strong ties contribute to the production of redundant information from multiple sources. Structural embeddedness in dense networks under certain conditions reduces adaptive capacities. On these grounds, Burt (1992) argues that sparsely connected networks provide a brokering advantage based on the ability of firms to arbitrage information that enhances overall efficiency. This is supported by the observation that in a network of firms linked by embedded ties the flow of new or innovative information decreases. Dense network structures tend to provoke isomorphic processes that lead to organisational inertia and override economic imperatives (Uzzi 1996: 684). The emerging organisational network aggravates economic performance due to lock-in effects into closed constellations and interactions.

The structural embeddedness perspective also examines causal mechanisms, conditions and reasons why network structures in terms of constellations and interactions affect economic agency and governance. With regard to the impact on economic performance, the debate on structural embeddedness comes to several conclusions. The general point of departure is that interfirm networks facilitate the creation of important economic outcomes. Positive impacts emerge through the institutionalisation of trust and reciprocity in dense network constellations that trigger thicker information exchange within closely-confined interactions. Organisational networks operating in this structurally-embedded logic of exchange improve performance by resource pooling, cooperation and coordinated adaptation. In contrast, Burt's empirical study (1992) proves true the opposite and suggests positive effects of sparse networks on economic performance due to novel information exchange in open actors' constellations.

Rowley et al (2000) examine the environmental conditions that determine the extent to which structurally embedded agency is advantageous. They argue that the effects of network structure on economic performance depend on the local context in which firms operate (Rowley et al 2000: 373). In uncertain environments firms seek to explore emerging innovations and changes while established environments provoke the strategy to exploit existing technologies, skills and information. This results from the different information requirements for exploitation and exploration and repeats discussions on the scope conditions for economic agency from the relational embeddedness perspective.

In the context of this dissertation, considering structural embeddedness of actors in network structures seems relevant as it acknowledges the realities of renewable energy governance in the multi-level system and goes beyond the public-private divide. This means that structural embeddedness will explore the interactions in evolving constellations between various actors embedded in networks and the effects on renewable energy governance.

New economic sociology debates fall short on integrating these two perspectives on actors' relations to provide more comprehensive understandings of socially-embedded action. This is due to the fact that the distinction between relational and structural embeddedness is conceptually ambiguous and not clearly differentiated. So far, only two major empirical studies attempt to integrate these concepts into the analysis of relational economic agency:

Rowley et al (2000) argue that the interrelation between structural and relational embeddedness constitutes an important explanatory variable for economic agency. They claim that the ways in which firms form strategic alliances through strong or weak ties are related to how they are structurally embedded in the network. This depends on the scope conditions since the appropriate network configuration depends on the degree of uncertainty and the required rate of innovation (Rowley et al 2000: 370).

Elaborating on that, the impact and requirement on relationally-embedded tie intensity and the structural interconnectedness within networks are both determined by the external environment. In a certain and competitive environment that demands for a high degree of exploitative behaviour to reap existing opportunities, strong ties, i.e. relational embeddedness, are important since they enable information exchanges, trust creation and joint problem solving (Rowley et al 2000: 384). In exploration contexts in which information sources might be few and redundant due to the novel and uncertain environment, high interconnectedness, i.e. structural embeddedness, is detrimental to finding niches to access economic opportunities. Their study derives at the general conclusion that under specific conditions, for specific purposes and at specific times both strong and weak ties might be beneficial for firms' performance (Hite & Hesterly 1999).

Simsek et al (2003) also make an attempt at integrating both embeddedness concepts in their empirical study that looks at the effects of structurally, relationally and cognitively embedded networks on entrepreneurial behaviour. In their model, structural embeddedness refers to the overall architecture of the network and also encompasses the properties of interfirm ties. Structural embeddedness is characterised by criteria such as closure, density, connectivity and hierarchy and measured in terms of absence or presence of ties between actors (Simsek et al 2003: 430). Relational embeddedness aims to describe the quality of exchange including the degree to which partners consider each other's needs and goals and the kind of behaviour towards each other. This is reflected in ties characterised by trust, norms, reputation, sanctions and obligations (Coleman 1990). As third element, cognitive embeddedness is drawn upon as ties characterised with regard to the similarity in the representations, interpretations and systems of meaning among firms (Nahapiet & Ghoshal 1998). As overall point of departure, the specific requirements of entrepreneurial behaviour impact on the ways in which economic actors and activities are socially embedded. Simsek et al distinguish two economic purposes or action orientation of firms. These are incremental organisational behaviour for efficiency improvements in well-established economic agency, and radical organisational behaviour for innovation-oriented agency in novel operations. The first observation is that incremental entrepreneurial orientations require similar cognitive contents and frames of meaning that facilitate communication and common understanding. Due to the prevalence of high levels of reciprocity and trust, a network of strong relations is beneficial for incremental entrepreneurial behaviour. The second observation refers to the fact that exploring new opportunities calls for radical entrepreneurial behaviour and demands low relational and cultural embeddedness. This is due to the requirements of informational diversity inherent in weaker ties as well as the limited reciprocity and social obligation. All in all, Simsek et al (2003: 438) derive at the conclusion that embeddedness perspectives provide a theory to explain how ties in networks settings shape firm behaviour.

In the context of this dissertation, the integration of relational and structural embeddedness perspectives to examine how actors and agency are embedded provides for richer understandings of what determines successful wind power deployment in India. In the light of governance debates, this allows for exploring the effects of emerging renewable energy stakeholder networks and dynamics on renewable energy governance.

How is Governed? - Actors' Relations in Stakeholder Networks

This section draws together new economic sociology debates in order to suggest a hypothesis on the linkages of actors' relations in stakeholder networks to effective renewable energy governance.

The consideration of how governance takes place with regard to wind power deployment in the light of governance perspective provides a promising entry into understanding conditions for effective renewable energy governance. This dissertation departs from the assumption that the overall economic performance in terms of effective renewable energy governance depends on how relationally embedded actors are in emerging arrangements and how the interactions evolve in network structures. New economic sociology debates propose the concept of embeddedness as relevant for exploring how relations in terms of arrangements and interactions between stakeholders shape and utilise opportunities for wind power deployment. Consequently, this dissertation will look at the impact of "actors' relations" as independent variable to explain the Indian wind energy puzzle.

Similar to organisational theories, new economic sociology departs from the conceptualisation of an actor in terms of social systems approach. Since economic activities are socially embedded, social systems or parts of the social system are the central units of analysis. These comprise private actors as well as organisations, public authorities, bureaucracies and organised or unorganised civil society. Several approaches were discussed that specify how relations and interactions between actors and within the relational network matter for economic agency.

Operationalising the variable "actors' relations" will give justice to the two central concepts of relational and structural embeddedness as well as consider relevant scope conditions for the impact on economic activities. Actors' relations integrate the two elements of dyadic ties as well as of interaction constellations between stakeholders involved in renewable energy governance in the wind sector. Dyadic ties are defined according to attributes of reputation, social capital and formality. The characterisation of dyadic ties constitutive of relations between stakeholders will enable the identification of typologies of actors' arrangements characterised by similar types of ties. Interactions in network constellations between stakeholders are acknowledged as the other feature characteristic and constitutive of networks. The following hypothesis proposes the effects of actors' relations in forms of specific arrangements and interactions within/of stakeholder networks on renewable energy governance in terms of wind energy deployment[40]:

Hypothesis 'Actors' Relations'

Successful wind energy deployment takes place if emerging local stakeholder networks comprise actors' arrangements embedded in dyadic ties of high quality and lower formality of relations. Additionally, the greater the structural openness of multi-

[40] As elaborated in Chapter I, this explanatory model and ensuing empirical analysis is set in temporal boundaries and confined to the year 2007 in the third phase of wind power deployment in India only.

stakeholder relational interactions within multilevel stakeholder networks, the more effective renewable energy governance

.

- *Scope Condition 1: Since the environmental context of the Indian wind market is characterised as certain; given the required information and asset exchange, successful wind power deployment is the effect of stakeholder networks characterised by arrangements of high quality and low formality of embedded ties.*
- *Scope Condition 2: Due to the wind market certainty wind sector activities are exploitation-oriented. Wind energy deployment is successful if stakeholder networks are characterised by open, cooperative and trustful interactions embedded in reciprocal and informal actors' arrangements.*

In summary, economic conditions emerge as relevant for renewable energy governance in India as far as relations and interactions between stakeholders in the context of multi-level stakeholder networks are concerned. New economic sociology elaborates on the factors and mechanisms how actors' relations matter for explaining the Indian wind puzzle.

Firstly, the relationally embedded agency between stakeholders defines how and what dyadic ties define the access to certain opportunity systems. In this regard, two implications for stakeholder agency in renewable energy governance are proposed. On the one hand, embedded ties, i.e. high relational embeddedness, positively contribute to renewable energy governance due to instilling cooperative and trustful behaviour given the existing high quality of relationships. On the other hand, effective renewable energy governance is inspired by disembedded weak ties since these allow for accessing novel information and enable exploitative learning in relatively uncertain settings. This dissertation argues that networks constituted mainly of relationally embedded actors' arrangements positively contribute to wind power deployment due to the importance of trust and mutual understanding.

Secondly, the structurally embedded agency between stakeholders determines the opportunities to engage in effective renewable energy governance in the context of network constellations and interactions. Regarding effects of structural embeddedness, Coleman (1988) argues that structurally closed network interactions based on a high number of embedded/strong ties lead to effective renewable energy governance due to mutually beneficial high-quality information exchange and reliability. In contrast, Burt's structural hole argument counters that open interaction constellations with a high number of arm's length/weak ties stimulate successful wind energy deployment due to broker and arbitrage advantages. This dissertation proposes that successful wind power deployment requires structurally open interactions constellations between relationally embedded actors given the importance of tacit exchange between multiple and multilevel stakeholders.

Thirdly, the impacts of actors' relations on effective renewable energy governance are discussed ambiguously in new economic sociology literature. This is not too surprising as new economic sociology stresses the importance of context and contingencies on how embeddedness influences economic agency. The first approach argues that given the certainty of the environmental context (Rowley et al 2003), embedded relations contribute to joint problem solving, trust and high-quality information. Considering the action orientation or economic purpose of business activities in the second approach, Simsek et al hold that disembedded ties are beneficial in situa-

tions of exploitation that require incremental improvements in existing procedures for the sake of market expansion and diversification. This dissertation departs from the premise of relatively certain environmental market conditions in which actors seek to harness existing opportunities. This requires stakeholder networks characterised by relationally embedded arrangements in structurally open interaction constellations.
Concluding on conceptual debates and empirical studies in new economic sociology, the following key points are of relevance for exploring actors' relations as explanatory variable in the theoretical model: The ways in which actors embed in the social context matters and are described in terms of dyadic ties and typologies, i.e. quality of relations and evolving arrangements between actors, and in terms of interactions, i.e. constellation between relationally embedded actors within the network. Apart from these two aspects, the scope conditions of environmental characteristics matter and imply certain agency and network requirements that determine the outcome of activities related to wind power deployment. This allows exploring how and with what means governance takes place to create favourable conditions for successful wind power deployment.

In a nutshell, two sociological theory debates suggest how characteristics and dynamics of stakeholder networks lead to effective renewable energy governance. Since the governance approach is helpful in analytical and interpretive terms, the following section will reflect on this debate. In addition, sociology-based network analysis is discussed as an insightful analytical tool to illustrate and operationalise who and how renewable energy is governed.

IV.3 Stakeholder Networks in the Governance Framework – Tackling the Puzzle

In the following two sections, the previous literature debate on determinants for effective renewable energy governance will be integrated into an explanatory model. For this purpose, the governance approach and stakeholder network analysis are discussed in order to provide the conceptual and analytical framework for explaining the Indian wind power puzzle.

This first section will venture into the governance debate and focus on discussions in the context of the research centre SFB 700 "Governance in Areas of Limited Statehood" (Risse & Lehmkuhl 2006). This specific focus is justified by the objective of the research centre to derive at an understanding of governance that goes beyond normative biases towards OECD-setting. Given the specific local context of the two Indian states, this approach is most appropriate to explore local governance arrangements as functional equivalents to provide public services like renewable electricity supply. For this purpose, actors and agency involved in governing these issues as well as structures setting the regulatory context are considered. The governance framework integrates the two aspects "who" governs and is governed, and "how" governance takes place, with what means and effects. The value added by this conceptual governance approach is that it provides scope to integrate both sociological theory perspectives to analyse characteristics and dynamics of governance arrangements. Governance debates provide theory guidance to understand and explain renewable energy deployment-related dynamics, mechanisms and governance structures in developing countries.

As point of departure, governance refers to *"institutionalized modes of coordinating societal activities towards collectively binding rules and/or the provision of collective goods for a defined group of people"* (Risse & Lehmkuhl 2007: 20). In other words, governance processes and structures intend to provide collectively binding rules, collective goods and services. This involves a variety of actors in so-called non-hierarchical modes of steering e.g. through incentives or horizontal coordination (Göhler 2007). Governance goes beyond traditional understandings of state-market relations that are characterised by and operate through hierarchical control-and-command. The governance debate refers back to earlier discussions in development politics and economics regarding the role of the state and public administration in less developed countries (Riggs 1964, Fuhr 1987, Fishlow 1990, Fuhr 1994). This debate has re-emerged in/due to the changing global scenario. The rise of global challenges such as climate change on the one hand goes beyond the territorial boundaries and sovereignty of nation states. On the other hand, globalisation sets limits to governments' capacities and capabilities to tackle complex issues and to ensure the delivery of basic public goods and services. This has contributed to the emergence of new actors like civil society or the private sector. In response to new global challenges, these actors engage in various forms of interactions in the multilevel context of international forums or at local level. They hence participate in previously traditional government tasks and affect what, how and to whom public goods and services are delivered.

Governance perspectives tackle questions about who is supposed to benefit from governance, towards whom are governance arrangements directed, and who actually benefits (Benecke, Branovic and Draude 2008). Additionally, governance perspectives consider how governance takes place and what means are employed to deliver public goods and services. Lastly, governance perspectives examine the effects of governance with regard to issues of effectiveness, legitimacy and power.

The governance concept does not only bear various meanings but has also travelled through numerous academic disciplines. Lately, the governance concept has been applied outside the OECD context (Risse 2007), which bears two conceptual problems: Firstly, the whole terminology relies on the notion of the modern (OECD) state and secondly, private and public spheres are separated (Draude 2007: 6; Risse 2007: 14). However, such differentiation is not compatible with the realities in most developing countries where borderlines between public and private or state and non-state actors are blurred (Reno 1998; Zürcher 2007). Consequently, this calls for the acknowledgement and identification of so-called functional equivalents to our Western understanding of governance, i.e. alternative actors and alternative means of providing governance services such as renewable energy deployment.

In addition, governance debates also discuss the interplay, dynamics and impacts of external stimuli on different contexts. This means that the governance framework considers a broad range of instruments, actors and mechanisms in interrelation to the overall objective of renewable energy deployment.

In summary, this dissertation will focus attention on the constitution, characteristics and dynamics of local governance arrangements looking at who governs and how. In the multilevel context of renewable energy deployment, local governance arrangements emerge as important for the delivery of public goods and services and the response to multilevel external stimuli.

This second section will suggest and elaborate on stakeholder network analysis as appropriate analytical tool to explore who governs and how governance takes place.

In addition, stakeholder network as conceptual approach will examine the explanatory power of characteristics and dynamics in renewable energy governance.
Results of the first inductive research on wind power deployment in India point towards the importance of actors' attributes as well as their relations and interactions in a multi-level network. The two literature and theory debates on organisational and new economic sociology specified the causal mechanisms in order to understand the research puzzle of spatially varying wind power deployment. On the one hand, actors' attributes are proposed as relevant for determining wind power deployment with regard to features such as capacities, interests and capabilities characterising all stakeholders. On the other hand, actors' relations in terms of the quality of dyadic ties in emerging actors' arrangements as well as interactions in structurally open network relations suggest insights into underlying conditions.
Since both aspects, the static actor-centric as well as dynamic relational features appear significant for a further exploration of the empirical context; an integrative framework is required to guide the analyses. The demands on such a framework are that it does not discriminate between the constitutive independent variables but allows for the identification of complementary effects hence incorporating static as well as dynamic features. Furthermore, it needs to acknowledge the historicity and multidimensionality of multi-level agency and provide the potential for rendering explanatory power to the impact and interplay of the independent variables. In the following, a short overview of network debates is provided to set the general contextual background. On this basis, stakeholder network analysis is elaborated as the analytical tool applied to this dissertation.

As point of departure, both organisational and new economic sociology utilise and integrate the network concept. On the one hand, new economic sociology debates point towards the descriptive potential for operationalising the embeddedness concept by social network analysis (Richter 2001: 9). This departs from the acknowledgement that economic actors coordinate interactions by informal systems, i.e. network governance, rather than relying on bureaucratic structures (Jones et al 1997: 911). Looking at networks and network governance advances the understanding of how economic agency is affected by dyadic relations as well as interaction constellations in structured network relations. On the other hand, different social science research traditions that include organisational sociology draw on different understandings and applications of the network concept. In the following, a cursory overview of network debates in the social sciences serves the purpose to derive at a concept suitable for understanding empirical realities and guiding the further research of this dissertation.

Generally speaking, social science approaches consider networks as new forms of social organisation (Börzel 1997). As lowest common denominator, networks are defined as a *"set of relatively stable, non-hierarchical and interdependent relationships linking various actors that share common interests which they pursue through cooperation"* (Börzel 1997: 1).
Babylonic diversity exists as concerns network terminologies and conceptualisations within social sciences. These range from conceptions of networks as metaphor, as analytical tool for analysing interactions, as method of social structure analysis or theory concept (Börzel 1997: 2). Specifically with regard to networks as a concept, the notion of networks is discussed ambiguously. One strand of literature looks at policy networks and network governance conceptualising alternative governance

mechanisms and their impacts (Bouteligier 2009: 2) in the course of changes of state-society relations. The other strand takes the network concept as a spatial metaphor and re-configures places, actors, activities in the context globally connected networks.

Social science debates on networks are similar to the extent that all of them engage in the key discussion of how to conceptualise networks in relation to state and hierarchy. Three main strands are distinguished that regard networks as clearly distinctive, or as hybrid or as integrative to state and hierarchy:
The first approach to networks defended by institutional economists proposes a clear distinction of networks both from state as well as from hierarchy (Podolny & Page 1998: 59). Networks are distinct from hierarchies as they do not posses clearly recognised authority to resolve disputes. Networks are distinct from markets since they integrate enduring relations. Institutional economics and sociological institutionalism consider networks as specific modes of coordinating action in contrast to other governance forms such as markets or hierarchies (Jansen 2002: 89).
The second stance on networks in the tradition of political sciences departs from the conceptualisation of networks as a *"new governance form situated in the continuum between hierarchies and markets"* (Mayntz 1993: 45). Networks combine elements characteristic of both as they integrate the autonomy of market and the capacities of hierarchies. Apart from conceptual theory debates, governance scholars also make use of policy networks as analytical concept. The overall purpose is to draw attention to the interactions of various actors coordinating activities through resources and interests. In this sense, resource exchange in governance networks is determined by the linkages or governance structures that differ according to intensity, standardisation and frequency of interactions (Börzel 1997).
The third perspective on networks based in sociology traditions constitutes the basis for social network analysis applied in this dissertation. The point of departure is that *"nearly everything constitutes a network"* (Bouteligier 2009: 4). In this sense, the only distinction is made between more hierarchical and more horizontal networks, i.e. regarding variations between hierarchy/market elements. This approach stresses the conceptualisation of networks as an analytical tool that encompasses and explains both markets and hierarchies as variations of network structures (Thompson 2003: 86).

In distinction to social and behavioural sciences, social network analysis (Hafner-Burton & Montgomery 2009: 6) serves most and foremost as a research methodology. Stakeholder network analysis is based on primary assumptions and formal tools and does not provide a unified set of theories. In this regard, stakeholder network analysis examines relationships defined by linkages among units (Hafner-Burton & Montgomery 2009: 6) and describes connections associating actors. Social network analysis departs from the principles that actors and their behaviour are mutually dependent, that relational ties serve as channels for resource diffusion and that persistent patterns of association create a social structure within which action takes place (Wassermann & Faust 1994).
However, the implications of social network analysis go way beyond descriptive purposes. Stakeholder networks analysis also allows for understanding wider impacts or power of social networks. This is based on the central argument made by Hafner-Burton & Montgomery (2009: 10) that networks inhere and exert social power since the relational ties enable the diffusion of resources and hence establish a lasting so-

cial structure. Material power is rendered less relevant since the coercive capabilities or power to get things done inherent in network actors is considered. Sociological perspectives go beyond discussions of stakeholder networks as methodological tool (Bouteligier 2009: 6). To start with, interdependent actors within structures constitute a network of relational ties through which resources are transferred. Networks of social relations explain socially-embedded economic behaviour and processes. Network analysis also integrates individual and group behaviour and bridges the micro-macro gap between individuals and institutions (Emirbayer 1994: 1415). Networks establish lasting patterns of interaction (Thompson 2003: 55) that provide opportunities as well as constrain individual action. Even though network analysis looks at constraining and enabling dimensions of patterned relations between actors within a system, it cautions against causal explanations. This is due to the fact that in order to allow for drawing implications on network effects, i.e. scaling up the notion of networks to a theory concept, theory perspectives need to be integrated. The network concept as such is not a meta-theory (Häußling 2005: 269). It requires the linkage with theory perspectives in order to allow for explanatory power of the procedural and structural patterns identified. In this sense, new economic sociology provides a methodological tool and a theory entry point to add explanatory power to stakeholder network analysis illustrated alongside two main approaches: Network analysis adopts a relational approach when behaviour or processes are explained through the density, strength symmetry, range of ties that bind (Emirbayer 1994: 1419). Additionally, positional approaches to network analysis focus on the nature of actors' ties not to one another but to third parties. They emphasise the importance of structural equivalence of relations between actors and look at the position or role that a set of actors occupies within the system as a whole (Emirbayer 1994: 1422).

In the context of this dissertation, stakeholder network analysis emerges as appropriate analytical tool to explore network characteristics and dynamics as conditions for effective renewable energy governance in the light of governance perspectives. In addition, stakeholder network analysis provides the potential to include conceptual theory debates about network mechanisms and effects on renewable energy governance. With regard to exploring actors' relations in context of new economic sociology, social network analysis is hence apt to operationalise the embeddedness concept (Richter 2001) to discuss how governance takes place. The analysis of social networks focus on the interaction of interdependent actors and the unit of analysis is based on actors as well as their relations (Thatcher 1998: 398). Considering actors' attributes in the context of organisational sociology, stakeholder network analysis focuses on the actor as constitutive element of networks. Characteristics in relation to other stakeholders and the overall context of renewable energy governance are examined to respond to the question who governs and is governed.

All in all, the main objective of stakeholder network analysis is to identify renewable energy stakeholder networks and to map their structures, characteristics and dynamics. This involves the identification and characterisation of key actors. In the context of this dissertation this is operationalised through the exploration of actors' attributes in relation to wind power deployment. Additionally, linkages and different characteristics of exchange relations are identified which allow for building typologies of actors' arrangements and interaction constellations. Lastly, studying the effects of networks requires looking at issues such as changes on policy processes, context of policy choices and path dependencies (Thatcher 1998: 406). Applied to this dissertation, the power of networks in exerting effects on the opportunity structures and the extent to which they are tapped in the context of renewable energy governance are discussed.

In summary, this section elaborated the analytical and conceptual framework that will structure the empirical analysis and interpretative discussion on the role of local stakeholder networks in renewable energy governance. In the following, the theory debates are condensed into an explanatory model in order to prepare the deductive empirical deep-dive into the puzzling realities of spatially varying wind power deployment.

IV.4 Explanatory Model for Wind Energy Deployment in India

This concluding section will draw together this chapter's literature debates that departed from the empirical findings of the first inductive research phase. Two aspects, namely actors' attributes and actors' relations require in-depth exploration as they hold explanatory power for effective renewable energy governance in developing countries (Figure XVI). The purpose of the analytical framework is to explore important governance questions about the extent to which characteristics and dynamics of local stakeholder networks constitute conditions for effective governance.

Figure XVI: Deductive Research Stage

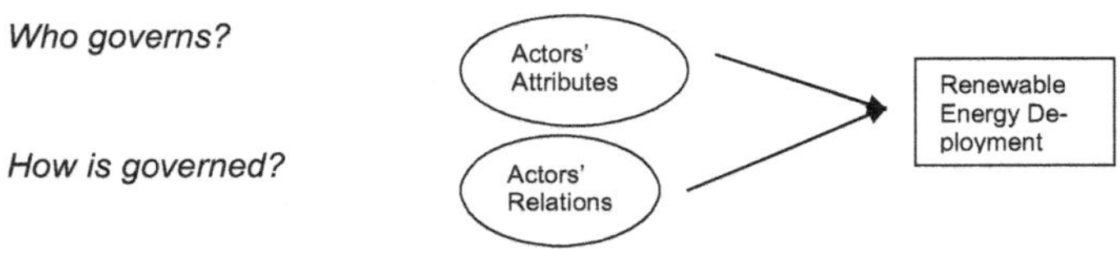

In the following, an explanatory model is developed to guide the ensuing in-depth deductive empirical research approach to the Indian wind energy puzzle. The explanatory model of wind energy deployment in India will integrate explanatory factors, elaborate hypotheses and scope conditions, and specify the (causal) mechanisms in a holistic framework. This suggests how, why and with what effects wind power deployment takes place differently in different local contexts in India.

The focus at this stage of this dissertation is on stakeholder networks in renewable energy governance in the case of the wind sector in Kerala and Tamil Nadu. The constitution and characterisation of stakeholder networks is empirically-derived as a result of an ethnographic, stakeholders' perception-based mapping exercise (cf. Chapter I).

Networks are formally characterised (Jansen 2002: 98) by network nodes and network boundaries. N*etwork nodes* describe stakeholders involved in renewable energy governance and integrate public, private and civil society actors. Organisational sociology and new economic sociology perspectives suggest interpretations as regards the effects of stakeholder characteristics on wind energy deployment. N*etwork boundaries* comprise the entire network of nodes, i.e. all stakeholders involved in wind energy deployment integrated in the multi-level system including the local, the national and the international level. The temporal boundaries of network analysis ini-

tially consider the third phase of wind power deployment and concretely 2007 as point of analysis.
The results and outcomes of ethnographic network analysis are depicted alongside two dimensions (Jansen 2002: 100). As for the first dimension, network nodes illustrate the characteristics of individual stakeholders and are discussed as the explanatory factor "actors' attributes" in the light of organisational sociology. Network nodes as the second explanatory dimension reflect the relational actors' arrangements and interaction constellations constituting renewable energy stakeholder networks as independent variable "actors' relations". New economic sociology suggests hypotheses on the effects on renewable energy governance with regard to the relations and emerging arrangements between stakeholders (Jansen 2002: 101).

For the sake of guiding deductive research on wind power deployment in the Indian case, the network concept is applied as analytical tool. In order to derive at explanations of how, why and with what effect network structures and processes are constituted the way they are, the integration of theory approaches is required. For this purpose, the dissertation mainly draws on the two theory perspectives of organisational sociology and new economic sociology for understanding stakeholder characteristics, i.e. actors' attributes, on the one hand and stakeholder relations and interactions, i.e. actors' relations, on the other hand. Reference to organisational sociology is justified by the insights provided into how wind energy deployment conceived as governance outcome related to stakeholder characteristics. The other important aspect identified by first inductive empirical research on renewable energy governance in India relates towards the importance of so-called economic conditions. Since specifically the relations and interactions between stakeholders in structured networks emerged as significant, new economic sociology was selected as theory approach.
These two explanatory variables that constitute the skeleton of the explanatory model will be contextualised in the governance framework and operationalised in the following:
The first condition for effective renewable energy governance lies in the question "who" governs and is governed that concern stakeholder characteristics. As regards the first independent variable, "actors' attributes" is defined as stakeholder characteristics constitutive of the respective organisational contexts. Based on organisational sociology literature, the following three criteria and indicators characterise this independent variable (Figure XVII). Firstly, *capacities* are considered that comprise the wind power budget, internal and external staff and its qualification, as well as autonomy in relation to resource use and independent decision making. Secondly, *interests* pertain to the motivation, attitudes and role of individuals and the organisation, attitudes of leaders, and the organisational reputation in terms of accessibility and efficiency. The third and last factor considers *capabilities*, i.e. the stakeholders' history and experience with regard to wind and renewable energy. As specified in Chapter I this independent variable will be described by various qualitative methods, sources and indicators.

Figure XVII: Actors' Attributes

Criteria	*Indicators*
Capabilities	History of experience and kind of experience with renewable energy deployment
Capacities	Human resources (staff size and qualification)
	Autonomy in resource use and decision making

	Financial Resources (internal & external budget for renewable energy)
Interests	Motivations and roles
	Leadership and attitudes
	Organisational reputation (accessibility, renewable energy, accessibility)

This independent variable seeks to explain wind power deployment as result of organisational agency that differs according to the role and contribution of stakeholders in renewable energy governance. The underlying premise is that organisational agency reflects actors' attributes. Understanding deviant organisational behaviour in terms of ineffective renewable energy governance requires the consideration of respective characteristics. The following hypothesis summarises the causal mechanisms that relate stakeholder characteristics, i.e. actors' attributes, to renewable energy governance:

Hypothesis on Actors' Attributes
Effective renewable energy governance in terms of successful wind energy deployment takes place if governance actors constituting local stakeholder networks are characterised by favourable attributes, specifically:

- *Firstly, actors' attributes are favourable of wind energy deployment if integrated in an institutionally embedded history of experience with renewable energy deployment as well as a supportive institutional environment.*
- *Secondly, actors' attributes are favourable of wind energy deployment if characterised by existing capacities in terms of resources including infrastructure, skills and other structural and procedural organisational features.*
- *Thirdly, actors' attributes are favourable of wind energy deployment if displaying cognitive values, interests and attitudes that embrace renewable energy. Organisational staff as well as leadership either establish or enforce routines that promote renewable energy activities.*

All in all, the main assumption is that stakeholders characterised by specific features gain the capacity to act in various ways shaping the policy dimension or engaging in the deployment process with effects on renewable energy governance. Hence the more favourable actors' attributes, the greater the chances for successful wind power deployment. This will be deductively explored in the following chapter and outcomes are represented in the illustration of network nodes constituting emerging stakeholder networks (Figure XVIII).

Figure XVIII: Actors' Attributes in Stakeholder Networks

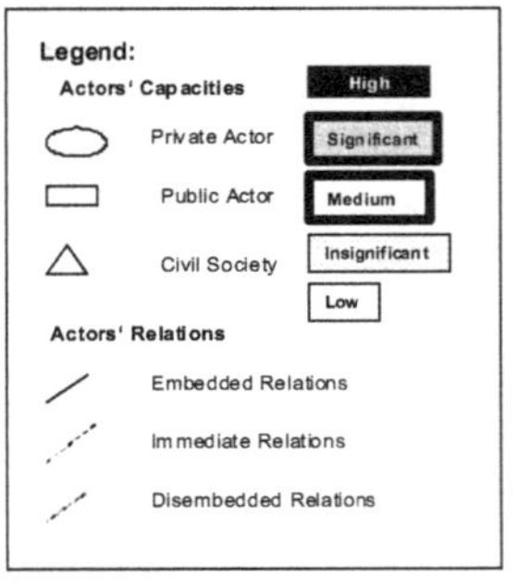

The second condition for effective renewable energy governance lies in the question "how" is governed and concerns dynamics between stakeholders. The second independent variable considers "actors' relations". Actors' relations are understood as relational arrangements and interactions between stakeholders that constitute a distinct stakeholder network. New economic sociology's embeddedness approaches are operationalised in order to specify the explanatory power of these concepts.

Before embarking upon the oprtationalisation of this explanatory factor, several issues need to be considered. First of all, nearly all of the empirical studies in the new economic sociology tradition to date are based in and focus on firms and economic agency taking place in the OECD world. However, economic activities and governance in areas of limited statehood without doubt encounter additional and/or different challenges particularly what regards the impact of general contextual conditions. Secondly, individual firms are taken as the main unit of analysis. In the case of renewable energy in India, however, public and private stakeholders are crucially involved in renewable energy deployment and governance and hence integrated into the analysis of this dissertation. As a last point, the differentiation between the different notions of embeddedness made in new economic sociology literature remains rather ambiguous in as far as the operationalisation, the conceptualisation and the effects on economic outcomes are concerned. The explanatory model to be developed will make adjustments to new economic sociology premises discussed above. The main objective is to render stakeholder network analysis as well as the underlying explanatory model as considerate of local circumstances as possible in order to understand renewable energy governance in India.

This dissertation departs from some basic assumptions on stakeholders, action orientation and environment oriented towards organisational and new economic sociology literature. Firstly, this dissertation focuses on multi-stakeholder networks that embrace for-profit private sector organisations, public institutions, non-profit organisations and local people involved in renewable energy governance with focus on wind energy deployment at multiple levels. Even though effective renewable energy governance is ultimately targeted at local people as governance addressees, stakeholder networks analysis will consider recipients and providers of governance services and disregard issues of in-/exclusion. Secondly, wind energy deployment comprises all economic and non-economic activities ranging from policy making with regard to e.g. wind tariff regulations to de facto wind power deployment. Wind energy deployment is considered as one instance of renewable energy governance. This dissertation makes the facilitating assumption that renewable energy governance is confined to the deployment of wind power. This omits, however, important dimensions of the energy governance process and is only one chain in the actual provision of electricity services. This takes place in an overall market environment characterised by relative certainty. Many empirical studies emphasise that the existing political barriers in the developing country context make the case for considering an uncertain and volatile environmental context. Yet, compared to other developing countries and other renewable energy technologies particularly the Indian wind energy sector constitutes a maturing and commercialised market. Thirdly, the logic of action characterising stakeholders in the wind energy undertaking differs according to the stakeholder and the theory perspective adopted. This means that private actors tend to pursue profit motives while public actors behave according to other interests, i.e. public welfare. In this sense, the theory perspectives employed allow considering several motivations such as rational calculus and intrinsic value orientation. With regard to entrepreneurial behaviour, wind sector engagement is oriented towards the purpose of exploitation in terms of market expansion, diversification and efficiency improvements. Again, one might argue that some local markets still await exploration; however, given the pre-existence of nation-wide demonstration activities and similar financial and political incentive conditions at the national level, the exploitation argument is deemed convincing.

Since renewable energy deployment is considered a socially-embedded activity, both, the opportunity structures provided by stakeholder networks per se as well as their accessibility is important. Examining stakeholder relations as well as dynamic interactions in network constellations draws upon structural and relational notions of embedded action. The application of stakeholder network analysis with focus on the second independent variable of actors' relations aims to identify relational arrangements and interaction constellations as constitutive and characteristic of specific renewable energy stakeholder networks. For this purpose the embeddedness approach is operationalised as follows with the objective to explore effects of stakeholder networks in terms of actors' relations on renewable energy governance:
The first constitutive feature of stakeholder networks relate to dyadic ties (Figure XIX). This means that ties, i.e. relations, between actors are considered and characterised according to three criteria. These are adopted from the new economic sociology literature and empirical studies available. Two of the criteria lean on the notion of relational embeddedness and are differentiated into reciprocity, i.e. the interaction intensity, and social capital, i.e. the interaction quality between stakeholders. *"Social capital"* refers to the relational quality between actors and describes their personal relations, knowledge and mutual trust. *"Reciprocity"* describes the relational intensity comprising the extent, frequency of exchanges as well as the degree of coordination and learning. The third criterion is partly drawn from literature about social capital while partly referring to debates about formal and informal governance. Called *"formality"* this criterion alludes to the sense of de jure and de facto obligation in ties and interactions between stakeholders and is characterised by resource dependency and autonomy. These three criteria of social capital, reciprocity and formality are selected as characteristic of dyadic relations due to interactions and network effects specified in new economic sociology debates above. As elaborated in Chapter II, several methods and sources of empirical investigation are drawn upon in order to gain knowledge about the different indicators and are complemented by a participant-based stakeholder mapping.

Figure XIX: Dyadic Network Relations

Factors	*Criteria*
Reciprocity	Extent & frequency of interaction
	Degree of coordination & learning
	Exchange of favours
Social Quality	Personal relations & knowledge
	Mutual understanding & trust
	Expectation certainty
Formality	Autonomy
	Legal obligations
	Resource dependency

The characterisation of actors' relations does not stop at dyadic ties. Based on Hite's study on the explanatory significance of different tie qualities, the objective is to identify new and emerging actors' arrangements characterised by similar dyadic relations. This dissertation aims to empirically derive at typologies of similar relational ties for acquiring knowledge about the kinds of evolving actors' arrangements. The classification of dyadic ties according to the three criteria gives rise to several typologies of 'relational embeddedness'. Figure XX summarises the nine possible actors' arrangements that might emerge according to the characterisation of the two significant attributes formality and social capital[41]. These typologies of relational strength in arrangements are represented and depicted in the illustrative stakeholder network mapping.

Figure XX: Actors' Relations in Stakeholder Networks

Tie Characteristics/ Typology	**"Disembedded Actors' Arrangement"**	**"Immediate Actors' Arrangement"**	**"Embedded Actors' Arrangement"**
Typical	High formality – Low social capital "distant – formal ties"	Medium formality – Medium social capital "immediate – formalised ties"	Low formality – High social capital "familiar – informal ties"
Mix	High formality – Medium social capital "distant – formalised ties"	High formality – High social capital "immediate – informal ties"	Low formality – Medium social capital "familiar – formalised ties"
Hybrid	Medium formality – Low social capital "hybrid – distant ties"	Low formality – Low social capital "hybrid – immediate ties"	Medium formality – High social capital "hybrid - familiar ties"

Assumption: Reciprocity held constant; considered not decisive for differentiating between tie quality

⟷ : "Embedded Actors' Relations

◄----► : "Immediate Actors' Relations

◄·········► : "Disembedded Actors' Relations"

In addition to analysing relational arrangements, a second element is drawn upon to describe and characterise actors' relations as explanatory variable. The analytical focus is on the emerging network constellations in the course of interactions. This refers to the notion of structural embeddedness and concerns the interaction constellations between relationally embedded stakeholders. The purpose of examining interaction constellations is to find out to what extent structural openness of constellations and tie characteristics affect network processes and outcomes, i.e. wind power deployment. As point of departure, a pre-existing differentiation of stakeholders accord-

[41] As elaborated previously, reciprocity is considered as less significant characterising dyadic ties and thus regarded a constant attribute. The individual attributes will be assessed according to the value range from high, medium to low.

ing to their so-called functional identity (public, private, civil society actor) often prevails stakeholder network concepts. This dissertation assumes that interactions go beyond pre-defined identities, i.e. public, private, civil society groups and evolve in constellations of differently characterised stakeholders. In the light of governance debates, the structural openness of interactions in respective constellations exerts effects on renewable energy governance. In the course of stakeholder network analysis, the first objective is to identify the relations between these pre-conceived groups, i.e. public-private, public-civil society and private-civil society actors and to other stakeholders. This gives rise to observations about the new modes of interactions and actors' constellations constituting the overall relational network structure. This ultimately allows drawing implications on the logic and functioning of entire multi-level stakeholder networks exploring effects on wind energy deployment.
The following hypothesis specifies the effect of actors' relations on wind power deployment:

Hypothesis 'Actors' Relations'
Successful wind energy deployment takes place if emerging local stakeholder networks comprise actors' arrangements embedded in dyadic ties of high quality and lower formality of relations. Additionally, the greater the structural openness of multi-stakeholder relational interactions within multilevel stakeholder networks, the more effective renewable energy governance.

- ➢ *Scope Condition 1: Since the environmental context of the Indian wind market is characterised as certain; given the required information and asset exchange, successful wind power deployment is the effect of stakeholder networks characterised by arrangements of high quality and low formality of embedded ties.*
- ➢ *Scope Condition 2: Due to the wind market certainty wind sector activities are exploitation-oriented. Wind energy deployment is successful if stakeholder networks are characterised by open, cooperative and trustful interactions embedded in reciprocal and informal actors' arrangements.*

All in all, the main theory-derived assumption proposes that the more actors' arrangements are embedded in terms of dyadic ties with high interaction quality and low formality the greater the chances for successful wind power deployment. Additionally, the greater the structural openness of interactions in networks of certain relational constellations, i.e. the greater the informal exchange between stakeholder groups, the higher the chances for wind energy deployment.

This dissertation sets out to explore the role of stakeholder networks for effective renewable energy governance in India in the case of wind power deployment in Tamil Nadu and Kerala. In the light of governance debates, this focuses on the characteristics and dynamics of renewable energy stakeholder networks as functionally equivalent governance arrangements. The main assumption is that local governance arrangements, i.e. actors' attributes and relations constituting renewable energy stakeholder networks, exert certain effects on renewable energy governance, i.e. wind power deployment.
What is lacking from the explanatory model so far is the specification of network effects arising from the characteristics and dynamics of renewable energy stakeholder

networks. This is due to research gaps in respective theory strands drawn upon for the elaboration of this model: New economic sociology, even though embarking upon relations, interactions and network configurations remains silent on actors' attributes. This is due to the implicit assumption that actor regardless of attributes act according to the logic of bounded rationality. Organisational sociology emphasises the individuality of stakeholders and allows for different and competing logics of action as a results of capacities and characteristics. Yet relations and interactions between stakeholders are not further specified and rather assumed to arise out of actors' attributes in concrete agency.

Table 25: Network Effects on Renewable Energy Governance

	Output	Outcome	Impact
Network Effects	Stakeholder agency in renewable energy governance – visible in policy and regulatory guidelines	Changing behaviour of stakeholders towards more respective agency – visible in different responses and approaches; implementation activities	Changes in the policy field with wider implications – visible in changing macro structures and principles, new opportunities

Table 25 stipulates that stakeholder networks exert certain effects alongside three dimensions that relate to renewable energy governance. The ambition of this dissertation is to identify stakeholder networks and respective causal mechanisms that explain the role of governance arrangements in renewable energy governance. However, the analytical and explanatory focus is in one renewable energy governance dimension, i.e. the outcome in terms of wind power deployment.

In summary, the theory-derived explanatory model examines the causal effects and mechanisms of renewable energy stakeholder networks on renewable energy governance in terms of wind power deployment. In this sense, it looks specifically at actors' attributes and relations in terms of stakeholder characteristics, relational actors' arrangements and interaction constellations. The added value of this explanatory framework is that it integrates these two aspects based in two important theory debates in order to explore multi-level governance issues such as wind energy deployment in local contexts of developing countries such as India. Different logics of action as well as scope conditions can thus be duly considered and various plausible and/or causal mechanisms and relations explored. The following chapter will embark upon the challenge to deductively test the explanatory model developed alongside the two case studies of wind power deployment in Tamil Nadu and Kerala.

CHAPTER V

Conditions for Wind Energy Development in Tamil Nadu and Kerala: Deep Dive into the Empirics Stage II

CHAPTER V
Conditions for Wind Energy Development in Tamil Nadu and Kerala: Deep Dive into the Empirics Stage II

"*Turning wind into power*" (Lincoln 1860) to tackle sustainable development challenges is conditional upon context-specific pre-requisites for effective renewable energy governance. This chapter will continue the second deductive research stage and empirically explore these conditions in the two cases of Tamil Nadu and Kerala guided by an explanatory model. This embraces theory discussions of the preceding chapter that pointed towards the importance of local renewable energy governance arrangements, i.e. stakeholder networks. The following empirical exploration is structured alongside the governance framework and tackles two main questions: "Who" governs and is governed and "how" does governance take place in the locally-specific cases of wind energy deployment? The purpose is to identify functional equivalents of providing governance services in the energy sector by analysing the characteristics and dynamics of local stakeholder networks.

At this stage, the concrete focus is on the empirical analyses of the wind power deployment case in the selected contexts of Tamil Nadu and Kerala. Interpretations and explanations of why and with what effects stakeholder networks relate to wind energy deployment are discussed in Chapter VI.
Wind energy deployment in Tamil Nadu and in Kerala are taken as most puzzling instances alongside which firstly, characteristics of governance actors and secondly, dynamics of governance in stakeholder networks are examined. Following from the descriptive-analytical stakeholder network analysis in the respective local contexts, comparisons between the local governance arrangements identified in Tamil Nadu and Kerala are drawn. Furthermore, the wider application of the explanatory model will be demonstrated tracing the process of wind sector development alongside three phases in the two states respectively.

V.I Wind Energy Deployment in Tamil Nadu - The Indian Success Story

Tamil Nadu is the model state for wind energy development in India. Out of an estimated wind power potential of 3050 MW, to date, Tamil Nadu has achieved an installed wind power capacity of 3492.7 MW (InWEA 2007)[42]. How come Tamil Nadu is that successful in harnessing natural resource potentials for wind power electricity generation?

The first inductive approach concluded with two preliminary observations that point towards the importance of local governance arrangements. Firstly, the continuous and stable policy and regulatory framework and their pro-active leadership emphasise that the characteristics of governance actors are relevant. Secondly, positive interaction effects between public and private stakeholders point towards the dynamics between actors involved in governing wind energy. The explanatory model elaborated in the previous chapter will structure the deductive in-depth exploration of how

[42] These counter-intuitive figures are explained by the insufficient and imprecise wind resource assessments (INT 25).

renewable energy governance arrangements are constituted and function to exert supportive effects on renewable energy governance.
This puts to debate the hypothesis that successful wind power deployment in Tamil Nadu emerges in the context of renewable energy governance arrangements characterised by favourable actors' attributes as well as embedded relations between stakeholders in open multi-level network constellations. Deductive research is guided by the governance framework and operationalised in the explanatory model and proceeds as follows: Firstly, the question who governs is addressed. On the one hand, the constitution of local governance actors is explored, i.e. the kind of actors involved. On the other hand, attention is turned towards qualitative aspects that characterise governance arrangements with regard to the attributes of individual stakeholders. Secondly, the question how wind energy is governed is explored. Attention is turned towards the identification of stakeholder arrangements in terms of relational ties and interactions within network structures.

V.1.1 Who Governs?: Actors' Attributes in Stakeholder Networks

The first approach to understand conditions for successful wind energy deployment in Tamil Nadu inherent in renewable energy governance arrangements is to look at who governs. This regards the configuration of stakeholder networks and the characteristics of governance actors, i.e. stakeholders' attributes. The explanatory model specifies how actors' attributes impact on wind power deployment. Borrowing from organisational sociology debates, actors' attributes comprise capacities, i.e. resources available to stakeholders, interests and attitudes towards renewable energy and lastly capabilities, i.e. experience in the area of renewable energy governance. This dissertation proposes that not the constitution but the characterisation of governance actors by favourable actors' attributes is a condition for effective renewable energy governance. This implies that public actors are more likely to engage in policy making and implementation that favours wind power. Private actors will more likely get involved in wind power entrepreneurial activities, and civil society actors will less likely oppose wind energy deployment. Actors' attributes bear implications on how transactions occur, and on how communication, information flows and practices evolve that influence the effectiveness of renewable energy governance.
In the case of Tamil Nadu, the following stakeholder network analysis will demonstrate that stakeholder attributes are favourable to wind power deployment. Reflecting on reduced transaction costs, uncertainties and more cooperation, this contributes to effective renewable energy governance. The following section will address the question who governs with the purpose to specify the characteristics of local governance arrangements. The equivalence functionalist analysis of renewable energy governance arrangements will firstly map the main stakeholders of wind energy deployment in Tamil Nadu. Secondly, stakeholders are characterised according to capacities, interests and capabilities. Causal mechanisms and explanations as specified in the explanatory model are discussed and preliminary conclusions on how actors' attributes characteristic of stakeholder networks might impact on the effectiveness of renewable energy governance are drawn.

Step I – Mapping the Constituents of the Local Stakeholder Network
In the pre-conditional first step, the actors engaged in wind power deployment are mapped.

Generally speaking, institutional and functional responsibilities and roles in renewable energy governance are clearly delineated in the Indian constitution and specific national as well as local legislations.
At the central level, responsibilities for the energy sector in which renewable energy and wind power are embedded as sub-sectors involve several ministries and government agencies. The Planning Commission is in charge of getting energy sector developments in line with other national objectives related to economic growth and development through the stipulation of Five-Year Plans. The development and administration of energy policy lie in the hands of several ministries such as the Ministry of Power, the Ministries of Coal, Oil and Gas and the Ministry of New and Renewable Energy Sources.
The most important player at the national level as regards the power sector in which renewable energy issues are embedded is the Ministry of Power (MoP). Staffed with 307 staff and accounting for a budget of Rs crore 143.399 (10th Five-Year-Plan period 2002-2007), its main responsibilities are to design and oversee the legislative and regulatory framework for power sector development in India[43]. Since according to the Indian constitution power is a concurrent subject, the ultimate responsibility for implementing power sector reforms and for steering the power sector lies with the federal states. In order to meet chronic problems of poor energy services and under-investment, the Government has ventured into a range of electricity market reforms. The most important piece of national legislation is the Electricity Act 2003 that aims to reform the power sector towards liberalisation, private sector initiatives, efficiency and coordination between Indian states – that ultimately promise greater scope for renewable energy. Even though no specific time frames are mentioned, the Electricity Act 2003 seeks to promote competition and to rationalise electricity tariffs in addition to clarifying the role of different organisations. The National Tariff Policy 2006 adds to the reform process in order to make the power sector commercially viable through cost-reflective tariffs and to provide nation-wide electricity access. In addition, the Integrated Energy Policy 2006 contributes specific recommendations to combat the energy challenges facing the country. Prominent features are ambitions to enhance renewables, to link incentives to energy generation, to enact promotional policies and to facilitate financing of renewable energy projects.
Responsibilities for renewable energy and wind sector are institutionalised with the Ministry for New and Renewable Energy Resources (MNRE). Its main functions are to issue policy guidelines and to provide support to the renewable energy sector through various programmes and subsidy schemes. Its history started with the institutionalisation of renewable energy issues in the Department for Non-Conventional Energy Sources (DNES) in 1981 under the Ministry of Science and Technology that was upgraded to a separate ministry in 1992. As of now, the MNRE employs 383 staff and accounts for a budget of Rs crore 7167 (10th Five-Year-Plan period 2002-2007). By the end of the 1980s the MNRE initiated the National Programme for Wind Energy Development. The designated objective of this programme was to induce private sector participation in this field and to render renewable power generation a commercially viable contribution to energy capacity addition (Rajsekhar et al 1999: 670). The programme included systematic wind resources assessments as well as demonstration projects starting in 1986. In 1993, the MNRE developed policy and tariff guidelines for state governments in order to promote renewable energy technologies (Pillai

[43] For this purpose, the Central Electricity Authority and the Central Electricity Regulatory Commission are set up as main executive bodies that coordinate state-level subsidiaries, i.e. State Electricity Boards and State Electricity Regulatory Commissions.

2006: 392). As part of the Indian-wide incentive scheme for the wind sector, central financial assistance was provided by the Ministry's financial arm, IREDA (Indian Renewable Energy Development Agency).

Other relevant stakeholders involved in renewable energy governance at national and international levels are finance institutions, international R&D business and companies working in the CDM business, e.g. CDM consultancies and Designated Operational Entities (DOE).

Power and renewable energy are concurrent subjects with the onus of policy making and implementation residing with the federal states. The central-level public actors are not included in the analysis of local governance arrangements since they provide similar opportunity structures in the energy sector across all local contexts. Private actors operating mainly at national and international level are also not directly included in local stakeholder network analyses as finance and research issues bear the same implications all across India. As elaborated in Chapter III, access and availability of services and financial support mechanisms through international R&D networks, FDI and CDM is not restricted to specific regions or stakeholders.

Narrowing the perspective to the local level enables the identification and selection of relevant stakeholders as result of ethnographic empirical research in Tamil Nadu (Figure XXI):

Figure XXI: Stakeholder Network in Tamil Nadu

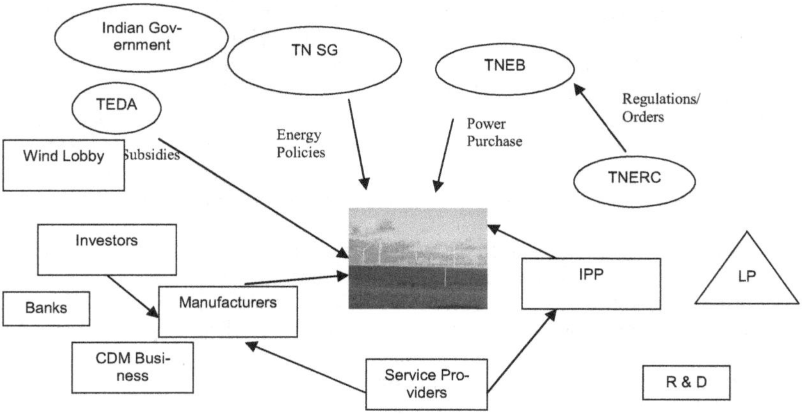

Legend:
TNSG: Tamil Nadu State Government
TEDA: Tamil Nadu Development Agency
TNEB: Tamil Nadu Electricity Board
TNERC: Tamil Nadu Electricity Regulatory Commission
IPP: Independent Power Producer
R&D: Research and Development Networks
LP: Local People

The most important public sector stakeholders for wind energy deployment in Tamil Nadu are state government, specifically the energy department that is responsible for policy making and implementation with regard to energy and wind energy issues. The Tamil Nadu Electricity Board (TNEB) as electricity utility is assigned with the tasks of grid management, electricity infrastructure, wind power purchase and transmission and distribution of wind power[44]. The Tamil Nadu Electricity Regulatory Commission (TNERC) was institutionalised under the Electricity Act 2003 and the Electricity Regulation Act 1998 to respond to the need for greater regulatory oversight and direction in the electricity sector. This semi-autonomous institution is assigned with the tasks to independently arbiter, fix and oversee the implementation of regulations for wind sector stakeholders, i.e. generators, suppliers, transmitters and users of these electricity services. As state nodal agency of the MNRE the Tamil Nadu Energy Development Agency (TEDA) promotes renewable energy deployment at the local level through channelling central-level subsidies, implementing demonstration projects and providing assistance to interested parties.

With regard to the private sector (cf. Chapter III), the most important stakeholders in wind energy deployment in the Indian context are manufacturers[45]. The wind turbine manufacturer Suzlon was selected as representative of this industry due to the fact that Suzlon is an indigenous actor and holds about 50 % of the market share. Increased power sector liberalisation opens avenues for new actors such as Independent Power Producers (IPP) to provide alternative business models in the wind energy segment through third party sale of wind power. IndoWind was selected as most representative IPP as only very few actors have so far engaged in the IPP-mode. Demand for wind power installations emerges to a large extent from investors that are not involved in the wind deployment procedure per se but invest for different reasons such as independent power supply or tax benefits[46]. Ashok Leyland represents a typical case of an investor that has its core business in the automobile sector. Service providers include a range of actors from knowledge brokers, suppliers of different technical parts for wind farm installations, legal and financial service providers to different types of consultants. The Centre for Wind Energy Technology (C-WET) was selected as a service provider that engages in turbine testing and a wide range of consulting activities with regard to wind farm installation procedures[47]. Even though wind lobbies represent and argue the case for wind turbine manufacturers and investors all across India, i.e. at national level, they were identified as particularly relevant to the local context of Tamil Nadu.

Lastly, local people constitute the civil society stakeholder group of the wind energy network of Tamil Nadu. Local people comprise the adjacent citizens that are not organised in any manner and do not have an official stake in the process but are nonetheless directly affected by wind farm installations. Regarding other civil society actors, NGOs and media representatives do not feel compelled to argue the case for wind energy or to defend the rights of affected communities.

In summary, the constitution of local governance arrangements does not have much explanatory power for the success of wind power deployment in Tamil Nadu. In the following, attention is turned to answer who governs in relation to the characteristics of governance actors. On the basis of the explanatory model derived from organisa-

---

[44] Furthermore, issues such as selling wind power to third parties or allowing wind power to be used for captive consumption also lie within the responsibility of the TNEB.

[45] As mentioned in the previous section due to the oligopoly structure of the wind energy market, manufacturers have quite significant stake and power in shaping market and business developments.

[46] Hence power requirements for internal consumption are immense and so are investment appetites as to avail of the tax benefits promised.

[47] Even though C-WET also operates on public funds, it functions as a quasi-private undertaking and most importantly extends its services across all India which is the reason for selection.

tional sociology, the three characteristic features of capabilities, capacities and interests are explored in turn with regard to the local governance actors.

Step II – Characterising Actors' Attributes: Capabilities
The first approach to characterise actors' attributes relates to their capabilities in renewable energy governance and looks at the experience of stakeholders. This describes the background and history of individual stakeholders with focus on wind energy deployment in the context of Tamil Nadu. According to organisational sociology, the history of experience entrenched in a specific societal culture (Meyer & Rowan 1977) is important for the outcome and perception of wind energy engagement. For example, negative experiences with failed wind farm installations can be turned positively into lessons learnt for future activities. Of course, negative experiences especially at early stages are likely to be perceived and evaluated in more general terms as failure of the entire wind energy undertaking. This leads to prejudices against as well as mental barriers and obstacles to further wind sector engagement. The history of experience explains how common knowledge, mutual understanding and relations between stakeholders come about and have evolved[48].
The following analysis will discuss governance actor characteristics with regard to capabilities, i.e. history of experience in renewable energy governance in Tamil Nadu.

(i) Highly-Favourable Wind Power Capabilities
The Tamil Nadu state government was very proactive in terms of wind energy promotion at quite an early stage. Tamil Nadu was one of the first states to adopt the non-obligatory MNRE wind power guidelines into state legislation and thus designed a favourable framework for wind power deployment (INT 1). Despite government changes over the last decade, annual policy making for the wind sector has remained stable and consistent. Consequently, state government capabilities steering wind energy development through the design and implementation of appropriate policies are characterised as high.

As counter-part to the government, the TNEB's capabilities with regard to the wind sector are also characterised as high. During the very early stages of wind sector development, the TNEB was rendered an important stakeholder in demonstration projects. As owner of wind projects and as buyer of the electricity generated, the TNEB gained first-hand experience with the realities and potentials of wind energy deployment. The TNEB has hence accumulated substantial knowledge as regards the operational business and ownership felt for wind energy. This goes beyond the usual role of utilities as buyer of power and signatory to power purchase agreements.

Manufacturers like Suzlon carry a wealth of experience not only with regard to their own organisational history but also looking back at its antecedents which refer to international joint ventures and mergers. Particularly in Tamil Nadu, as one of the first states Suzlon commenced its business, operations are diversified through backward integration alongside the wind turbine production chain. Suzlon did not only internalise the marketing of wind farm installations but also the manufacturing as well as indigenous production of individual components particularly in locations in Tamil Nadu.

[48] The implicit historic dimension related to the acquisition of capabilities and evolution of stakeholder networks is subject to analysis in the last section of this chapter.

In response to emerging CDM opportunities, Suzlon by now outsourced an own consultancy business, Senergi Global.

(ii) Significant Wind Power Capabilities

Service providers such as C-WET are characterised by significant capabilities in the wind energy sector in Tamil Nadu. This is due to their set-up during the initiation stage of wind energy development in response to the demand for better quality and servicing of wind turbine machinery. Going beyond that, C-WET's relation to the international R&D community contributed to the transfer of many fruitful lessons learnt. C-WET does not only participate actively in the wind energy development process but also accumulates a common pool of experiences and knowledge across public and private sectors.

(iii) Medium Wind Power Capabilities

Ashok Leyland as investor in wind power installations has its main business operations in the automobile sector in which it looks back at a success story in Tamil Nadu. Both its core business and its wind sector engagement have some historical background. Investors experienced changes in regulations pertaining to the subsidy and taxation regime as well as through economic crisis such as the Asian Crisis that also affected the wind business.

TEDA was an important participant in the first wind energy demonstration project at the very early stages of wind energy development in Tamil Nadu only. Even though, TEDA constitutes a useful vehicle for channelling subsidies, its involvement in the wind sector today is negligible. Due to the insignificance of its mandate and role in wind power deployment, its capabilities regarding wind energy deployment are characterised as average.

(iv) Insignificant Wind Power Capabilities

Over the last couple of years, independent power producers (IPP) such as Indowind have emerged as new actors in the Indian wind scenario. This is predominantly the case in states like Tamil Nadu where regulatory framework, policy implementation and utility attitudes are not overly averse to third party sale. Ever since IndoWind entered the stage, a new knowledge base on how to provide an alternative model of doing business in the wind sector was created. However, IndoWind's activity in the Tamil Nadu wind energy scenario is relatively short-lived and hence capabilities with wind energy development in Tamil Nadu are still insignificant.

As the other newcomer to the Indian power sector scenario, the Tamil Nadu Electricity Regulatory Commission (TNERC) entered the stage in 2002[49]. TNERC was institutionalised relatively late, arguably due to existing stable policy directions and purchase policies as well as vested interests in the status quo. Due to the late entry into the wind energy scenario at a stage where many stakeholders and processes were well established, capabilities with wind energy deployment are rather insignificant.

[49] In contrast to other Indian state, Tamil Nadu has not been among the latest actors to set up the regulatory commission.

(v) Low Wind Power Capabilities

Without doubt, local people are the stakeholders the most directly affected right from the inception of this business. Local people are characterised by hardly any experience in wind energy deployment as they neither have official stakes in the deployment process nor are actively engaged in any form of protest.

Figure XXII: Actors' Capabilities in Tamil Nadu

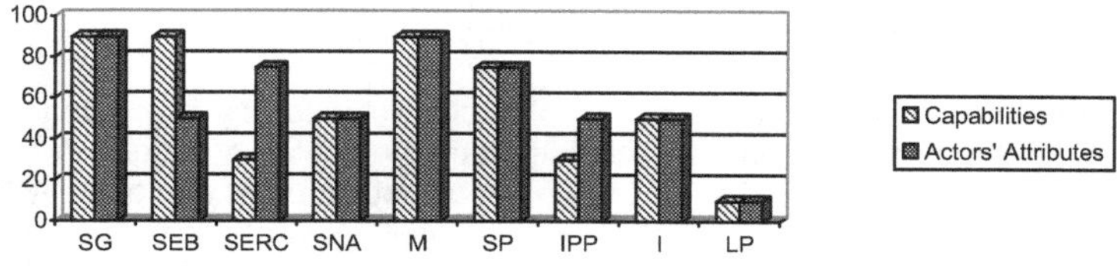

In summary (Figure XXII) the Tamil Nadu stakeholder network is characterised by actors' capabilities that range between remarkably high for public actors and Suzlon to relatively insignificant as regards new forms of public and private engagement, e.g. IPP, SERCs. In the light of organisational sociology literature, actors' capabilities reflect and have over time mutually constituted a societal environment in which wind power is perceived and discussed in a rather favourable manner. This translates into organisational cultures and norms supportive of wind power that promote functional organisational behaviour.

Step III – Characterising Actors' Attributes: Capacities

The second characteristic feature of actors' attributes concerns stakeholders' capacities and resources. Capacities comprise the stakeholders' endowment with internal and external budgets earmarked for wind energy, the staff available for wind energy issues as well as its qualification, and lastly autonomy and independence in resources use. These capacities are derived from certain legal, political and institutional sources. According to organisational sociology (Vaughan 1999), capacities are part of governance actor characteristics and shape organisational structures, practices and tasks. These determine how and with what effects governance actors are empowered to embark upon renewable energy governance.

(i) High Wind Power Capacities

The Tamil Nadu state government is characterised by high capacities. As regards resources, Tamil Nadu state government allocated Rs crore[50] 29.65 for renewable energy out of Rs crore 8028.65 for the entire energy sector during the 10th Five-Year-Plan period (2002-2007). This means that indeed 20 % of the 10th Five-Year-Plan is dedicated to the energy budget (Tamil Nadu State Planning Commission 2006). In comparison to other Indian states this is quite significant. However, one needs to take into consideration that due to electoral promises of free power for farmers the Tamil Nadu power sector also faces high expenditures. With regard to human resources, the state government assigns a distinct department with the responsibilities for energy and renewable energy issues. High-level staff working in this department on a long-term basis is characterised as well qualified and knowledgeable in the field of renewable energy (INT 3). On the one hand, this is due the background of the energy

[50] Rupees: Rs lakh 1 = Rs 100.000; Rs crore 1 = Rs lakh 100 = Rs 10.000.000

secretary in the Indian Administrative Service (IAS) who plays an important role in energy-related policy making[51]. In the state context of Tamil Nadu, an IAS background of state officials is positively associated with an efficient and integer personality. These posts are not only high in demand but also require substantial qualification (INT 2). State officials such as the energy secretary also benefited from exchanges with external consultants and industrialists for example through site visits to Germany that *"engendered the potential of renewable energy and specifically wind options into our minds"* (INT 3). In addition, internal training programmes render government staff not only aware but also educated on wind energy issues.

The wind turbine manufacturer Suzlon is also characterised by high capacities with regard to wind energy deployment. This is reflected not only in budgetary resources recurred from operations in India ($ 208 Mio in domestic orders) and globally ($ 4,304 Mio orders abroad) but also in the staff development. With a significant staff size of about 9500 employees from 12 different nationalities, regular internal and external trainings are offered (Suzlon 2007). This qualification potential, the job prospects and the promising business projections render Suzlon a very attractive and competitive employer in the wind energy sector. Apart from wind farm installations Suzlon operates several manufacturing plants and training centres in Tamil Nadu.

(ii) Significant Wind Power Capacities

C-WET as service provider possesses quite significant resources. Its budget recurs not only on internal resources from government (Rs crore 4 in 2006) but is also added to by international assistance and own earnings. C-WET disposes of quite a lot of staff of around 40 people (C-WET 2008). A significant number of professionals experience intensive training and regular exchange with international knowledge hubs, e.g. UNEP Risoe, and R&D partners. This renders them highly knowledgeable not only with regard to national circumstances but also with reference to the international context. C-WET is also involved in hard and soft technology transfer. Undoubtedly, this contributes to C-WET's comparative advantage and autonomy vis-à-vis other public and private stakeholders.

With regard to the group of investors, Ashok Leyland's budgetary position and staff size is very substantial (Ashok Leyland 2007). However, only very few people are actually involved in wind energy operations as none of its day-to-day operations concern wind power with exception of procurement issues. In spite of limited resources in this regard, one might argue that investors have significant capacities due to their financial power. Resources invested in wind energy installations are accrued independent from the whims of financial institutions given its financially stable core business. As well-reputed customers with established operations in other business segments such as cement, textile, automobile or other industries and with known balance sheets to financial institutions, obtaining finances is not an issue.

As one of the first IPPs to enter the wind energy business in Tamil Nadu IndoWind's capacities are surprisingly significant. Financial capacities are based on internal capital resources in addition to their ability to find investors or obtain finances from the market and the early engagement in international CDM business. Despite this newcomer status, a high number of well-qualified staff is designated to diverse opera-

[51] More so than the energy minister that functions as representative or at the most signatory.

tional tasks. These range from the installation of wind farms, marketing wind farms as complete products to customers, i.e. investors, or negotiating power procurement options with public actors to catering for power purchase and qualifying for CDM status.

TNERC as public actor is characterised by significant capacities in the wind energy sector in Tamil Nadu. This is not related to its quantitative capacities as the national electricity legislation delineates constitution, composition and resources of regulatory agencies in order to render them comparable across India. The TNERC disposes of a staff of about 21 people and a budget of around Rs lakhs 113.60 commissioned by the state government in 2006 (TNERC 2006). Differences between regulators across India emerge in the composition of the commission, staff qualification, regulatory output in terms of regulations and petitions as well as attitudes towards renewable energy. With regard to the TNERC, most of the commission staff has a background in the Indian Administrative Service System (IAS). Given the Tamil Nadu context, this contributes to its good reputation as non-judgemental and well-qualified. Staff quality is upheld by regular training programmes and workshops as well as by the constitution of consultative committees calling upon expert staff as regards tricky issues regarding the wind sector. The high number of petitions against rulings and orders reflects TNERC's autonomous and independent decision-making. This speaks for the regulator's impartiality since parties with varying stakes and interests in regulatory processes are not necessarily effective in applying informal means, i.e. drawing back on hard or soft persuasion, to argue their case.

(iii) Average Wind Power Capacities

The fact that the TNEB is characterised by average actors' capacities in terms of resources and capabilities seems at first sight astounding as generally speaking, the TNEB is perceived a powerful player. Its staff size of about 77.647 people is relatively high in comparison to other Indian states (TNEB 2006). As implementing agency of the state government's free power policies the TNEB is assigned with the crucial policy-relevant tasks. The TNEB has a separate division for non-conventional energy sources with designated budgets and staff that experience regular trainings held internally or sponsored by development assistance. Yet, staff qualification has often come under dispute (INT 4) particularly as regards issues of grid management, maintenance and power planning. Recently, the more or less healthy financial condition of the TNEB is reversing[52]. The weak point concerns the TNEB's autonomy and independence in decision making from state government. Even though execution of power purchase and clearance procedures for wind farm installations are in the hand of the TNEB, general wind policy directions are clearly determined and implemented by state government.

The other public actor characterised by medium capacities with regard to resource endowment and capabilities is the state nodal agency TEDA. Due to the interest of foreign investors from Denmark and DANIDA the internal budget is not only composed of MNRE resources, which are Rs crore 20 of 2006, but also added to by external funding. However, in total, renewable energy are only assigned a share of 0.4 % of the total energy budget (TEDA 2006). In comparison to other Indian states, this figure is not breathtakingly high at all. One might speculate that the reason for this insignificant budget lies in the fact that renewable energies specifically in the wind

[52] Revenue accounts deficit in 2006: - Rs lakh 13.5521 (TNEB 2006)

sector are better established and on the route to commercialisation. In addition to the limited budget, the staff size of 30 people is also not impressive. However, due to exchange with international partners, staff qualification is characterised as experienced and knowledgeable.

(iv) Low Wind Power Capacities

The only stakeholder characterised by low capacities are local people. They are relevant only in the very early stages of the wind energy deployment process as concerns issues of land acquisition. Since the Tamil Nadu legislation does not require environmental impact assessments for ordinary wind power installations that render public consultations mandatory it is up to individual wind entrepreneurs to negotiate construction details and land purchase with local people. Local people construe the weakest link in the wind sector chain since in the Tamil Nadu context the rural population living at the windy sites is not well educated. Neither do local people possess any financial resources to organise their interests. No environmental lobby or local media feels called upon to argue a potential case. Even though local people in theory dispose of substantial bargaining power due to the dependence of government on their votes, in reality they do not or cannot make use of it because of lacking awareness or skills to organise their interests. Furthermore, local representatives are easily acquiesced by the wind business.

Figure XXIII: Actors' Capacities in Tamil Nadu

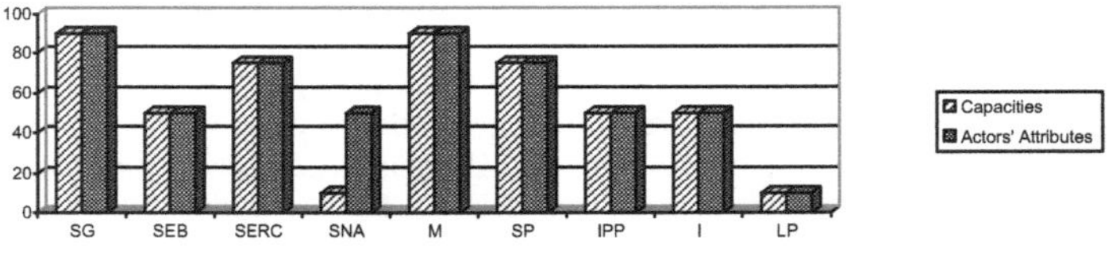

In summary (Figure XXIII), most of the private and public stakeholders are characterised by relatively significant capacities oriented towards renewable energy deployment. According to Organisational sociology, organisational structures, practices and tasks assigned to the wind business express support to wind power due to resources invested and capacities acquired (Vaughan 1999). Firstly, the better equipped the staff and the more flexible the procedures and structures, the more information exchange is facilitated and transaction costs are reduced. Secondly, well-informed and integrated staff also contributes to the generation of trust among organisational members that leads to more productive and constructive functioning and outcomes. This incurs internally low transaction costs, facilitated information exchange and trust in the wind energy undertaking. Consequently, relatively high capacities contribute to shaping the characteristics of governance actors in favour of effective renewable energy governance.

Step IV – Characterising Actors' Attributes: Interests

The third approach to characterise actors' attributes in stakeholder networks is to look at the interests and attitudes prevailing at individual basis as well as collectively within organisations. Organisational sociology takes up the importance of individual as well as collective cognition entrenched in the organisational culture, leadership or political will (Vaughan 1999). Causal arguments stipulate that internal characteristics

and processes that display interests and attitudes influence organisational activities and their effects on wind power deployment. Examining the interests and attitudes characteristic of actors' attributes tackles the following questions: Do actors promote wind energy deployment because it is simply their job or are they motivated by more than that and pursue other agendas? And, how do stakeholder attributes in terms of interests and attitudes impact on wind power deployment? Interests reflect on the one hand, the official roles and functions of stakeholders and on the other hand, the actor's intrinsic motivation to get involved in the wind power business. Attitudes describe both the reputation of an individual leader as well as that of an organisation. Individual attitudes are captured by the proactiveness and initiatives started by a high-level official within an organisational context. Organisational attitudes are less tangible as they refer to the accessibility to and the efficiency with which an organisation works regarding wind power deployment.
First inductive empirical research points towards the fact that in many cases individual and organisational attitudes and hence reputation diverge, which influences renewable energy governance (INT 5).

(i) High Wind Power Interests

The Tamil Nadu state government is characterised by a consistent and positive mindset that reflects high interests in wind energy (INT 26). Wind energy is perceived a valuable asset in the Tamil Nadu energy mix. Furthermore, the reputation not only of the energy minister and secretary but also of the Tamil Nadu Chief Minister is favourable of wind energy (INT 13). This is partly triggered by the industrial demand for power vis-à-vis the actual power supply situation. State government's main interest is to respond to the demands of its constituency and to cater for the welfare of Tamil Nadu state. Wind power has a stake in the efforts to meet power shortages, and industry demands, and stick to electoral promises of free power. State government also increases its economic and international profile establishing wind energy as a thriving business in Tamil Nadu. The personal relations between officials at the MNRE being of Tamil Nadu origin and state government officials contributed to more openness towards new ventures into the wind business. MNRE officials put significant amount of resources and efforts into persuading state government officials to opt for the wind energy route (INT 6). Yet, once successful not only individual leaders but also the organisation, i.e. state government, as a whole obtained a reputation of being proactive and favourable in terms of wind energy promotion.

Suzlon as private actor has high interests in wind energy deployment. Suzlon's main business is to manufacture wind turbines and cater for the entire wind power process – providing all-around services to their customers. Suzlon's interests are most and foremost profit-oriented towards increasing the returns on investment. Suzlon wears with pride its status as the first and the only Indian indigenous wind turbine manufacturer. Beyond its ambitions to expand business activities globally, Suzlon seeks to venture even further into the Indian market. In the case of Tamil Nadu, this means expanding operational activities and offering new kinds of services such as CDM investment to their clients. Attitudes of individual leaders and organisation converge along similar lines. "*Pro-activeness in promoting wind energy is written in capital letters not merely as a profitable business but also based on other interests, last but not least 'making India big' on the global market*" (INT 7).

As state nodal agencies of the MNRE the Tamil Nadu Energy Development Agency (TEDA) is as a matter of fact supportive of wind energy deployment since its existence is justified on the grounds of renewable energy promotion. TEDA's role as partner for the wind sector has changed since TEDA nowadays features quite insignificantly as facilitator of wind energy deployment. TEDA provides assistance in wind resources assessments, and linking interested parties still holds strong interests in wind energy promotion. Surprisingly, TEDA's chairman has the reputation of nonchalance and laissez-faire leadership not in the driver seat in terms to steer the organisation and give it a special place or role in the wind energy deployment process. The lack of individual leadership clashes with the organisational culture, i.e. the proactive organisational lead in wind power. Organisational routines are still characterised as efficient and supportive of wind power, and the organisational reputation is positive not only due to assigned functions but also due to the prevailing working culture.

(ii) Significant Wind Power Interests

Interests and attitudes of the regulator, the TNERC, are characterised as significant. So far, regulations pertaining to the wind sector have translated existing policies into a stable and consistent framework favourable of wind energy deployment. The level of arbitration in rule setting leaves the organisation with a controversial reputation. While the private sector complains about a potential bias towards government bodies, the high number of petitions allow for assuming that its operations are highly contested yet thoroughly exercised. On the whole, the TNERC is perceived to operate according to the principles of fairness and to display an inherent interest to align the different viewpoints on wind energy regulations (INT 14).

C-WET as service provider in the wind energy sector reveals significant interests in wind energy deployment. C-WET came into being on the initiative of the MNRE given the pressure by the wind industry for reliable information, knowledge and services related to the siting of wind projects as well as the adequate fit of turbine technology to the chosen location. C-WET's interests in wind energy are of commercial and of altruistic manner. On the one hand, C-WET earns profit, diversifies its business in the wind energy sector and engages in market expansion activities. On the other hand, C-WET caters for the growth of the Indian wind energy sector in general and contributes to international technology transfer. Despite the lack of sufficient indigenous research and qualification, C-WET staff has a favourable reputation and is held in high esteem also due to its international linkages.

(iii) Medium Wind Power Interests

Investors are characterised by average interests and favourable attitudes towards wind energy. Ashok Leyland as a typical case of an investor in wind farm installation follows two main interests: This first one is to secure power supply for the own highly electricity-consumptive manufacturing plants, which is independent of the occasionally unpredictable and unstable grid power. Wind power as an alternative to grid electricity holds some promises. It circumvents not only problems and past ambivalent experiences with the TNEB that at times forces them to shut down operations due to power shortages but also pre-empts aggravating power scenarios. The second reason is to avail of the tax benefits offered on investment in the wind business. This is due to renewable energy policy making by central government that aims to encourage investment. However, as this form of policy support is highly contested and criticised for its inefficiency, it seems a matter of time that incentive structures will be changed.

Independent power producers are characterised by average levels of interests towards wind energy. IPPs such as Indowind specialise in the business of wind farm installation and sell the image of a 'green power producer' as marketing feature and comparative advantage to their clients that are private or public power consumers. The main interests are, on the one hand, to gain market access and make profits exploring a market niche. On the other hand, deliberately selling a green image, IndoWind embraces social constructivist rationales about the influence of norms and values. Due to the uniqueness and early timing of its activities, an organisational reputation that is taken seriously and shared by other stakeholders is still lacking. Many private actors (INT 8) regard IndoWind as a newcomer with a business concept still to be proven viable. However, its ability to survive and establish a niche so far has received some acknowledgement and recognition as innovative and courageous player.

Generally speaking, utilities such as the Tamil Nadu Electricity Board (TNEB) are suspicious of wind energy due to the different issues experienced in wind electricity generation (cf. Chapter III) – some of them perception-based, some of them de facto technical-operational, others due to regulatory specificities. The TNEB constitutes one of the very few exceptions to the generalised hostility to wind energy displaying average interests in wind energy deployment. The wind power facility of the TNEB was construed as a single-window clearance system which means that the entire application and clearance procedure for wind farm installations go through one single institution. In addition to the fact that it is the only buyer of wind-generated electricity, this renders the TNEB extremely important. The TNEB has commercial interests to support wind energy deployment as it earns from procuring wind power. This is not necessarily forced upon them as renewable portfolio standards for the energy mix are by far overachieved in Tamil Nadu. Rationality is one key driver as the TNEB is responsible for satisfying the increasing power demand while fulfilling the promise of free power. Individual and organisational attitudes are also perceived as rather favourable to wind energy. This is reflected in the rapid and efficient clearance procedures and power purchase negotiations.

(iv) Low Wind Power Interests

Lastly, local people display both low interests as well as unfavourable attitudes towards wind energy deployment. In fact, they do not earn much from wind energy installations as land purchases take place informally and without them being in a powerful position to bargain – if at all – for a fair price. Since they do not benefit from wind power installation and suffer from (un)intended consequences such as noise and pollution, interests in wind power are low. The fact that no violent protest or organised action has so far emerged from the local level can only be attributed to the lack of resources and the dominance of the pro-wind stakeholders.

Figure XXIV: Actors' Interests in Tamil Nadu

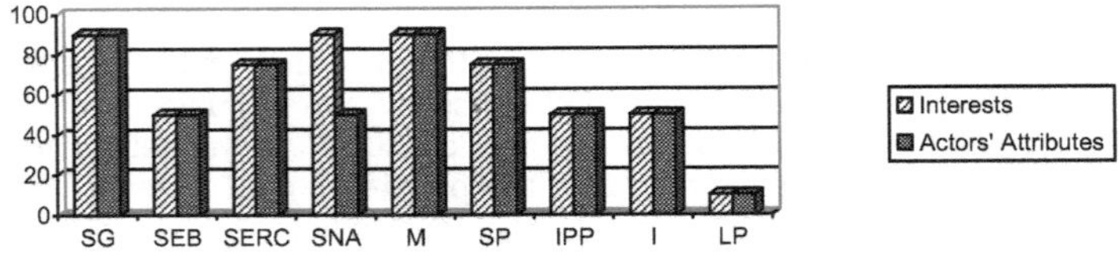

All in all, interests and attitudes of stakeholders, particularly public and private actors, that dominate the wind business are characterised as rather favourable of wind power. This reflects not only proactive leadership at the top and at other organisational levels but also implies the internalisation of organisational routines that support wind power governance. Organisational sociology supports the observation that if stakeholders are characterised by certain interests in promoting wind energy and individual as well as organisation attitudes are favourable of wind energy, deployment activities will take of.

Step V – "Actors' Attributes" in Tamil Nadu's Stakeholder Network

As summarised in Table 26, public and private actors' attributes constituting the Tamil Nadu stakeholder network are on the whole characterised as favourable to wind power.

Table 26: Actors' Attributes in the Tamil Nadu Renewable Energy Stakeholder Network[53]

	State Government	*TNEB*	*TNERC*	*TEDA*	*Manufacturer*	*Service Provider*	*IPP*	*Investor*	*Local People*
Capacities	High	Medium	Significant	Low	High	Significant	Significant	Significant	Low
Interests	High	Medium	Significant	High	High	Significant	Medium	Medium	Low
Capabilities	High	High	Insignificant	Medium	High	Significant	Insignificant	Medium	Low
Actors' Attributes	High	Medium	Significant	Medium	High	Significant	Medium	Medium	Low

This supports the proposition that successful wind power deployment in Tamil Nadu is explained by a stakeholder network constituted of actors characterised by favourable organisational contexts with high interests in, capacities and capabilities for wind power (Figure XXV):

[53] Indicators and threshold values are elaborated in the light of organisational sociology in Chapter IV.

Figure XXV: Actors' Attributes in the Tamil Nadu Stakeholder Network

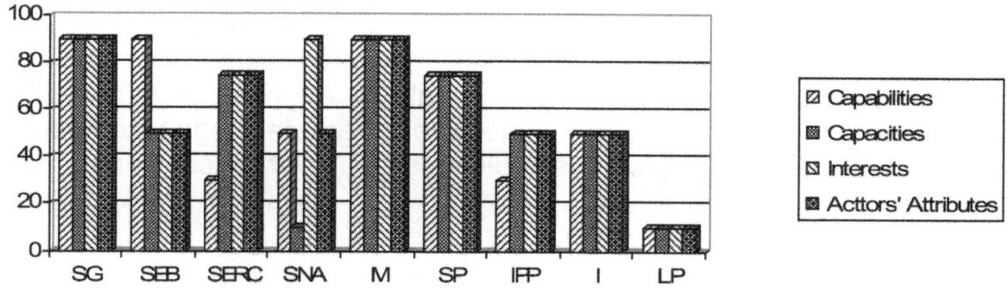

Characterising the local governance arrangement in Tamil Nadu reveals three outstanding features: Firstly, private actors dispose of substantial resources and also experience in wind energy deployment. This leads to an active and persistent investment and engagement in the wind business that has evolved into a highly attractive market. Secondly, characteristics relevant for effective renewable energy governance regarding public institutions are staff allocation and more specifically staff qualification and political will, individual and collective attitudes and interests. Thirdly, local people feature as outliers implying the lack of voice and importance.

In the light of organisational sociology, characteristics of who governs bear three implications for the requirements for effective renewable energy governance: Firstly, a societal environment favourable of wind energy reflects on stakeholders who establish certain cultural identities such as wind power entrepreneurs or forerunners. This is particularly the case for public stakeholder like state governance and private actors like Suzlon. This supports the argument that high capabilities in terms of historic experiences contribute to wind engagement in the first place. Secondly, stakeholders' resource characteristics emerge as relevant not necessarily in quantitative but qualitative terms. This aspect turns out to be important for public authorities since it translates into reduced transaction costs and emerging trust in interactions supportive of wind energy deployment. Thirdly, interests and attitudes favouring wind power are embedded in organisational routines and leadership particularly of private stakeholders. These are also shared by public stakeholders and lead to non-deviant organisational behaviour. Consequently, implementation practices and business engagement are more likely favourable for wind energy deployment activities.
The effects of stakeholder characteristics as regards who governs trigger down to the following mechanisms. Well-equipped and –experienced public sector staff contributes to the streamlining of wind farm installation procedures which reduces administrative hurdles. Political stability and credibility are the result of reliable and well-established procedures, policy making and implementation within the public sector and shape a supportive investment climate. This stimulates wind power deployment and as multiplication effect attracts even more private sector players into the wind business. Adding to the private actors' resource endowment and experience, this contributes to the diversification, commercialisation and professionalisation of renewable energy stakeholder networks in Tamil Nadu. Immediate implications are reductions in access costs for new types of business such as IPPs and CDM as well as available innovation opportunities. In the long run, this will contribute to the survival and transformation of wind business adapting to changing global circumstances as well as securing energy supply in the state of Tamil Nadu.

This section analysed characteristics of who governs, i.e. actors' attributes of the Tamil Nadu stakeholder network in the wind sector. Two aspects remain subject to interpretation in Chapter VI, and concern how stakeholder characteristics trigger certain network effects, i.e. causal mechanisms, and why actors' attributes are constituted as identified by stakeholder network analysis. All in all, the Tamil Nadu case study demonstrated that the favourable attributes characterising stakeholders in local networks contribute to effective renewable energy governance that translates into successful wind power deployment.

V.1.2 How is Renewable Energy Governed?: Actors' Relations in Stakeholder Networks

This section will address the question how renewable energy is governed. Relations and interactions in stakeholder networks are explored as conditions for effective renewable energy governance. The objective is to look at how public, private and non-governmental actors are embedded in dynamic relations in the local context in order to understand the effects of stakeholder networks on wind power deployment[54]:
Firstly, the role of relational arrangements between governance actors characterised by dyadic ties between individual actors is considered, i.e. how stakeholder relations display elements of reciprocity, social capital and formality. As specified in Chapter IV, three criteria are applied to assess how stakeholders are embedded in relations in the local context. Reciprocity comprises the extent, effort and quality of interactions, social capital describes the mutuality and personal knowledge involved in relations. Lastly, the qualitative strength of stakeholder embeddedness is characterised on the basis of formality inherent in relations. The analysis of dyadic ties between various stakeholders in a network gives rise to the identification of actors' arrangements on the basis of relational typologies, i.e. stakeholders that are embedded in this network through similarly characterised ties.
Secondly, on this basis, the interaction constellations between embedded stakeholders in the renewable energy network are explored. This focuses on the extent to which interactions are structurally open or restricted to certain groups. The conventional way to cluster stakeholders in renewable energy networks considers their functional characteristics as public, private or civil society actors groups. In this dissertation, interaction constellations, i.e. certain cliques or groups, are identified on the basis of similarly embedded ties that go beyond pre-conceived group identities. This allows for characterising the structural openness of renewable stakeholder networks to multi-stakeholder interactions. In addition, empirical observations on renewable energy deployment in developing countries point towards the importance of public-private interactions beyond hierarchical governance and market boundaries. The impact of network structures and dynamics on economic activities such as renewable energy deployment is controversially discussed in new economic sociology literature[55] (Chapter IV). This dissertation argues that open network structures and dynam-

[54] The analysis and interpretation of stakeholder relations is based on qualitative interviews, participant observations and ethnographic stakeholder mapping techniques.
[55] Structural closure of stakeholder networks applicable to closed public actor constellations implies negative impacts on renewable energy deployment. Due to the inaccessibility to private stakeholders, their needs cannot adequately be considered in policy making. Yet, among private actors, closed networks and interaction constellations might be positive for renewable energy deployment as business transactions are pursued much more efficiently due to the greater trust and confidence established.

ics within interaction constellations emerging between public and private stakeholders are required for triggering wind energy deployment.
Thirdly, certain scope conditions that influence the extent and way in which stakeholder networks impact on effective renewable energy governance are discussed. These comprise the environmental context on the one hand, and the focus of renewable energy governance activities on the other hand[56].

All in all, this gives rise to implications on how governance takes place through relational interactions in the overall governance network with effect on wind power deployment. As discussed in the previous chapter IV, new economic sociology (Svedberg 1997) basically argues that renewable energy stakeholder networks characterised by strongly embedded relations and structurally open interactions do not only create conducive investment climate but also insinuate trust in proper process and procedures.

Step I – Dyadic Ties in Emerging Actors' Arrangements
In the following, typologies of emerging actors' arrangements characterised by similar dyadic ties between the public, private and civil society actors are identified alongside the criteria of reciprocity, social capital and formality of relations. Categories of actors' relations with similar tie characteristics are established on the basis of considering variances in the criteria formality and social capital. Justified by empirical observations and studies, the reciprocity aspect is held constant as quantitative measures of stakeholder relations are less important. In the case of Tamil Nadu, three main types of actors' arrangements are identified (Table 27) which will now be looked at in more detail:

Table 27: Stakeholder Arrangements in the Tamil Nadu Stakeholder Network

Tie Characteristics/ Typology	**"Disembedded Actors' Arrangements"**	**"Immediate Actors' Arrangements"**	**"Embedded Actors' Arrangements"**
Typical	High formality – Low social capital "distant – formal ties"	Medium formality – Medium social capital **"immediate – formalised ties"** 24% dyadic ties	Low formality – High social capital "familiar – informal ties"
Mix	High formality – Medium social capital "distant – formalised ties"	High formality – High social capital "immediate – informal ties"	Low formality – Medium social capital **"familiar – formalised ties"** 29% dyadic ties
Hybrid	Medium formality – Low social capital "hybrid – distant ties"	Low formality – Low social capital **"hybrid – immediate ties"** 43% dyadic ties	Medium formality – High social capital "hybrid - familiar ties"

[56] As discussed in Chapter IV, the environmental conditions of the Indian wind energy market are argued as relatively certain entrepreneurial behaviour thus exploitation-oriented.

Tamil Nadu Case I – Immediate actors' arrangements characterised by "hybrid-immediate relational ties"

The majority of stakeholders are involved in relations of low interaction intensity and quality as well as low elements of formality. This means that 43.3 % (i.e. 16 out of 38 stakeholder relations) of emerging actors' arrangements in the renewable energy network in Tamil Nadu are characterised by "hybrid-immediate ties". In the following, this aspect is discussed with regard to six characteristic examples:

Firstly, due to Tamil Nadu's relatively stable and self-sufficient power supply position and due to the fact that energy is in the hands of state legislators, relations between TNEB and MoP are quite distant.

Secondly, as mentioned previously, the role of finance institutions for wind energy deployment in Tamil Nadu is not of outstanding importance; hence the low levels of quality and intensity characterising relations to the private sector. Even though financial service providers are stakeholders in renewable energy governance, private as well as public actors (state government) still draw mostly on internal funds or sources of income.

Thirdly, due to its status as a newcomer to the wind energy scenario in Tamil Nadu, the IPP features low quality and intensity of relations with the TNEB in contrast to established actors such as Suzlon. Even though moves towards a common understanding with regard to tripartite agreements for power purchase arrangements were mentioned (INT 15), this is still on shaky grounds and far from becoming an established and reliable relation.

Fourthly, since local people possess little capacity or voice to interfere with business activities, investing in local stakeholder relations is not considered worthwhile on the parts of private actors. Some elements of reciprocity characterise relations between state government and local people that are based on the exchange of favours; i.e. votes for security of power supply. Since re-election depends on the people's votes, interactions with the local population are considered of greater importance. However, since the local constituency is not particularly vociferous and demanding, relations on the whole are characterised by low interaction quality.

The fifth point is that arrangements between public and private actors particularly with regard not only to lobby groups and IPPs but also to government actors such as TEDA and the TNEB integrate some amount of social capital. This implies that trust is put into the integrity of policy making in terms of setting up regulatory and tariff regimes. In this sense, the high number of petitions against TNERC's rulings might simply be part of the game as some degree of mutual understanding and reliability concern TNERC's decisions and role. Due to the budgetary dependence and the nomination procedure that is determined by state government, relations between regulator and state government reveal more formal elements. The case of the manufacturer constitutes an outlier as relations are of low formality and low reciprocity but involve high social capital. This means that despite lacking mutuality in actual exchange, a high level of understanding and trust prevails in relations to TNERC. The regulator is recognised as neutral arbiter that has the role and responsibility to mitigate different interests but can be relied upon to make decisions that do not disrupt the stable policy and regulatory environment.

The sixth feature describes actors' arrangements between investors and state government that are not based on a great deal of reciprocity but characterised by some trust and mutual understanding. This implies that due to the relative independence of the state economy from the industries engaged in the wind energy sector, exchange of favours happens not on a frequent basis. Even though textile, automobile or cement industries feature significantly in the Tamil Nadu economy, they have reached a status of maturity which renders them independent from government protectionism. Generally speaking, the lowest degree of reciprocity and social capital prevails in relations between the financial sector and private as well as between public stakeholders; in relations between public stakeholders such as the TNEB and the federal level ministries, state government and regulators; and in relations between local people and private actors.

Tamil Nadu Case II – Embedded actors' arrangements characterised by "familiar – formalised ties"

In the second largest typology of actors' arrangements, dyadic ties are characterised by high reciprocity, social capital and low formality. This typology that comprises 29.7 % (i.e. 11 out of 38 relations) of emerging arrangements is characterised by "familiar-formalised ties" and was initially expected to be the most prevalent in Tamil Nadu. Four examples of this type of actors' arrangement are shortly identified:

Firstly, the only actors' arrangement featuring relations with higher levels of formality evolves between TNEB and state government. This is due to the institutional set up and ownership structure that establishes legal bindings and official procedures between these public institutions. Due to the prospect of looming electricity sector liberalisation[57] and due to the relative autonomy of both state government and TNEB, relations are not symbiotic and characterised by overt formalism and bureaucracy. Without doubt, interaction quality and intensity of relations between these institutions is high. Mutual understanding in terms of power sector and wind energy development exists that goes hand in hand with coordination and learning from shared experiences.

Secondly, the most striking feature characterising the embeddedness of private stakeholders in emerging arrangements is the focus on the manufacturer. Manufacturers not only historically uphold intensive and frequent interactions with service providers and investors. They also invest a considerable amount of loyalties and shared resources in these arrangements upon which they base various expectations. In fact, relations to service providers are characterised by high interaction quality and intensity not only concerning manufacturers but also IPPs. Importance is attached to the fact that service providers do not only extend frequent and mutual relations to private sector stakeholders. They also reach beyond national boundaries and are tied in actors' arrangements of informal but consistent manner to the international R&D community. This is not only a one-way-street but characterises mutual network processes that involve a high degree of learning and trust. A common element of private stakeholder relations is the entanglement with the wind lobby that goes beyond the realm of Tamil Nadu to national and international levels. Relations to the wind lobby originate from the lobby groups and are characterised as formal and of medium-frequency. The wind lobby also extends sporadic and informal relations to other national-level stakeholders such as the ministries, banks and beyond that the CDM business.

Thirdly, high reciprocity and average degrees of social capital are characteristic of relational arrangements between manufacturers and investors and IPP. Even though coordination and exchange of favours in terms of tacit information transfer takes place, relations are not as intense with regard to social capital. As for the IPP, this is

[57] This is rendered mandatory by the Electricity Act 2003.

explained by the natural competition for installing wind farms and customer groups that render too great a deal of trust and reliability impossible. Despite some shared experiences and common learning, investors are not regarded as highly reliable. Given potential changes in fiscal incentives structures or the global economy, investment interests in the wind business are not unlikely to fluctuate.

The fourth and final point is that actors' arrangements of high social capital are noticeable between national-level institutions such as the MNRE and public actors such as TEDA and state government as well as service providers. These are based on mutual understanding and trust rather than on the actual exchange of favours.

Generally speaking, as for actors' arrangements involving private actors, reciprocity in relations matters immensely whereas high social capital is important especially for public participants and in the relation with private stakeholders.

Tamil Nadu Case III – Immediate actors' arrangements characterised by "immediate-formalised ties"

Lastly, the third typology of actors' arrangements is characterised by relations of average social capital and reciprocity as well as medium formality. This group comprise roughly 24.3 % (i.e. 10 out of 38 relations) of stakeholder arrangements constitutive of "immediate-formalised ties" as three examples will illustrate:

Firstly, with regard to public stakeholders, relations between state government, TNEB, TEDA and TNERC are intensive reflected in the number of interactions and the frequency of exchange. Actors' arrangements involve relations of relatively higher degrees of formality that refer to the official administrative structure, i.e. TEDA being the state nodal agency of the MNRE. Reciprocity or social capital also features significantly in interactions between these stakeholders. This is due to the fact that TEDA considers its role to support state government in its ambitions to scale up wind energy in Tamil Nadu. Relations between state government and TNEB as well as state government, TEDA and MNRE are based on a shared history of wind energy deployment. These are characterised by formal and informal elements of coordination and occur on a reciprocal basis. The interactive quality reflects relatively high mutual trust, shared learning experiences, similar reputation, and strategic information transfer. On a more general level, this means that the role of wind energy is acknowledged and agreement exists that a stable and credible policy and regulatory framework is required to encourage private participation in the wind business. Furthermore, processes and procedures for wind farm installation, e.g. clearances, purchase agreements etc., need to be facilitated- This requires service-oriented and constructive action and coordination on the parts of the public stakeholders involved, i.e. the TNEB, TNERC, TEDA.

Secondly, due to IndoWind's pioneering status actors' arrangements between IPP and service providers also fall into this category. Even though actors' arrangements are of some relational quality and interaction intensity, ultimately, sufficient time is required for significant trust and reciprocity to embed into respective relations.

Thirdly, actors' arrangements that symbolise the global reach of private sector emerge in relations to the CDM business. Interactions slowly start to emerge and originate from private sector stakeholders in Tamil Nadu which is why the quality of dyadic ties is still characterised by lower social capital. In this context, IPPs, manufacturers and service providers engage in relations of medium levels of reciprocity and social capital. Mutual understanding, coordination and learning are the most important aspects of these relationships. Due to the lacking proximity and the lengthy reach of business transactions, not the same amount of trust and exchange is likely to evolve as with stakeholders operating in the local context (INT 16).

In summary, the majority of actors' arrangements constituting the renewable energy stakeholder network are characterised by lower levels of interaction quality and intensity and lower to medium levels of formality. With regard to the formal element inherent in relations, either legal obligations or some kind of semi-official dependence due to membership in and nomination of executive boards exist. This finding implies that the local stakeholder network is characterised predominantly by actors' arrangements of immediately-embedded relations.

Step II –Interaction Constellations in the Stakeholder Network
After identifying actors' arrangements according to the intensity and quality of relations between stakeholder, attention is turned to structural and dynamic aspects in stakeholder networks. This refers concretely to interactions within the structure of network relations. In this respect, the extent to which stakeholders embedded in certain arrangements extend relations beyond that of the established public, private sectors or civil society groups is examined. This aspect captures the degree to which public and private divides are actually crossed. The underlying assumption is that the greater the interactions between public and private groups, the more the accessibility of and exchange between actors is facilitated which ultimately contributes to effective renewable energy governance. In the light of the governance debate, this serves to explore how wind energy is governed. This considers the emergence and role of functionally equivalent governance arrangements to provide for electricity service delivery through the successful deployment of wind energy[58]. In the following, three actors' constellations are identified analysing the interactions between public and private stakeholders in renewable energy stakeholder networks:

Public-Public Clique Constellations
Public stakeholders are clustered in relational constellations characterised by informality and high degrees of loyalties, resources, learning experience and coordination. Due to a shared history and learning, sufficient trust and expectation certainty is established. Strong mutual dependencies and formalistic ties do not constitute the characteristic features of public-public relations. Still, relations between state government, TEDA and TNEB are characterised by structural closure which means that more than half of the interactions take place within the group of public actors. This emphasises the centrality of interactions and policy-making among the public sector constituents. TEDA exhibits the most confined network position among all the public actors. On the other end of the spectrum, regulators display relatively open network structures which supports the argument of their impartiality and openness.

Private-Private Clique Constellations
The great intensity of relations among private actor constellations in addition to the structurally open interactions with private actors crossing state-level boundaries is striking. Interactions in constellations of crucial private sector stakeholders occur on a relational basis involving high levels of trust, resource and information sharing and

[58] Cf. assumption of governance framework that successful deployment automatically implies the effective electricity service delivery. This is a simplifying restriction that does not meet Indian realities since transmission and distribution are crucial intervening factors but is required for research feasibility reasons.

mutuality. Particularly manufacturers and service providers maintain very open interactions in this network structure that extend not only to other private but also to public actors. As regards IPPs and investors, actors' arrangements are more closed and interactions take place mainly with other private actors. Investors are one exception since they establish closed networks that mostly involve private sector participants. Since the manufacturer functions as central figure for the engagement of investors in the wind business, maintaining relations and engaging in interactions that go beyond the private actor constellation is not necessarily required. Due to its status as a relative newcomer to the wind energy scenario, IPPs put greater emphasis in establishing relations and interactions in arrangements with private sector participants, i.e. securing suppliers, consultants etc.
All in all, interactions between private stakeholders are intense and the respective actors' constellation is characterised by relations of high intensity and also quality reflecting the mutual dependencies in the Indian business model. Going global also requires private actors to extend network relations and interactions to a less intense extent beyond state-level boundaries in the Tamil Nadu case.

Public – Private Clique Constellations
The striking feature of the Tamil Nadu renewable energy stakeholder network is the structural openness of interactions and network relations between public and private stakeholders. As briefly discussed alongside some examples, evolving constellations are characterised by relations of some degrees of informality and a considerable amount of trust and expectation certainty.
With the central focus on the manufacturer, private stakeholder constellations engage in medium- to lower-intensive interactions with public stakeholders specifically the TNERC and TNEB but to less an extent also state government. These network constellations are characterised by a lower number of interactions, some degrees of informality and mutual understanding. As for state government, relations to manufacturers are important and suggest the openness and access of policy making to external input. Apart from manufacturers, also IPPs gain access to the TNEB. Relations to government actors such as state government, TNERC and TEDA dominate its network of relations. The insignificant role of TEDA in wind energy deployment processes is reflected in its limited interactions with private sector stakeholders. In contrast, TNERC is characterised by structural openness of relations which reflects its position as arbiter between the interests of the diverse range of public, private and non-governmental.
Even though interaction degrees among respective public as well as private actors' arrangements vary, a strict differentiation between private and public stakeholder groups cannot be maintained in the Tamil Nadu renewable energy network. Without doubt, tendencies for internal clique-building exist as concerns public stakeholders, i.e. state government, TNEB and TEDA, and also private stakeholder, i.e. investors. All in all, interactions between actors in the renewable energy network in Tamil Nadu are characterised by relative openness of differently embedded relational arrangements. This implies that interactions and transactions between stakeholders, i.e. private as well as public parties, are facilitated. With regard to the interaction effects, reduced access and information costs in addition to instilled trust and expectation certainty across private/public relations promotes wind energy deployment.

Step III: Renewable Energy Governance in the Tamil Nadu Stakeholder Network

The aim of this section is to discuss the mechanisms how "actors' relations" in emerging stakeholder networks relate to wind power deployment.

The empirical analysis concludes that actors' arrangements are characterised by relations of relatively high stakeholder embeddedness which contributes to wind energy deployment in Tamil Nadu. The fact that the majority of actors' arrangements put high emphasis on reciprocity and social capital in relations supports the argument about the importance of trust. Confidence and mutual understanding shape a common ground for exploiting opportunities of wind power. This goes beyond the logics of rational decision making based on repeated interactions and factual exchange of securities. The finding that most of the actors' relations also feature low to medium levels of formality and less qualitative aspects than expected seems surprising but is explained by two phenomena: Firstly, a great number of actors extend their relations beyond the local context of Tamil Nadu entering into arrangements with actors either at the national or at the international level. This implies that intensity and quality of interactions cannot be maintained at the same level as in the case with actors operating in closer proximity. Geography matters at least as concerns the way in which relations are established and arrangements are shaped at the initial stages of business activities. Secondly, this observation supports one stream of debate in relational economic geography (Chapter IV) that predicts less intense and lower quality of relations with an increasing maturity and commercialisation of the market (Simsek et al 2003). One might argue that in the case of Tamil Nadu, wind energy embarked upon a development stage that requires stakeholders to be less embedded in networks of relations than in previous phases.

With regard to interactions in stakeholder networks, the analysis of renewable energy stakeholder networks concludes that interactions take place in the context of immediately-embedded ties which promote wind power deployment. The renewable energy network in Tamil Nadu features relatively open structures with many interactions extending to actors other than those of the same public, private or civil society group. This implies greater mutual reliance and the use of informal mechanisms and elements in network processes. Generally speaking, relatively openly-structured actors' arrangements enable interactions across public-private divides and hence promote renewable energy deployment. Network relations among public actors remain sparse with the exception of TNERC that is most open to all stakeholders. Interactions among private stakeholder with the exception of the IPP take place in structurally open network constellations. For successful wind power deployment, the cartelisation of and restricted interactions in private constellations seem to be less advantageous than establishing trustful relations and engaging in interactions with public and non-governmental stakeholders. New economic sociology literature argues that structurally open networks facilitate the exchange of information and assets that reduce transaction costs in doing business. Repeated interactions on this basis lead to the creation of trust and mutual understanding that promotes engagement in wind power deployment activities.

In summary, governance in the Tamil Nadu stakeholder network takes place in relationally-embedded multi-actor arrangements characterised by the tendency towards structurally open multilevel interactions.

V.1.3 Governing Renewable Energy: Stakeholder Networks in Tamil Nadu

In order to draw together the findings of the Tamil Nadu wind power case, the concluding section will summarise the key results and discuss conditions for effective renewable energy governance on the basis of who governs how in the local stakeholder network.

The deductive empirical analysis of wind energy deployment in Tamil Nadu guided by the explanatory model derives at several conclusions on the effects and role of stakeholder networks for renewable energy governance. In the light of the governance debate, stakeholder network analysis identified and characterised stakeholder networks with regard to actors' attributes, arrangements and interactions:
Firstly, with respect to who governs, stakeholder network analysis captured the characteristics of stakeholders. Actors' attributes are characterised as highly favourable of wind deployment due to supportive organisational features fostering agency that facilitates effective renewable energy governance. Empirical analyses confirm the propositions on positive effects of stakeholder characteristics related to renewable energy governance. Firstly, the majority of stakeholders look back at a history of wind power engagement and institutionalised norms and culture that entrench wind power support. Secondly, organisational characteristics, structures and processes feature significant resource endowment and capabilities that in terms of independent and autonomous agency promote wind power. Thirdly, attitudes and interests positively-oriented towards wind power are entrenched in organisational routines and noticeable in organisational leadership. The analytical results amount not only to key private actors such as manufacturers but also to public stakeholders such as government and TNEB that play important roles in stakeholder networks. As key findings, political will as well as quality of capacities (e.g. staff qualification) are crucial attributes required for public stakeholders to promote wind power.

Secondly, in response to how wind energy is governed, stakeholder network analysis captured actors' relations in renewable energy governance arrangements. This regards on the one hand, actors' arrangements emerging on the basis of similarly embedded network relations and on the other hand, interactions in structures of network relations. As proposed by the explanatory model, successful wind energy deployment in Tamil Nadu is traced back to a stakeholder network constituted predominantly of actors' arrangements characterised by immediately-embedded relations. Informal and qualitative dyadic ties embed most of the stakeholders in trustful, cooperative and informal arrangements. This instils trust and renders interactions and transactions less formalistic but oriented towards the common objective of wind power deployment. Since the majority of emerging actors' arrangements are on the whole characterised by immediately-embedded relations, wind power deployment is successful.
As regards interactions in evolving networks constellations, most of the private actors engage in relatively open interactions constellations going beyond local level and group boundaries. While the interactions between public actors are characterised by slightly more structural closure, interactive processes between public and private actors are the constitutive characteristics of renewable energy governance arrangements in Tamil Nadu.
As a word of caution, empirical findings also point towards the surprising outcome that despite successful wind power deployment, actors' arrangements are not em-

bedded in highly qualitative relations and entirely open interaction constellations. Interpreted in the light of the scope conditions, this implies that agency is not fully exploitation-oriented as the environment still features slight uncertainties.

The following stakeholder network mapping will summarise and integrated the results of the descriptive-analytical exploration of the local renewable energy governance arrangement in Tamil Nadu:

Map E: Tamil Nadu Renewable Energy Stakeholder Network

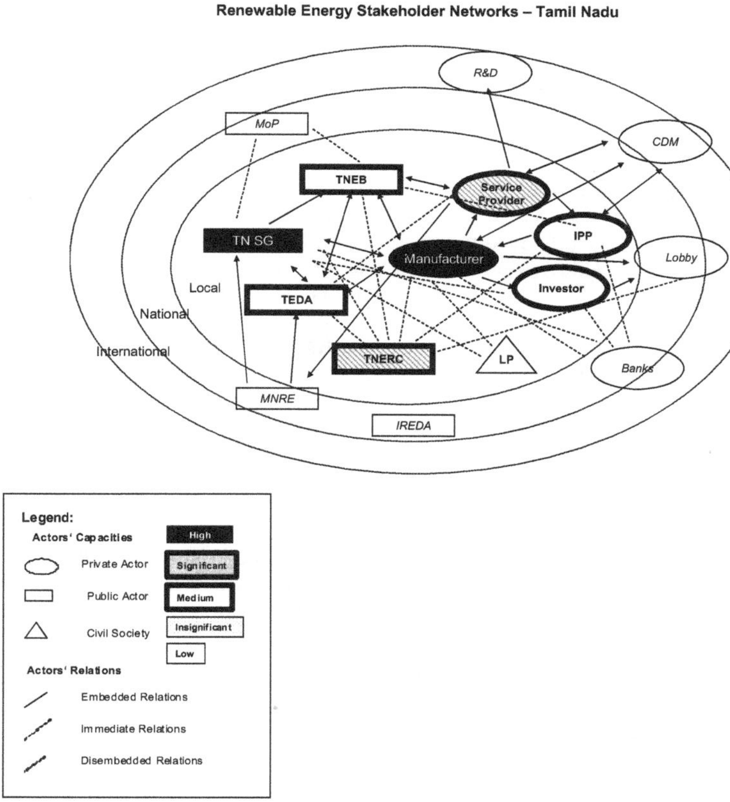

This illustration summarises the conditions for effective renewable energy governance as regards the characteristics of who governs and how governance takes place in local stakeholder networks. Stakeholder networks as explanatory factor integrate the interrelation between actors' attributes and relations in the context of Tamil Nadu. This specific governance arrangement is identified on the basis of ethnographic research that involved participant-based mental mapping of stakeholder network configurations.

The Tamil Nadu renewable energy stakeholder network exerts certain effects on renewable energy governance. The local wind energy stakeholder network that promotes wind deployment engages favourably-characterised actors in embedded relations within evolving multi-stakeholder constellations of structural openness. In order to specify the effects of Tamil Nadu renewable energy governance arrangements, organisational and new economic sociology theories are discussed. The main purpose is to suggest entry points for further interpretative discussions in Chapter VI.
Even though new economic and organisational sociology provide different perspectives, this dissertation argues that both approaches are required for a comprehensive understanding of stakeholder network effects on renewable energy governance. On the one hand, new economic sociology theory assists the identification of how wind energy is governed in the context of actors' arrangements. This departs from actor-centric assumptions of agency driven by bounded rationality that leaves actors' attributes out of the analysis and focuses on relations, ties and networks. On the other hand, organisational sociology addresses who governs and conducts rather static analyses of actors' attributes. It remains silent as regards the explicit impacts on relations and interactions. The empirical results call for the integration of both aspects and their interplay in order to understand stakeholder network processes and effects on renewable energy governance.
The main argument with regard to the Tamil Nadu stakeholder network is that favourable attributes particularly in terms of interests, i.e. political will, and capacities, i.e. staff qualification, create the possibility for emerging actors' arrangements to go beyond formal and superficial arms-length relations. Stakeholders require a certain amount of trust and expectation certainty in emerging relations to embed in actors' arrangements with intense and quality ties. Mutual understandings and common goals are only realistic and constitutive of actors' arrangements given that stakeholder interests and attitudes are characterised by common goals and are equipped with necessary capacities to render credible a mutual engagement in wind energy deployment. This gives rise to the conclusion that both, actors' attributes and actors' relations, characterise specific stakeholder networks and exert effects on how wind power deployment takes place.

Summarising the first case study of wind power deployment in Tamil Nadu, for effective renewable energy governance, stakeholder networks need to be characterised by favourable actors' attributes and actors' arrangements embedded in structurally open and interactive constellations under the conditions of market certainty and exploitation-oriented agency. Two aspects remain subject to interpretive discussions that will be embarked upon in Chapter VI. These regard firstly, how renewable energy networks work in terms of causal effects and mechanisms towards effective renewable energy governance and secondly, why networks are constituted and have emerged the way they have.

V.2 Wind Energy Deployment in Kerala – India's Lost Case?

Diplomatically put, the wind sector in Kerala constitutes an emerging market: Out of a wind resource potential of 800 MW capacity, only 2 MW have been installed and commissioned as of 2007 (InWEA 2007). This empirical case study will highlight the importance of local governance arrangements for effective renewable energy gov-

ernance. This will show concretely that the absence of supportive stakeholder networks is detrimental to wind power deployment.
The following sections will analyse renewable energy governance arrangements similarly to the deductive analytical procedure applied in the Tamil Nadu case study. Firstly, the governance question who governs wind power in Kerala will be explored by stakeholder network analysis with regard to the governance actors and their characteristics. Secondly, the governance question how governance takes place is examined mapping emerging actors' arrangements and interactions. Thirdly, conclusions on the effects of the local stakeholder network on wind power deployment in Kerala are drawn.

V.2.1 Who Governs?: Actors' Attributes in Stakeholder Networks

The first approach to understand why wind power deployment in Kerala has not taken place is to look at who governs. In the light of governance debates, the purpose is to identify and characterise new actors in governance arrangements that constitute functional equivalents to renewable energy service provision. This relates to the organisational context characterising stakeholders that constitute the renewable energy network. The organisational context describes the attributes of stakeholders involved in wind energy deployment in terms of capacities, interests and capabilities.
The following section will examine to what extent stakeholder characteristics featuring these actors' attributes provide explanatory value for wind power deployment in Kerala. This will commence with a mapping of the stakeholders identified as relevant for renewable energy governance in Kerala by ethnographic research. Following from that, features characterising governance actors are analysed with focus on the Kerala stakeholder network.

Step I – Mapping the Constituents of the Local Stakeholder Network
As discussed in the Tamil Nadu case study, the stakeholders involved in renewable energy governance at the national and international levels are similar and provide the same kind of opportunity structures. With regard to the central level, the two ministries, MNRE and MoP are mentioned as relevant due to the policy directions, subsidies and wind power programmes. These stakeholders in addition to private actors such as financiers are not directly involved in wind energy deployment in Kerala and hence not considered in the following analysis.
Narrowing the focus, the stakeholders engaged in renewable energy governance in Kerala comprise a significantly less number of actors, particularly as concerns private stakeholders. Service providers, lobby groups and to some extent also investors lack local presence. This does not mean that these actors are not involved in the wind energy sector in Kerala but that their activities are very limited and that their offices are located in other parts of India. IPPs to date are still not active in Kerala due to the general newcomer status to the wind sector. Since the wind energy sector is characterised by private sector participation, this feature is already symbolic for the negligence experienced in Kerala in spite of existing and promising wind power potential. Consequently, the governance actors constituting Kerala's wind energy stakeholder networks are predominantly public stakeholders. The following illustration (Figure XXVI) depicts the most important stakeholders involved in renewable energy governance with focus on the wind sector in Kerala as identified in a participative manner by the local actors:

Figure XXVI: Stakeholder Network in Kerala

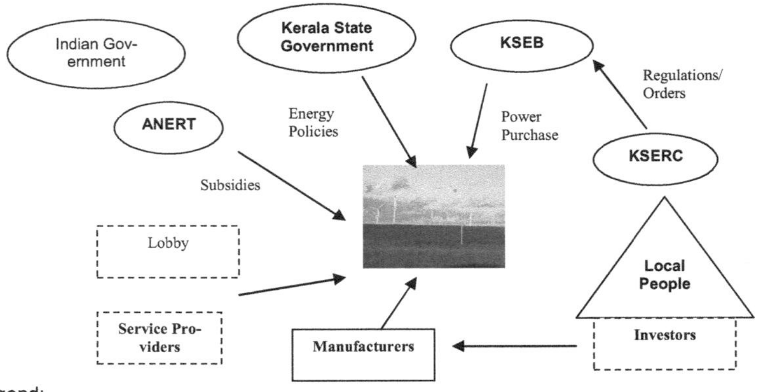

Legend:
ANERT: Agency of Non-Conventional Energy Technologies
KSERC: Kerala State Electricity Regulatory Commission
KSEB: Kerala State Electricity Board

The most important public actors at the local level include Kerala state government, Kerala State Electricity Board (KSEB), Kerala State Electricity Regulatory Commission (KSERC) and the Agency of Non-Conventional Energy and Rural Technology (ANERT) as state nodal agency. In comparison to the stakeholder setting in Tamil Nadu, this reflects a fairly similar picture. Since the Indian Constitution renders the issue of power and electricity a concurrent subject, not only responsibilities but also organisational structures constituting the public domain are similar in the Indian states. As stakeholder network analysis will demonstrate despite the same public sector constitution, great variations pertain with regard to stakeholder characteristics that impacts on how wind power is governed.

State government is one of the most significant public stakeholders in the entire power sector policy cycle and energy governance in the case of Kerala. Responsibilities for renewable energy issues are divided between the legal wind of the power department under the head of the minister of law and the administrative department headed by the minister for electricity. This implies potential bottlenecks to effective renewable energy governance inherent in an unclear division of tasks, responsibilities and labour. Additionally, the Kerala State Electricity Board (KSEB) constitutes an important player responsible for power generation, transmission and distribution. Similarly to the Tamil Nadu case, the KSEB has so far avoided power sector reforms[59]. With regard to the KSERC, the 2003 Electricity Act stipulated the same provisions and procedures concerning set up, responsibilities and tasks for all Indian states. The only difference to the TNERC is the fact that the institutionalisation of the KSERC took place some time later. Lastly, the state nodal agency of the MNRE, ANERT (Agency for New and Renewable Energy and Rural Technology), is involved in renewable energy promotion through channelling subsidies and implementing pro-

[59] The only provisional step in terms of power sector reforms was the functional unbundling of the KSEB into three competency centres; however, this structural change did not impact on the operations and functions of the KSEB.

grammes of the MNRE. In the case of Kerala, ANERT has not undertaken any kind of own initiative such as the wind resource assessment by TEDA.
As mentioned above, the direct participation of private stakeholders involved in wind energy deployment within the boundaries of Kerala state are limited to the manufacturers Suzlon and Vestas. Suzlon is selected as representative since to date Suzlon is the only manufacturer that has opened an office in Kerala, i.e. operates in the local context. Despite many prejudices, interviews confirmed that investors slowly start to voice some interest in wind energy deployment and are either based in Kerala or located elsewhere in India or abroad (INT 15). As elaborated in the Chapter III, even though the industrial structure of Kerala state is different from Tamil Nadu, investment appetite for the wind power exists in the commercial or service sector. Recently, the Kerala power sector scenario started to shift towards greater supply restrictions and is generally known for the high tariffs for commercial and industrial consumers. Service industries and the commercial enterprises operating in Kerala voice an interest in wind power investment in order to become independent from erratic and expensive grid electricity. Additionally, private parties with significant monetary resources either residing within or outside the state are keen to reap the tax benefits available for wind power installations. Yet, as no individual investor is identified that represents this diverse community, investors are analysed as one stakeholder group. Similar to the case of Tamil Nadu, service providers establish important connections and play crucial roles in the entire production chain of the wind power business. These actors have so far not started to open local offices and mostly operate from other states such as Tamil Nadu. For the sake of analysis, C-WET is considered the most representative actor. Even though it operates from the local context of Tamil Nadu, interviews and observations confirmed its role and function as most important service provider (INT 16).
International R&D companies as well as the CDM business are left out of the analysis. Firstly, they do not play any significant role in the wind energy business in Kerala to date. Secondly, they operate at national and international levels and hence provide similar opportunities all across India. This is also true for lobby groups that entertain relations to local manufacturers but have so far not ventured much into the Kerala context. This also illustrates the early stages of wind energy deployment in Kerala and implies the lack of business savvy and skills to establish connections to stakeholders already experienced and present in the wind business elsewhere.
Lastly, in contrast to the Tamil Nadu case, local people constitute significant stakeholders in renewable energy governance. Similarly to Tamil Nadu, civil society is not organised and represented by NGOs and their concerns are not picked up by media. As mentioned in Chapter III, education levels in Kerala are among the highest in India. In addition, environmental awareness is significant and finds expression in protests, e.g. accompanying hydro sector activities in Kerala. Hence civil society engagement and organisation with regard to NGOs and interest groups around environmental issues is quite remarkable. Even though wind energy is not an established field of activity, local people are perceived and taken seriously as a potentially vociferous and significant stakeholder (INT 17).

This stakeholder mapping is followed by the analysis of governance actors involved in wind energy deployment in Kerala alongside characteristics such as capacities, interests and capabilities. Similarly to the case study on wind energy deployment in Tamil Nadu, the guiding question is who governs wind power in Kerala. The explanatory model will operationalise governance actors and their characteristics as well as specify mechanisms through which actors' attributes impact on wind energy deployment. The purpose is to explore the hypothesis that local stakeholder networks characterised by favourable actors' attributes of fairly high capacities, capabilities and interests promote effective renewable energy governance in terms of wind power deployment.

Step II – Characterising Actors' Attributes: Capabilities

The first approach characterises governance actors according to their capabilities reflected in the history of experience within the organisational environment. Experience considers the extent to which stakeholders have already been engaged in wind energy activities and look back at a history of past activities in this specific local context. As argued previously, not the kind of experience but what actors make of it matters and has consequences for subsequent actions. This implies that even negative experiences of failed demonstration projects might translate into valuable lessons learnt according to which procedures and processes are adjusted to render wind energy deployment successful. Yet, in the Kerala case, the consequence of a history of failed wind power projects was the manifestation of reservations against wind energy. Apart from these normative implications of past experiences, history matters as regards the length of activities, accumulated knowledge in this field and engagement in the local setting.

(i) Average Wind Power Capabilities

The story of wind power development commenced with the first demonstration project installed in Kerala by the KSEB, ANERT and state government in 1986. This was part of the nation-wide MNRE-led initiative to promote wind power on a large scale. However, this undertaking suffered from technical and operational difficulties. The ensuing failure of the wind demonstration project in the early 1990s right from the start reflected a certain reluctance of the utility and government. Wind energy deployment encountered largely perception-based but fundamental objections particularly by the KSEB. The KSEB argumentatively utilised project failures in the past to point towards the techno-operational infeasibility of wind power installations in the Kerala setting. Power purchase agreements and infrastructure facilities for evacuating wind power were nearly impossible to obtain. Despite the fact that the KSEB as one of the first actors engaged in wind power deployment and holds substantial experiences, the ensuing negatively pre-conceived perceptions render actors' attributes only averagely-favourable of wind energy.

Over the last decade, Kerala state government has been torn between promoting and withdrawing support to wind energy development: In contrast to Tamil Nadu, MNRE guidelines or central government support of wind demonstration activities were not taken up. Renewable energy policies exist since the late 1990s and hydro resources contribute significantly to electricity addition. Until recently these policies have not specifically mentioned wind power. Government changes contributed to the shifting attitudes towards wind power and prevented the realisation of wind sector governance. With a new government coming into power in 2001, old policy drafts were revised and translated into a piece of policy legislation. On the one hand, this implies quite a long history and to some extent also an interest in wind sector development in Kerala. On the other hand, the instability and uncertainty in policy making reflects frequent government changes and the fact that no cross-fractional coalition emerged to promote wind energy. The value of wind energy resources for electricity capacity addition has not diffused the political arena and government bureaucracy.

Suzlon as wind turbine manufacturer was the first private sector entity to open an office in Kerala in 2001. Not only Suzlon but also Vestas voiced interest in the wind business in Kerala and entered into negotiations for land, clearances, permission, power purchase and evacuation details with the respective public stakeholders. In the context of Kerala, Suzlon has a comparative advantage as first mover learning from early engagement in wind energy business within this distinct local setting. Given the fact that staff employed at the office did not work in the wind sector previously, experience is less compared to other local offices. Notwithstanding, staff is qualified with regard to engagement in the renewable energy field in the local context of Kerala. Pre-existing local knowledge and network base is institutionalised locally. Suzlon also draws on an organisational memory of manifold successful and negative experiences within the wind energy business all across the country.

(ii) Insignificant Wind Power Capabilities

Even though investors have raised strong interests and revealed their potential to engage in wind sector activities in Kerala, actors' attributes in terms of experience are characterised as insignificantly-favourable. Exceptions are those investors recently acquired by Suzlon as well as private parties residing in other Indian states. This assessment refers both to experience in wind power investment as well as to investment in this specific local context. Speaking about private sector investment in Kerala often encounters the prejudices that Kerala constitutes a hostile investment context due to the Communist government. In reality, this is not confirmed as de facto regulations and policies as well as local interviews with private sector stakeholders describe Kerala as private investment-friendly context (INT 17).

ANERT constitutes another public sector stakeholder characterised by insignificantly-favourable actors' attributes as regards experience in wind energy deployment in Kerala. This is due to the fact that ANERT became involved in the wind energy business only recently. Notwithstanding, nowadays it is assigned with crucial tasks and functions that relate to the different clearances and procedures required for wind energy installations.

Another stakeholder of insignificantly-favourable actors' attributes in this respect is the group of service providers taking the example of C-WET. Without doubt, particularly C-WET obtains a wealth of general experience in the wind sector as revealed in the Tamil Nadu case. However, due to the lack of local presence and the only recent involvement in Kerala, a locally-specific organisational culture did not evolve.

Capabilities of local people with regard to wind deployment are also characterised as insignificant with regard to the wind sector. This is due to the fact that until today only some protest has been voiced against wind energy installations due to the limited affectedness given the low level of deployment activities. However, lobby activities on environmental issues are substantial and Kerala has seen a history of environmental protests against hydro projects. This renders local people potentially powerful since not only wind entrepreneurs but also public stakeholders have to fear vociferous objections to wind energy-related undertakings.

(iii) Low Wind Power Capabilities

In the case of Kerala, the regulatory authority KSERC became operational relatively late as of 2004. This renders its experience with policy making and arbitration in the field of wind energy rather low. The background of the IAS officials and bureaucrats employed in the KSERC[60] does not contribute to the establishment of a common knowledge base of shared experiences in similar sectors or activities, i.e. renewable energy governance. Furthermore, lagging activities in the wind energy sector have so far not provided much scope for learning.

Figure XXVII: Actors' Capabilities in Kerala

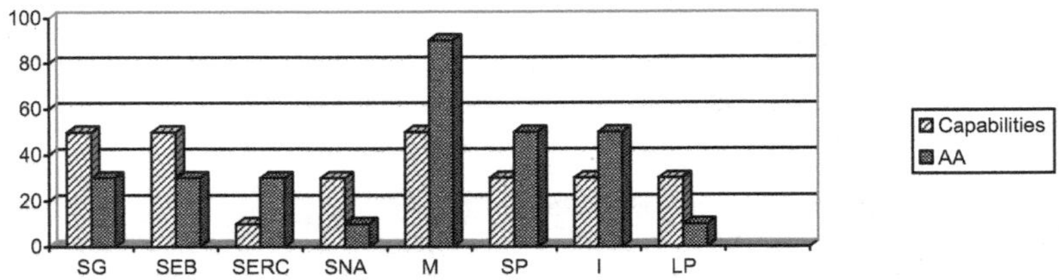

In conclusion, capabilities with regard to the history of experience that characterise actors' attributes tend to be less favourable in terms of wind power support. While for investors and service providers this refers to the short duration of engagement in wind business in this local context, local people as well as ANERT are influenced by the negative normative implications of past attempts. Manufacturers look back at not too long operational activities but dispose of an overall organisational memory of success as well as failure stories in terms of wind energy installations in other local contexts. As for public actors, both the long history of wind energy deployment as well as the negatively perceived experiences count towards the characterisation of actors' attributes as average in terms of wind power support. Organisational sociology argues that deviant behaviour such as declined support of and limited engagement in wind power arises out of the lessons learnt and experiences made and perceived. Despite the fact that some stakeholders draw on a rich pool of experiences other actors, e.g. public stakeholders, do not share similar histories and hence reflect different organisational environments. As for Kerala, this implies a societal context that seemingly questions the value of wind power in which concrete demands for wind power capacity additions have only emerged recently.

In summary, this section contributed mixed results with regard to stakeholder capabilities in terms of history as well as normative implications pertaining to wind power activities in Kerala. Actors in the Kerala stakeholder network are hence characterised by average levels of capabilities in wind power deployment.

Step III – Examining Actors' Attributes: Capacities

Organisational sociology points to the importance of resources and capacities in shaping the characteristics of an organisation. These determine not only structures but also processes and tasks that lead to functional or deviant organisational behav-

[60] Interview confirmed that in contrast to Tamil Nadu, the IAS background of officials in Kerala is not held in high esteem. This is due to the unfavourable reputation of Kerala and respective local positions (INT 27).

iour (Vaughan 1999). Stakeholder capacities contribute to reducing transaction costs in wind sector activities and to creating trust in a common purpose. Actors' capacities comprise the resources inherent in the financial budgetary situation accrued from internal sources as well as from external sources that is dedicated to renewable energy investments specifically to wind sector development. The autonomy and independence of decision making within an organisation closely relates to the resources aspect. For example, if an organisation depends on the other for the specification of the budget and the designation of staff, decision making is not entirely independent. Resources also include the human resource aspect with regard to the number as well as the qualification of staff assigned to renewable energy issues. As the following analysis will demonstrate, this distinction proves essential particularly in the case of Kerala as a quantitatively high staff level does not necessarily correspond to qualitative aspects related to the efficiency, effectiveness and productivity of work.

(i) Significant Wind Power Capacities

Within the Kerala stakeholder network, the manufacturer is characterised by the highest capacities regarding these organisational characteristics. In comparison to Suzlon's operation elsewhere in India, the Kerala office in terms of staff and budget is small. Yet, this office is embedded in the greater organisational network and its internal procedures and budget are to some extent independent of the success of business operations in the Kerala context. Suzlon draws on external assistance from established service providers outside the state of Kerala and Suzlon's autonomy in terms of business decisions is high. Furthermore, the staff working in the Kerala office is not only qualified but also no newcomer to the Kerala context and the renewable energy scenario.

(ii) Average Wind Power Capacities

In contrast to the situation in Tamil Nadu, civil society stakeholders are characterised by average degrees of capacities. This is due to the fact that even in remote areas local people are well educated and highly aware of sensitive issues pertaining to the environment and energy. They are capable of autonomous decision-making and action and they defend their own interest vis-à-vis public and private stakeholders independent of representatives to speak for them. Since in some instances this has resulted in vociferous protest, local peoples' soft power and voice is taken seriously due to (perceived) repercussions on electoral outcomes and impediments to business operations.

Private actors' groups constitutive of service providers as well as investors are characterised by average levels of capacities. As analysed in the Tamil Nadu case, in general both stakeholders avail of substantial internal financial resources or, as regards C-WET, employ highly-qualified staff. Due to their restricted activities and local presence in Kerala, resources assigned for the specific purpose of wind energy deployment in this local setting are limited. Independent and autonomous decision-making in terms of wind sector engagement is more restricted given the local absence, limited historic involvement and insignificant stake in Kerala's renewable energy governance scenario.

(iii) Insignificant Wind Power Capacities

The KSEB as a public sector entity is a highly important employer even though the number of staff assigned to deal with wind energy and renewable energy issues is relatively low in comparison to other Indian states (INT 17). Responsibilities are not institutionally located in a specific division but wind energy is dealt with in the electricity generation unit. Recently, KSEB staff has benefited from not only technical but also managerial training. On the basis of Canadian development assistance, the KSEB experienced not only functional restructuring within the organisation but moreover additional financial and technical training. Until recently, the KSEB has been a profit-making entity as Kerala even sold power to other states. Additionally, the KSEB received project support in the hydro sector from the Chinese business. The budgetary position in addition to the high staff number renders the KSEB on the one hand relatively superior and independent in decision making from state government[61]. On the other hand, this apparent autonomy is questionable taking into consideration that the high-level staff assignment and the power sector decisions are taken by state government.

In contrast to other Indian states, the KSERC was established rather late. The staffing of 21 persons and budget (Rs lakh 46,3 for 2005/06) of the KSERC similarly to other states is relatively small (KSERC 2007). However, in the case of Kerala no training or qualification programmes of its personnel are implemented and staff qualification is rather low. Furthermore, no external consultancy is drawn upon when it comes to technical matters or regulatory issues pertaining to wind energy. Interviews confirm that regarding decisions on wind tariffs and regulations the KSERC's decisions are slanted towards the utility's and government's interests and objectives (INT 20).

(iv) Low Wind Power Capacities

At first sight, the low assessment of state government's capacities in terms of resource endowment might appear surprising due to the still comfortable power position. However, responsibilities for the wind sector are divided between two departmental units within state government bureaucracy. Even though both fractions are to some extent involved with renewable energy governance, neither section of Kerala state government power department avails of specialised staff qualified in this regard. The budget under the 10th Five-Year-Plan (2002-2007) spreads financial resources for the power sector into Rs crore 3425 for the KSEB and Rs crore 70 for ANERT, i.e. renewable energy (Kerala Planning Board 2007). In comparison to other Indian states, this features quite lowly. Training programmes or any other staff qualification do not take place or are not officially stated. External consultancy or budget contribution from international development assistance or other sources is also not mentioned. The independence and autonomy in decision-making are questionable due to the dominant position of the KSEB. This arises out of the power of organised labour force in the KSEB, i.e. unions, its significant government budgetary contribution through power sale and the general social obligation of government to provide power services to its constituency. In conclusion, due to the fact that staff and budget within state government assigned specifically to renewable and wind energy is negligible, actors' capacities are low.

---

[61] In the Kerala case, data was impossible to obtain or access was restricted.

Despite the important tasks ANERT is charged with in the wind power deployment process, resources and capacities are characterised as low. At first sight, this might be surprising since ANERT disposes of a very substantial number of 60-70 staff (INT 19) compared to other Indian states. In addition, ANERT frequently calls upon external assistance such as service providers and consultancies from other Indian states as for example C-WET. Yet, ANERT faces serious problem as regards staff qualification and continuity. For the past ten years, ANERT has not had a director besides temporary deputations from other organisations or from Indian Administrative Service (IAS) staff in retirement or leave processes. ANERT suffers from a bad reputation as its staff turned out incapable and unqualified to meet job requirements. This point is often emphasised since it reflects a dramatic dilemma: ANERT is assigned crucial tasks in wind power governance but lacks implementation capacities and capabilities. ANERT obtains no autonomy in decision-making and is strongly dependent and orientated towards the KSEB. No institutional vision, no human resource policy and no continuity in procedures related to the organisation and wind power are established within the organisation (INT 18). Some interviews also remarked that ANERT is negatively biased against wind power and supports other renewable energy options. This renders ANERT an institutional monolith that is overstaffed but badly qualified and incapable of meeting crucial functions in the wind power deployment process. As human resource issues are state government's responsibilities, an interesting question is to what extent this situation is intended to be like that in order to increase invisible hurdles to wind power development.

Figure XXVIII: Actors' Capacities in Kerala

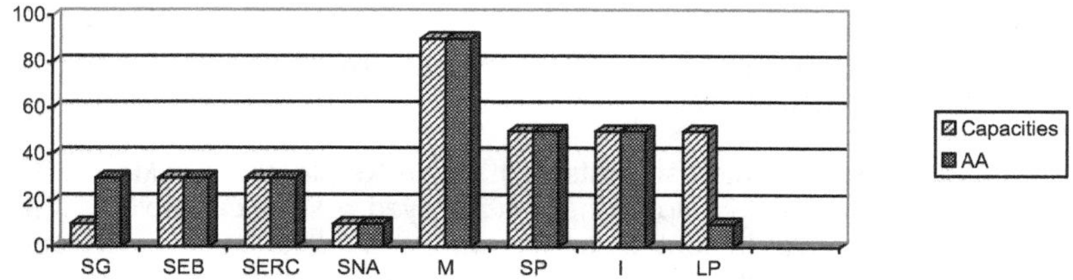

In summary, stakeholder capacities on the whole are characterised as less favourable to the wind energy business. The analysis demonstrates that particularly with regard to staff qualification, public institutions specifically ANERT are badly equipped. Lacking qualification cannot mitigate high staff and budgetary size and hence negatively affects performance and effectiveness. This also feeds back negatively on organisational reputation and impedes procedures and interactions with other stakeholders. In contrast, private actors are better positioned with regard to resource endowment since they draw on affiliations beyond state boundaries. In addition to greater resource capacities, their autonomy to act in favour of wind energy deployment is also enhanced. Lacking capacities particularly on the parts of public stakeholders as regards qualification translates into organisational practices and tasks that inhibit transactions related to wind deployment. Given the mutual interdependence and perceived obligations between stakeholders, autonomy in decision-making is severely impeded. As proposed by organisational sociology, dysfunctional organisational behaviour in terms of ineffective renewable energy governance arises out of capacity-related characteristics that pose impediments to wind power deployment.

Step IV – Characterising Actors' Attributes: Interests

The third aspect that characterises governance actors with regard to engagement in renewable energy governance focuses on interests and attitudes.

This dissertation draws on organisational sociology that provides an integrative view on how interests and attitudes reflected in organisational features explain renewable energy governance. Two aspects are differentiated regarding the attitudes of stakeholders: While the one issue concerns the reputation established by the organisational leadership, the other issue touches upon the attitudes reflected in organisational behaviour. As remarked previously, both aspects do not necessarily go hand in hand but might diverge. In the case of the Uttar Pradesh Non-Conventional Energy Agency (NEDA), a very proactive leader held in high esteem for his engagement in promoting renewable energy in other assignments within the Indian administrative system was not able to stimulate efficient and effective performance in the organisation that had the reputation of being averse and procrastinated in terms renewable energy support (INT 21). Organisational sociology perspectives emphasise that interests and attitudes are shaped by individual as well as collective rules, routines and leadership. As proposed in the explanatory model, wind power deployment depends on the extent to which leaders embrace this undertaking, and the value of wind power is entrenched in organisational routines.

(i) Significant Wind Power Interests

As manufacturer, Suzlon has substantial interests in wind sector development in Kerala. This is not only due to the company's expansive business strategy in India but also refers to self-understanding and perception of being the indigenous pioneer of the Indian wind energy market. Most and foremost this relates to profit motives but also to gaining comparative advantages to competitors as first mover in a new context. The inside knowledge of and the locational proximity to the Kerala context put Suzlon in an advantaged position.

Substantial investment interest exists both within Kerala and outside state boundaries, even though investors so far have not played a significant role in wind sector development in Kerala. In contrast to other Indian wind states, investors emerge not predominantly from power-intensive industries but from commercial or service sectors. Investors pursue not only business motives such as securing power supply for companies and factories or availing of tax benefits. Interviews also revealed the desire to contribute to economic development in the home state (INT 9) as investment interests emerge from Keralites working abroad or in other Indian states.

C-WET also has substantial interests and favourable attitudes in terms of wind power deployment in general as well as concerning the engagement in the local setting of Kerala in particular. This is not only due to altruistic reasons of promoting renewable energy across India but also relates to business interests in expanding activities on a nation-wide basis.

(ii) Medium Interests

State nodal agencies like ANERT have an inherent interest to promote renewable energy as a matter of functional responsibility and self-perception. Responsibilities assigned to the Agency for Non-Conventional Energy and Rural Technology (ANERT) as the state nodal agency for renewable energy and wind issues go beyond the channelling of central government subsidies or brokering information. In contrast to

Tamil Nadu, Kerala state government assigned ANERT with the task to schedule and verify certifications and clearances required for wind farm installations. The fact that Kerala did not establish a single-window-clearance system with the KSEB but demands of entrepreneurs to fight the battle through different bureaucracies on the one hand increases the administrative hurdles. On the other hand, this can be interpreted as a more transparent procedure in which voices of all affected stakeholders particularly those of local people are heard. Without doubt, ANERT's official position is supportive and in favour of renewable energy and also wind energy deployment. However, one might argue that internal characteristics such staff qualification render questionable the extent to which ANERT functions as the intended support hub or whether these intricacies imply hidden aversions to wind energy.

Interests of the Kerala State Regulatory Commission (KSERC) in the wind sector are ambiguous with regard to its support to deployment activities. Even though not openly and outspokenly averse of wind energy, the KSERC is subject to significantly high influence and insinuation by state government and the KSEB. This is not only mentioned during interviews but is also reflected in the very low number of court cases and petitions (INT 20). Since this is rather exceptional for such highly contested issues such as tariffs one might argue that public sector influence is quite substantial. This renders support for wind energy relatively low. De jure, the KSERC is not supposed to display interests or attitudes of its own as regards wind sector development. As an autonomous organisation KSERC has to mitigate different interests concerning wind energy issues. De facto, attitudes and interests are influenced by private sector lobbying on the one hand but more substantially by the human resource policies and implicit state government lobbying on the other hand. Similarly as for ANERT, director and chairman are designated by state government. As interests and attitudes are biased towards state government objectives, the KSERC's attributes reflect average support towards wind energy.

(iii) Insignificant Wind Power Interests

On the whole, state government showed limited interest in wind energy. Awareness levels among public officials on wind energy issues vary widely. Interviews during the empirical field research gave rise to the observation that the higher the status and power, the lower the attention to and interest in the potentials offered by wind energy (INT 28). This leads to the conclusion that not only autonomy and independent decision making within state government is limited, but also interests in engagement in new ventures such as wind energy are low. Decisions hinge on the individual power calculi of individual politicians and bureaucrats. Their attitudes towards wind energy are determined by several factors such as the interest in re-election. State government seeks to avoid potentially tricky issues and aims to maintain good and consistent relations to the local population. The relatively comfortable power sector position in Kerala has so far not put any pressure on the state to harness additional energy resources. Additionally, alternative energy options such as hydro power have been embraced for quite some time which increases entry barriers to newcomers such as wind energy. These factors render the mindsets and attitudes of public officials and civil servants in state government rather hostile to wind energy. Additionally, government changes and ensuing internal power struggles turned policy making for wind energy development unstable and unpredictable.

The organisational context characterising the Kerala State Electricity Board (KSEB) is slightly more favourable of renewable energy than that of the state government. Wind energy is predominantly perceived as a business benefiting less society but more as an undertaking catering to business interests and profits. The main interests of the KSEB are availing of the cheapest power possible since the provision of affordable and stable electricity supply is considered a social obligation and KSEB's central objective. KSEB argues that constituting a people's enterprise it is obliged to service delivery that serves the people's interests (INT 10). Negative perceptions on wind energy cause operational and technical difficulties, result in the lack of infrastructure facilities and the resistance to come up with any kind of burden sharing with wind energy investors and manufacturers. With regard to renewable energy deployment, alternative paths such as hydro power are emphasised within the KSEB and reflect government's agenda. Even though this attitude has not significantly changed, the KSEB's mindsets started to move towards becoming more of a private-sector undertaking (INT 23). Interviews confirmed that attitudes slowly move towards considering wind energy a potential option for electricity addition.

(iv) Low Wind Power Interests

Local people have a relatively unfavourable mindset towards renewable energy and particularly wind energy in Kerala. Environmental education and awareness in general are quite significant in Kerala. Local people even in remote villages are conscious of potentially negative effects of wind power such as noise pollution and land destruction in other parts of the country through media coverage and word of mouth (INT 2). Due to the fact that local representative are forced to respond to local people given the traditionally high influence of unions, getting organised is an easier undertaking than in Tamil Nadu. Local people do not shy away from organising protest on their own as demonstrated in marching against the manufacturer's office (INT 11). Local people are also considered powerful in the sense that they constitute the voters' base, which renders state government and public officials highly interested in not upsetting them and endangering their re-election. As a result, procedures for land purchase and project installation are more complexly stipulated in order to ensure the least local objections possible.

Figure XXIX: Actors' Interests in Kerala

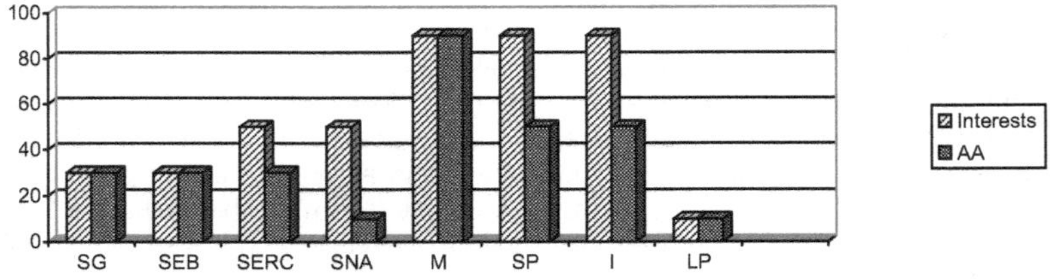

In summary, attitudes characterising governance actors bear implications on renewable energy governance (Figure XXIX). Private actors have significant interests in wind energy due to the anticipated values inherent in wind business engagement. Interests and attitudes of state government and KSEB are ambivalent since no common norm embracing wind power as valuable and attractive pervades the organisational context. This reflects their status-quo state of mind with regard to wind power.

Lastly, local people have no interest in wind energy and object developments in this sector.
These observations demonstrate that by far not all stakeholders display dysfunctional behaviour. Private actors for example hold high stakes in wind power. Given the interdependence of dynamics and structures in the stakeholder network, a critical amount of organisational deviance by some actors seems to be sufficient to impede the entire wind energy undertaking. This implies that effective renewable energy governance depends on favourable actors' attributes characterising a critical mass of stakeholders in the renewable energy stakeholder network.

Step V – Discussing Actors' Attributes related to Wind Power Deployment in Kerala
The first step to discuss stakeholder networks as condition for effective renewable energy governance looks at who governs in terms of specific actors' characteristics. Stakeholder network analysis explored governance actors with regard to their attributes in Kerala's renewable energy stakeholder network and revealed the following results: Local governance arrangements in Kerala are characterised as less favourable with regard to actors' attributes in relation to wind power deployment (Figure XXX):

Figure XXX: Actors' Attributes in the Kerala Stakeholder Network

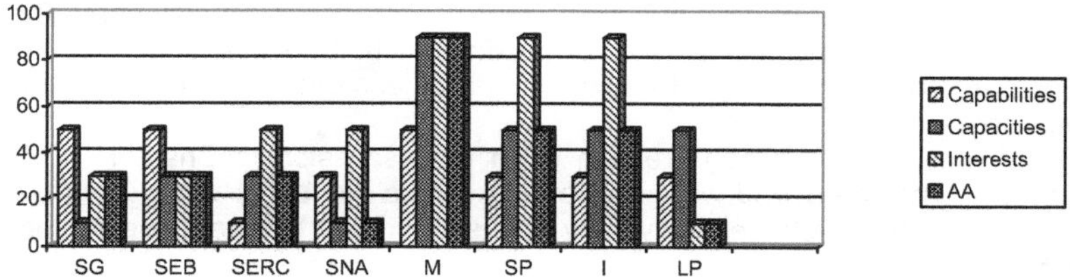

Another striking feature of the Kerala setting is the asymmetric characterisation of actors' attributes in the local renewable energy stakeholder network (Table 28).

Table 28: Actors' Attributes in the Kerala y Stakeholder Network

	State Govern ment	*KSEB*	*KSERC*	*ANERT*	*Manu facturer*	*Service Provider*	*Investor*	*Local People*
Capacities	Low	Insignifi-cant	Insignifi-cant	Low	Significant	Medium	Medium	Medium
Interests	Insignificant	Insignifi-cant	Medium	Medium	Significant	Significant	Significant	Low
Capabili-ties	Medium	Medium	Low	Insignifi-cant	Medium	Insignificant	Insignificant	Insignifi-cant
Attributes	insignificant	Insignifi-cant	insignifi-cant	Low	Significant-medium	medium	Medium	Medium-insignifi-cant

While public stakeholders and local people are characterised by insignificant to low attributes as regards wind energy deployment, private actors display medium to significant orientations towards the wind sector business. In the case of public stakeholders, this is explained by the lack of political will reflected in adverse attitudes and interests at both organisational and individual levels and in bad human resource qualification. One might speculate that these two aspects are interrelated since the more adverse the interest, the less emphasis is put on staff qualification. Human resources policies are applied as political tool to hide the lack of genuine interest in wind energy development behind the veil of officially high staff number and budget contribution. As a result, organisations lack substance and teeth to render support to wind power an effective activity. The case of private stakeholders is less debatable as without doubt interests and attitudes correspond to adequate resources and capabilities. However, in the context of the stakeholder network this cannot compensate for their lack of experience and unfavourable public stakeholder characteristics in the local context of Kerala.

In the light of organisational sociology, the characteristics of governance actors bear implications on transaction costs and trust, i.e. the main mechanisms through which wind power deployment is encouraged. Transaction costs are high due to low interests in facilitating procedures and insignificant resources invested particularly in terms of staff quality. Limited levels of trust prevent transactions and interactions in the stakeholder network pertaining to renewable energy governance and the wind sector to establish and become routinised. A critical mass of stakeholders, referring to the public sector, characterised by negative attributes is sufficient for creating obstacles to successful wind energy deployment in Kerala. The Kerala case supports the hypothesis that stakeholder networks in wind power are characterised by governance actors with adverse orientation towards wind energy deployment. The by-and-large unfavourable characteristics of public stakeholders constitute stakeholder networks that on the whole will likely contribute to less effective renewable energy governance.

In summary, two questions remain open for discussion in Chapter VI. These are firstly, in what ways and with what effects actors' attributes relate to renewable energy governance; and secondly, why actors' attributes feature these distinctive characteristics.

V.2.2 How is Renewable Energy Governed?: Actors' Relations in Stakeholder Networks

Renewable energy governance arrangements in Kerala were so far descriptively analysed to tackle the question who governs alongside the characteristics of respective stakeholders. Attention at this stage is turned towards the question how wind energy is governed. Stakeholder network analysis will explore emerging actors' arrangements and interactions constituting local stakeholder networks. The purpose is to draw conclusions on the effects of the embeddedness of stakeholders in relational interactions on wind energy deployment. Similar to stakeholder network analysis in the case of Tamil Nadu, the exploration of actors' relations is guided by the explanatory model based on new economic sociology and governance literature. In the following, the analysis first and foremost starts with the characterisation of actors' arrangements with respect to the dyadic ties between stakeholders. Summarising emerging constellations of actors' relations of similar dyadic ties then gives rise to the identification of certain typologies, i.e. actors' arrangements. The second step in the analysis of actors' relations is to explore interaction constellations between stakeholders in the network structure. The purpose is to assess in how far actors engage in interactions and extend relations to similar stakeholders or go beyond the public-private divide. This allows for statements about the openness or closure of renewable

energy networks which bears implications with regard to public-private dynamics and how wind power is governed. New economic sociology argues that for industries at early development stages closed networks constituted of tightly-bound cliques are necessary in order to establish mutuality and trust in exchange procedures (Rowley et al 2000). With the maturing of industries no real need exists to entertain tightly-confined exchange relations due to different requirements at this development stage. This dissertation departs from the assumption that wind power markets are certain (cf. Chapter I) which requires stakeholder networks to be characterised by structural openness, i.e. interactions going beyond public/private divides. In this section, stakeholder network analysis will discuss the extent to which the lack of structural openness in Kerala's renewable energy stakeholder network impedes effective renewable energy governance. For this purpose, interaction constellations between public and private as well as among the respective groups are examined according to on the one hand, interaction density and on the other hand, the quality of relations.
In the third analytical step, the entire renewable energy stakeholder network in Kerala is explored. The overall purpose is to pay attention to interrelations and implications of who and how wind power is governed. New economic sociology suggests on the one hand, that intensive interactions constellations characterised by embedded network relations promote wind energy deployment. This is due to reduced transaction costs relating to deployment procedures and shared norms or common understandings of the benefits of wind energy deployment. On the other hand, fewer interactions in embedded networks are suggested as pre-requisites for effective renewable energy governance. Since procedures are institutionalised anyway, successful wind energy deployment depends more on actors' attributes than on interactions within actors' arrangements.
The following three steps will discuss the hypothesis that wind energy deployment in Kerala has so far not taken off due to the facts that stakeholders are not well embedded in governance arrangements that feature closed structures and interactions.

Step I – Dyadic Ties in Emerging Actors' Arrangements

The purpose of this section is to characterise actors' relations and to identify new actors' arrangements based on ethnographic research. In contrast to the Tamil Nadu case, a slightly higher number of actors' arrangements emerge in the context of Kerala. The embeddedness of public and private entities in relations within governance arrangements is categorised as follows (Table 29) and will be discussed in the four cases identified:

Table 29: Actors' Relations in the Kerala Stakeholder Network (Chapter V, p)

Tie Characteristics/ Typology	*"Disembedded Actors' Arrangements"*	*"Immediate Actors' Arrangements"*	*"Embedded Actors' Arrangements"*
Typical	High formality – Low social capital **"distant – formal ties"** 29% of ties	Medium formality – Medium social capital **"immediate – formalised ties"** 11% of ties	Low formality – High social capital "familiar – informal ties"
Mix	High formality – Medium social capital "distant – formalised ties"	High formality – High social capital "immediate – informal ties"	Low formality – Medium social capital **"familiar – formalised ties"** 17% of ties
Hybrid	Medium formality – Low	Low formality – Low so-	Medium formality –

	social capital **"hybrid – distant ties"** 43% of ties	cial capital "hybrid – immediate ties"	High social capital "hybrid - familiar ties"

Kerala Case I – Disembedded actors' arrangements characterised by "hybrid – distant ties"

This first kind of actors' arrangement comprise the majority of actors with a stake in wind energy deployment in Kerala (43 % of dyadic ties) and is constituted of stakeholders that engage in relations with low social capital and reciprocity. Aspects of formality are not strongly integrated but nevertheless prevail as characteristic feature as exemplified in four instances:

Firstly, the KSEB extends the most frequent and intense relation to the Ministry of Power but also holds regular contact to Suzlon and ANERT. Interactions take place on a day-to-day, formal level as stipulated in the procedures for wind power development. The fact that little social capital and reciprocity exists illustrates the KSEB's hesitant interests and attitudes to wind power. It also shows the difficulty to come by attitudinal differences through establishing mutual understanding and cooperation with manufacturers. Interactions between these stakeholders remain focused on formalistic procedures. This in turn impedes smoother, more informal transactions that facilitate wind power deployment in other states such as Tamil Nadu. The relations between ANERT and manufacturer are of institutionalised and formal manner reflected in the wind energy deployment process regarding power purchase and procurement.

The second case concerns the even more formalistic relations between the state government and the KSERC. This implies the strong shadow of state government hierarchy apparent in decisions about budget and staffing of the KSERC. This is upheld as to guarantee that regulatory decision making does not contravene state interests. Indirect control and dependence is established even though state government does not have an official say in KSERC's agency.

Thirdly, manufacturers are involved in more formal relations to local people. No common understanding, trust or social capital is inherent in relations. Interests and attitudes of local people clash with the business and profit orientation of manufacturers. This is aggravated by the anticipated consequences of harmful environmental impacts of wind power that drive at times the aggressively-perceived actions on the part of civil society actors.

Lastly, the fourth example depicts relations between the MNRE, KSERC and ANERT that are either non-existent or very weak. This suggests that apart from subsidy transfer and implementation of central schemes, no attention is paid to the additional guidance and orientation offered. Interactions reflecting these kinds of relations are of low intensity and also reveal a low quality.

According to new economic sociology, the observation that actors' arrangement are predominantly characterised by disembedded relations between actors in the local stakeholder networks bears several implications. Transactions regarding policy and business issue in wind power deployment are oriented towards highly formalistic and standardised procedures of little flexibility that consider no case-to-case specificities. Since wind power deployment procedures are institutionally managed in a very complex and complicated manner that involves many administrative layers and procedural stages, transactions costs are very high. Finally, the low levels of trust and social capital that characterise dyadic ties between public and private actors impede smooth interactions oriented towards common objectives of successful wind power deployment.

Kerala Case II – Disembedded actors' arrangements characterised by "distant - formal ties"

This second actors' arrangement comprising 29 % of actors' relations is characterised by disembedded dyadic ties that are highly formal but display some reciprocity while social capital is negligible. Actors' arrangements of this typology comprise stakeholder constellations among government institutions as well as between government and private actors, which the following three cases will illustrate.

Firstly, with regard to state government, very tight reciprocal relations are entertained with both the KSEB and ANERT. Number and frequency of interactions between these public institutions are high. Decision-making by far does not take place autonomously from each other. This is due to the fact that state government has an important say in budgetary, human resource and policy direction decisions. On the whole, certain vested interests in terms of party politics, electoral objectives and labour issues render all of the organisations highly dependent on each other. Interviews confirmed the regular transfer of high-level staff among organisations, which also includes the KSERC, as well as the membership on the respective boards (INT 17). Policy decisions are not taken without unofficial as well as formal consultation of the affected public sector stakeholders. Relations between ANERT and state government are of a rather top-down nature as ANERT does not hold much stake in policy and economic processes. However, ANERT is of significance in terms of shielding state government against aversions or resentments by local people. This is reflected in the institutionalisation of wind power procedures in the organisational responsibility of ANERT. All in all, little trust and social capital characterise these dyadic ties. Mutual objectives exist but are overshadowed by the emphasis on procedures and formalities. This is partly due to the local constituency's demand for "proper process" which might be one reason why corruption rates are the lowest compared to all other Indian states.

The second observation refers to another set of very tight relations between manufacturers and ANERT. Manufacturers are highly dependent on ANERT due to required procedural formalities. Interactions between both parties involve little shared understanding or reciprocity. This is reflected in the frequent complaints by manufacturers about the lacking efficiency of ANERT. Manufacturers uphold regular relations characterised in a similar manner with the KSERC. Particularly with regard to the KSERC, manufacturers attempt at a high degree of informal communication in order to move towards a tacit and common understanding. So far, this has not produced any effective results in terms of "convincing" the KSERC of the concerns of the wind power business community which would translate into more favourable regulatory decisions.

Lastly, relations between local people and state government also fall into this typology. State government is highly concerned about people's consent with regard to repercussions on electoral outcomes (INT 22). In order to allow for as many safety nets as possible, provisions to inform and include the public into policy making and implementation processes are provided. Wind energy procedures are just one example of state government shedding responsibilities to ANERT as well as the KSEB. In contrast to Tamil Nadu, environmental impact assessments and additional procedures are legislative requirements for wind power deployment. This serves to protect the official integrity of wind power activities and to avoid being at the centre of responsibilities and hence potential protest. Some prevailing elements of social capital refer to the exchange of favours – votes - in return for attending to people's interests and concerns as regards environmental integrity and power supply.

Kerala Case III – Embedded actors' arrangements characterised by "familiar – formalised ties"

Actors' arrangements in the third typology comprise embedded actors with relations of high interaction quality and/or reciprocity as well as low formality constituting 17 % of dyadic ties. Relations feature high reciprocity and significant social capital while formal elements play an insignificant role. Remarkably, these actors' arrangements emerge between private stakeholders only.

Shifting perspectives to the private sector, the manufacturer, i.e. Suzlon, constitutes the stakeholder with the greatest number and density of relations. Interrelations with other private stakeholders such as service providers and investors are very intense since manufacturers rely and depend on their input. In contrast to Tamil Nadu, this is not reflected in a high number and high frequency of day-to-day and face-to-face contacts given the fact that most of the suppliers operate from other local contexts. Since service providers and investors are not institutionally located inside the state of Kerala, this requires greater trust and mutuality of the dyadic ties in order to facilitate smooth business transactions. This also proves that geographical proximity is not a pre-condition for relational quality to evolve in stakeholder arrangements. Interactions between these business stakeholders benefit from previous experiences in other states, i.e. drawing on a collective inter-organisational memory. Relations are characterised by some mutual understanding and tacit knowledge of each others' requirements. This relational quality of dyadic ties seems to some extent make up for the possibility of high intensity of face-to-face interactions.

As one of the few stakeholders located at the national level that matter in the local context of stakeholder network in Kerala, wind lobbies hold similarly embedded relations to manufacturers and investors, as in Tamil Nadu. However, as they do not venture into the public domain their influence remains insignificant till date.

Two main implications emerge in the light of new economic sociology debates. Firstly, the transaction cost argument does not seem to hold. Since stakeholders are located in quite some distance from each other, communication and other costs involved in transactions and exchanges are high. Still, embedded relations exist and profit-oriented exchange and interactions take place. Secondly, this observation supports arguments emphasising the importance of trust and mutual understanding inherent in embedded ties that mitigate these issues and facilitate interactions involved in wind power deployment.

Kerala Case IV – Immediate actors' arrangements characterised by "immediate – formalised ties"

The last actors' arrangements emerging in Kerala's renewable energy stakeholder network comprises 11 % of actors with relations of some degrees of formality as well social capital and reciprocity. These immediately-embedded relations illustrate dyadic ties between public organisations and service providers in two cases:

Firstly, since the KSEB has specific interests in how the power purchase regulations are shaped by the KSERC, relations go beyond the formalistic level into the more tacit realm of taking influence on regulatory decisions. The relatively high quality of relations in terms of social capital invested is based particularly on the high density of these ties.

Secondly, ANERT upholds similar kinds of relations to service providers that are characterised by low degrees of formality and high trust and reciprocity. As ANERT depends on service providers such as C-WET for assistance in wind energy processes due to the lack of own expertise, more effort is invested in coordination and learning that translates into greater interaction intensity. Through repeated and frequent interactions some level of trust and personal knowledge has evolved that depicts higher interaction quality.

In the light of new economic sociology debates public/ private actor' arrangements in this typology put less emphasis on formal ties. The efficiency of interactions is higher given the level of trust, even though this not sufficient for promoting wind energy deployment due to the limited effects on reduced transaction costs.

All in all, the characteristics of dyadic ties in relations between stakeholders engaged in wind power deployment in the context of Kerala led to the identification of four main actors' arrangements. The most striking feature is that the majority of relations are of high formality and low social capital. This pertains mainly to actors' arrangements between public and private actors. According to new economic sociology theory, the high formality inherent in relations also among public stakeholders increases transaction costs of doing business and governing renewable energy. The embeddedness of relations emerging in actors' arrangements between private actors does not seem to be sufficient to overcome the disembedded relations characterising constellations between public and private actors. Low levels of trust and higher transaction costs hence impact negatively on processes and procedures involved in wind energy deployment.

Step II – Interaction Constellations in Stakeholder Networks
The purpose of this section is to characterise stakeholder interactions and to identify interaction constellations going beyond public/private divides. After the analysis of emerging actors' arrangements with regard to how strongly embedded actors are, the extent to which relationships extend beyond the boundaries of established groups, i.e. private sector, public sector, civil society, is examined. This relates both to the quantity as well as to the quality of interactions and evolving interaction constellations. The overall objective is to examine the dynamics of interactions, i.e. how governance takes place. In this regard, structural openness expresses the extent to which actors maintain interactive relations with other actors beyond public/private divides. This allows for identifying cliques or groups of actors with similar kinds of relational structures engaged in interactions.
One main interest of this dissertation is to look at the public-private distinction characterising multilevel renewable energy governance arrangements in the distinct local context. This is justified on the grounds of the empirical observation that the interrelation between public and private actors presumably promotes wind energy deployment. The assumption is that the more closed constellations particularly among public actors, the less feasible interactions within the stakeholder network, i.e. between public and private actors. This impedes exchange and coordination, and reduces the propensity of renewable energy governance to gain the necessary support and clout required for wind deployment to take place. The analytical outcome of this examination and discussion is the characterisation of the renewable energy stakeholder network in Kerala with effect on wind power deployment. New economic sociology suggests the positive impact of structurally open interactions on effective renewable en-

ergy governance. In the light of governance debates, the added value of this analysis is the identification of functionally equivalent interaction arrangements with what means wind power is governed.
In the following, observations on the stakeholder network structure and dynamics will consider the evolving interaction constellations among public, among private and between public and private actors.

Public – Public Clique Constellations
Interaction constellations among the public stakeholders state government, KSEB and ANERT are characterised by the frequency of interactions and high reciprocity. The relational quality is reflected in frequent information exchange, formal cooperation and mutual understanding featuring very dense formalistic ties. This implies that decision making is dependent on agreement among all the stakeholders and indicates a high degree of common objectives and orientation towards each other. This leads to the fact that access to any of these organisations by non-public stakeholders is rendered more difficult. This is particularly apparent when talking about the objectives of gaining a more tacit agreement based on mutual commitments, shared information and goals, and a certain degree of informal exchange.

Private – Private Clique Constellations
Interaction constellations among private stakeholders specifically manufacturer, service provider and investors are characterised by very high intensity. Given the geographical distance, the frequency of interaction is not as important as featured in the public stakeholder cluster. The relational quality of ties in interaction constellations is characterised by a long history of shared experiences, mutual commitments and understandings and high degrees of also informal contacts. Although transaction costs might increase due to the fact that supply chain partners are not located in closer proximity, interactions are characterised by elements of trust and mutuality inherited from previous or similar relations. This in turn provides the basis for successful wind energy deployment as private sector requirements are considered. Remarkably, interaction constellations and relational ties between private stakeholders are the only ones that go beyond the state boundaries of Kerala.

Public – Private Clique Constellations
Network dynamics between public and private stakeholders are of less intensive nature. Interactions are characterised by high formality and reflect low social capital. Undoubtedly, manufacturers hold the most intense interactions particularly with ANERT due to the administrative requirements of the wind installation procedure. The quality of relations featuring in interactions is quite a one-way-street and not based on mutual understanding or cooperative behaviour beneficial to both parties. Interaction processes between private and public parties, even though they might be quite frequent, take place on a unilateral basis. They mostly emerge on the initiative of private stakeholders and are not reciprocated in a similar fashion and with the same degree of informality. While private stakeholders attempt at certain degrees of informal contact or trying to gain common ground and to increase levels of trust, public stakeholders particularly KSEB and state government shy away from a similar kind of relational quality in interactions.

Step III: Renewable Energy Governance in the Kerala Stakeholder Network

This section will specify the effects on wind power deployment as suggested by the explanatory model for wind power deployment in India. In the Tamil Nadu case, a positive impact of structurally open interactions particularly between public and private stakeholders proves favourable for wind power deployment. As for the Kerala case, the hypothesis also seems to hold true since relations in interaction constellations are characterised as rather close. Even though private actors try to establish dyadic ties translating into interactions and respective constellations with public actors, public authorities restrict the dense web of closed relations to their own group. In this context, structural closure of network relations impedes interactions between public and private parties that turns out detrimental for wind power development.

New economic sociology suggests that the characterisation of public-private relations in terms of overall network constellation matters. Business activities depend on and are interrelated with policy-related activities ranging from power purchase arrangements, tariff regulation to clearance issuance. In Kerala, structural closure of interactions between public and private stakeholder constellations implies that transactions take longer and are more costly. This is also related to the lacking sense of common purpose or confidence in the wind energy undertaking.

With regard to the conditions under which stakeholder networks exert different kind of effects on renewable energy governance, the Kerala case provides scope for debate. Firstly, market characteristics particularly compared to Tamil Nadu might not be considered as certain due to the relatively early stage of wind power deployment and respective governance activities. However, this dissertation makes the case that due to similar historic trajectories, the overall environment for wind power is similar across India and hence characterised as certain. Secondly, wind sector activities at first sight appear not exploitation-oriented due to the limited private and public agency. Yet, as private actors obtain a wealth of experience in other contexts while public engagement has existed to some extent, wind sector agency in Kerala is also focused on exploiting available resource potential. In conclusion, the Kerala case confirms the hypothesis about the effects of governance arrangements on renewable energy governance.

V.2.3 Renewable Energy Governance: Stakeholder Networks in Kerala

Alongside the ethnographically-derived stakeholder network mapping below (Map F), the main observations of the network effects in relation to renewable energy governance in Kerala are now discussed. This concluding section aims to integrate the constitutive elements of Kerala's renewable energy stakeholder network, i.e. stakeholders and their specific characteristics, actors' arrangements and interactions, in order to understand who governs how. While this approach to exploring Kerala's renewable energy governance stakeholder network remains at the analytical level, Chapter VI will go into depth exploring causal mechanisms and underlying rationales.

Map F: Kerala Renewable Energy Stakeholder Network

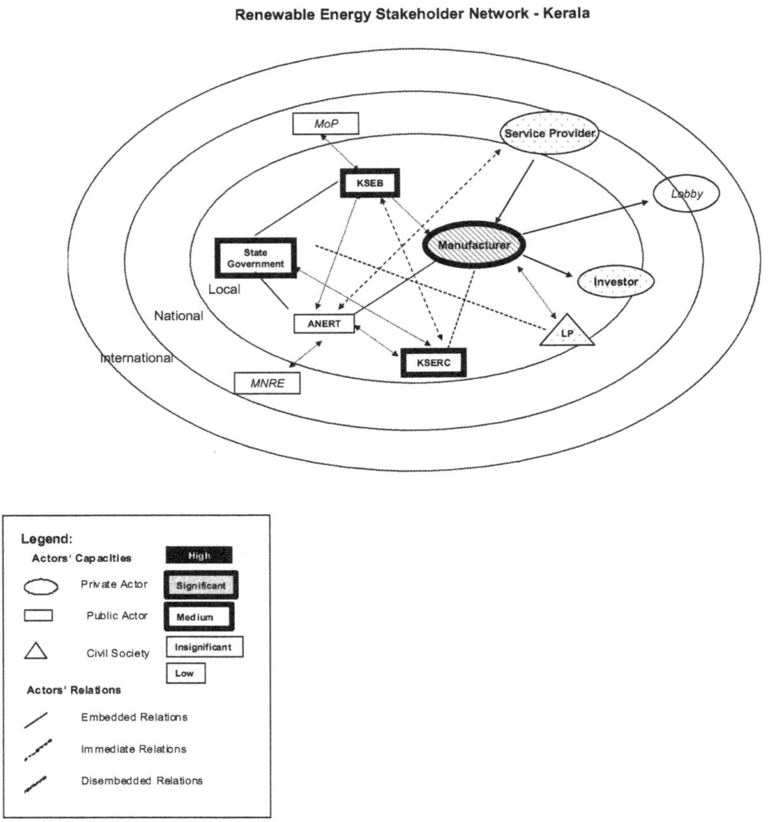

In the light of governance debates, this theory-guided empirical venture into the realities of wind power deployment in Kerala derived at the following conclusions.

With regard to who governs, actors' arrangements are characterised by formalistic and less trustful relations. This goes beyond the formal institutionalisation of certain renewable energy procedures but extends to the interdependence and obligation of decision making with regard to wind power deployment. Another aspect refers to the limited amount of social capital inherent in relations that renders the accessibility and contribution of stakeholders to renewable energy governance more difficult. Concretely, private actors have a difficult time to establish relations based on reliability, common understanding, coordination and social capital with public sector stakeholders. The effects are the lack of information transfer and exchange of assets, and reduced levels of trust and expectation certainty that impede private actors to benefit from wind energy investment. Renewable energy governance is dominated by public stakeholders. The fact that private actors are excluded from emerging coalitions of common interests – a de facto renewable energy network – turns out unfavourable for investment activities in the wind energy sector in Kerala. In the light of new eco-

nomic sociology, the variety of actors' arrangements with relatively small qualitative elements characterising relations imply obstacles to smooth transactions. Lack of trust and mutual understanding poses tacit barriers and invisible transaction costs.
With regard to how wind energy is governed, interactions in Kerala's renewable energy stakeholder networks reflect the structural closure of public stakeholder constellations. This confirms the proposition on the negative effects of network closure on wind power deployment. Governance arrangements are characterised by limited stakeholder embeddedness and little interaction density and quality between public and private stakeholders which contributes to insignificant wind power deployment. While interactions between private actors feature relational quality, public stakeholder interactions are characterised by high degrees of formality that are based on tacit understanding and mutual control over agency. Despite high interaction density and formal relational quality constituting respective private and public interaction constellations, wind energy deployment is not taking place. This is due to the fact that interactions within the entire stakeholder network take place in structurally closed constellations. In other words, the structural closure of relations entrenched in public stakeholder constellations that do not share social capital and reciprocity in ties with private actors turns out detrimental for wind energy deployment. Structurally close public constellations might be considered as more transparent and less corrupt as not prone to lobby influence. However, the structural closure of public actor constellations might also be interpreted as an institutionalised corruption since access and participation of affected parties is de facto inhibited. A last observation refers to the fact that network dynamics with the rare exceptions mentioned do not venture beyond the border of the Kerala state. In summary, with regard to the explanatory model, unsuccessful wind energy deployment in Kerala is due to the local stakeholder network featuring unfavourable characteristics of actors in disembedded relational arrangements and structurally close interaction constellations.

Wind energy deployment takes place in the context of multi-level stakeholder networks integrating the dynamic interplay of stakeholder attributes and relations in specific network interaction constellations. These renewable energy governance arrangements are expected to exert certain effects on renewable energy governance given that a critical time dimension has passed in order for renewable energy stakeholder networks to evolve. With regard to the Kerala case study, the empirical results interpret the interrelations between actors' attributes and actors' relations in the local stakeholder network as follows: Low interests in wind power by public actors relate to the reluctance to establish more than impersonal business relations with private stakeholders. The lack of experience in the local context renders more difficult for private actors to find right entry points to establish trustful relations with public actors. The generally high formalism characteristic of relations in actors' arrangements also reflects low capabilities and autonomy of actors in decision-making.
One critical question points to the ways through which the stakeholder network characterised by specific actors' attributes and relations exerts effects on wind energy deployment[62]. Integrating new economic and organisational sociology debates allows for identifying two central mechanisms that link attributes and relations constitutive of stakeholder networks. Firstly, according to the transaction cost argument, low human resource endowment and interests on the parts of public authorities puts unneces-

[62] As discussed previously, the inherent difficulty with interdisciplinary approaches is the explicit focus on either attributes – organisational sociology – or relations – new economic sociology.

sary obstacles to wind power deployment through prolonging processes and duration of required procedures. Clearances that are not required in Tamil Nadu are obligatory to obtain for entrepreneurs investing in wind power in Kerala. Additionally, the only source of authority is ANERT that is insufficiently equipped for this task since characterised by unfavourable attributes in terms of staff qualification and attitudes. Secondly, according to arguments about trust and common understanding, the lack of local knowledge and experience on the parts of private actors renders the establishment of embedded relations and the engagement in structurally-open interactions more difficult. Public authorities such as the TNEB view with suspicion emerging new actors such as IPPs and are hence reluctant to sign PPAs.

In summary of the second case study of wind power deployment in Kerala, stakeholder network analysis revealed two main results: Firstly, the Kerala case shows that the less favourable actors' attributes featuring the local stakeholder network are one explanatory factor for unsuccessful wind power deployment. The failure to harness the wind energy potential is specifically amounted to the resistance and stickiness of relevant policy and regulatory institutions and lack of experiences by private actors. Secondly, this case study demonstrates the extent to which the characteristics of actors' arrangements that lack the embeddedness of dyadic ties in stakeholder network relations impede wind power deployment. Moreover, interactions within the network are characterised by structural closure impeding governance among private, public and the local actors. On this basis of this analysis, Chapter VI will embark upon the theory-guided exploration of causal mechanisms and underlying rationales. The following concluding section will contrast the deductive empirical findings on renewable energy networks in wind power deployment in Tamil Nadu and Kerala and discuss the validity of the explanatory model through process tracing.

V.3 Wind Energy Deployment in India – Results and Implications

In the course of this chapter, the explanatory model for wind energy deployment in India was deductively tested in the two local contexts of Tamil Nadu and Kerala. The following section will summarise the main results and firstly, contrast the empirical-analytical findings of the two settings. Secondly, a process tracing of wind power deployment in Tamil Nadu and Kerala is conducted in order to prove the validity of the explanatory model.

V.3.1 Contrasting Determinants of Wind Energy Deployment in Tamil Nadu and Kerala

The case study analysis of Tamil Nadu and Kerala confirms the hypotheses of the explanatory model that the local stakeholder networks exert favourable or unfavourable effects on the outcome of renewable energy governance. Analysing the extent to which stakeholder network characteristics and dynamics explain wind energy deployment concludes on three key observations (Table 30):

Table 30: Renewable Energy Stakeholder Networks in Tamil Nadu and Kerala

	Actors' Attributes	Actors' Relations
Tamil Nadu	- Highly-favourable actors' attributes with regard to wind symmetrically characterising the organisational context of the ma-	- Actors' arrangements emerging are characterised by fairly embedded ties of actors in local networks multilevel system

	jority of stakeholders - Actors' attributes of public actors reflect high interests and staff quality imminent in organisational routines and leadership - Actors' attributes of private actors display high experience and resources - Civil society actors' attributes are low given absence of resources and organisation	- Structurally open interactions include private actors into more trustful relations constellations with public stakeholders
Kerala	- Less favourable actors' attributes for wind energy deployment asymmetrically characterising different stakeholders - Actors' attributes of public actors reflect low interests and staff quality in organisational routines and cultures despite individual leadership efforts - Actors' attributes of private actors suffer from low experience - Civil society attributes reveal some capabilities but negative attitudes to wind energy	- Disembedded ties characterise the variety of public, private & public actors' arrangements - Structurally close constellations and interactions between private and public actors

First of all, differences emerge as regards who governs that do not refer to the constitution but to the characteristics of governance actors (Figure XXXI). In the Tamil Nadu case, a strong orientation towards renewable energy deployment on the parts of public actors with regard to human resource aspects and political will translates into highly-favourable actors' attributes. Private actors are willing to invest resources and avail of experience with regard to wind power deployment.

In contrast, the Kerala case shows that actors' attributes unfavourably constituted with regard to wind power prevent effective renewable energy governance. Public actors avail of resources and experience but lack interests, staff qualification and political will. Despite individual leadership, organisational culture and characteristics remain unfavourable of wind energy. Private actors have little experience with the local context. Hence stakeholder networks are by and large characterised by unfavourable actors' attributes for wind energy deployment. Surprisingly, attributes of civil society actors are more pronounced than in Tamil Nadu due to existing experiences and resources but remain unfavourable of wind energy in relation to attitudes.

Figure XXXI: Actors' Attributes in Tamil Nadu and Kerala

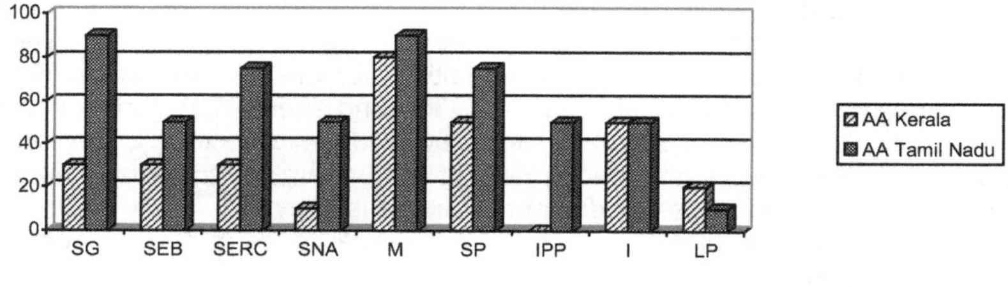

Secondly, differences emerge in terms of how and with what means governance takes place (Table 31). In the Tamil Nadu case, emerging actors' arrangements are characterised as such that public and private actors are relatively embedded in relations that feature low degrees of formality and significant amounts of interaction intensity and quality. As surprising outcome, the majority of actors' arrangements reflect dyadic ties with less emphasis on factors such as social capital and reciprocity. This is explained by the scope of actors' relations and interactions that extend to multi-level networks. This implies that space, locality and also time are important for embedded ties to emerge in actors' arrangements. Interactions take place in networks structured in quite an open manner particularly on the parts of key private sector players.

In contrast, the Kerala case shows a greater variety of emerging actors' arrangements predominated by relationally disembedded ties of low quality and high formality. This is particularly true for actors' arrangements involving relations between public and private stakeholders. Public actor arrangements are characterised by formalistic elements enhancing mutual dependencies and the focus on proper process that allows for little flexibility and access. Emerging actors' arrangements between private actors go beyond state-level boundaries and feature embedded ties. This implies that interactions in structurally closed constellations are detrimental for wind energy deployment. Particularly the fact that interaction constellations between public actors are constituted of formalistic ties impedes exchange with and access to private stakeholders.

Table 31: Stakeholder Relations in Renewable Energy Networks

	Tamil Nadu	***Kerala***
"Disembedded AR"		**"distant – formal ties"** 29% of ties **"hybrid – distant ties"** 43% of ties
"Immediate AR"	**"hybrid – immediate ties"** 43% dyadic ties **"immediate – formalised ties"** 24% daydic ties	**"immediate – formalised ties"** 11% of ties
"Embedded AR"	**"familiar – formalised ties"** 29% dyadic ties	**"familiar – formalised ties"** 17% of ties

In overall conclusion, wind energy deployment takes place given that stakeholder networks are characterised by and constituted of actors with wind-supportive attributes that are embedded in structurally-open interaction arrangements. The deductively-guided empirical analysis of the conditions for wind deployment in Tamil Nadu and Kerala rested on an explanatory model for wind energy deployment in India. Derived from organisational sociology and new economic sociology, the hypotheses proposed that the successful deployment of wind energy is conditional upon renewable energy networks constituted of and characterised by:

i. Favourable attributes characterising a critical mass of stakeholders that participate in renewable energy deployment
ii. Actors' arrangements with embedded in dyadic relations of high social capital, reciprocity and low formalism in structurally open multi-stakeholder interactions

iii. Given a certain market environment in which the exploitation of new opportunities in emerging sectors such as wind energy is required.

V.3.2 Tracing the Process of Wind Energy Deployment in Tamil Nadu and Kerala

The following section will look at the past of wind power deployment and trace respective processes. The explanatory model has so far been applied to analyse the current context of wind power deployment in Tamil Nadu and in Kerala. As emphasised particularly by historical institutionalism and organisational sociology theories, history matters immensely in shaping organisational experiences and environment. First inductive empirical evidence confirmed the importance of path-dependent development. Apart from the ambition to bring history back in, the purpose of the following process tracing is to establish the validity of explanations through increasing the number of (temporal) case studies.

Figure XXXII: Wind Sector Development in India

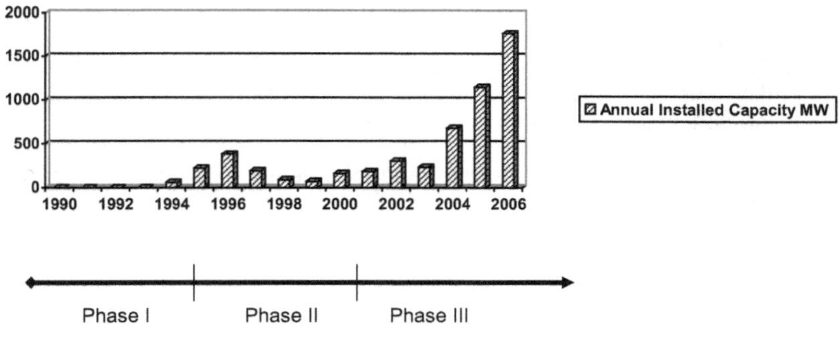

For this purpose, the history of wind energy deployment in Tamil Nadu and Kerala is explored in three different phases, i.e. as distinct case studies, identified in Chapter I (Figure XXXII). The aim is to prove whether the explanatory model can account for these additional cases ensuring the circumstances, i.e. local context of Tamil Nadu and Kerala, are similar. The analysis is based on qualitative interviews and hence differs in empirical richness. This is due to the absence of historic data or statistics to corroborate evidence on the history of wind power development and the respective explanatory factors[63].

Phase I – Initiation Stage
Generally speaking, during the initiation stage, wind sector development in India experienced a slow take off. Capacities increased to roughly 400 MW over approximately a decade. Geographically speaking, the majority of wind power growth took place in Tamil Nadu. The most important events characterising this phase were the liberalisation process that set in around 1991, the design of policy guidelines by the MNRE, the launch of the wind energy programme and the set-up of IREDA .

[63] The process tracing is largely based on interviews since consistent and historic data for the two cases was not available.

In the case of *Tamil Nadu*, the moderate but continuous increase in installed wind energy capacity is explained on the grounds of the emerging stakeholder network.
Firstly, looking at who governs at this stage allows for the conclusion that the stakeholder orientation towards wind energy deployment was quite significant. Outstanding features are the pro-activeness of state government officials in launching a wind energy policy according to MNRE guidelines. Attitudes of the TNEB reflected the willingness to participate in wind farm installations and to support wind energy deployment. TEDA's attributes at this stage are characterised as favourable due to the important role with regard to providing expertise and directly engaging in wind power procedures. Despite the lack of capacities and knowledge about wind energy, private sector stakeholders showed keenness and investment savvy. In summary, despite the lack of experience and human and financial resources, actors' attributes reflected a significant orientation towards wind energy deployment.
Secondly, with regard to how governance took place, stakeholders started to develop embedded ties. Relations between public actors reflected a common understanding and the joint effort to promote wind energy. Relations to private actors slowly evolved and were initially characterised by greater degrees of formality. Reciprocity or trust was not established also as no shared history existed. Interactions at this stage were relatively closely-confined to public/private sector constellations due to the fact that the public sector was in the driving seat as regards initial wind energy activities. Private participants were engaged in establishing business relations among themselves and mainly concerned to get the crucial actors on board.

In the case of *Kerala*, some wind energy deployment took place with the launch of the first demonstration project in 1992. This implies that to some extent the renewable energy stakeholder network was emerging.
Firstly, with regard to who governs, actors' attributes were shaped in a manner as to allow the entry of wind power as potential local energy option. This means that some interest in wind energy deployment and cautiously curious if not proactive attitudes of stakeholders such as the KSEB and state government existed. This might be due to the incentives promised by the MNRE or the interest in exploring additional options for power generation. Private stakeholders availed of less favourable attributes than in the Tamil Nadu context. Even though business interest in wind power existed; at this time, picking the low-hanging fruits in the more accessible Tamil Nadu wind context was preferred. Wind power business did not rush to venture in relatively more uncertain environments characterised by a perceived more complex wind regime.
Secondly, with regard to how governance took place, actors' arrangements were not relationally embedded in dyadic ties. In the case of private stakeholders, dyadic ties to other private actors were less formalistic and characterised more by social capital than between public and private stakeholders. Employing private stakeholders for the installations of the wind power project implies that relational arrangements involving private and public stakeholders began to take some shape. Compared to no-wind-business scenario, the stakeholder network started to become more favourable for private and public interactions and interrelations to develop. This is due to the fact that public stakeholders proclaimed willingness to embark upon new options such as wind energy deployment. However, uncertainty prevailed about the conditions and the way private stakeholders were to be included in such a new undertaking.
In conclusion, as regards the initiation stage, wind energy deployment is explained most and foremost by favourable actors' attributes in the Tamil Nadu case, specifically political will. Stakeholders were at this point not deeply embedded in actors' ar-

rangements and interaction in structurally closed networks. However, due to the fact that private actors and of course public actors are not newcomers to the Tamil Nadu context, a certain amount of mutual understanding, reciprocity and also trust was engrained in relations. These observations are in line with new economic sociology literature (Chapter IV) that emphasises the importance of dense ties and network closure at the initiation stage of business activities. In order to establish sufficient amounts of trust to other parties and to facilitate information and knowledge transfer among similar actors, close network structures are required.

Phase II – Development Stage

During the development stage of the Indian wind sector, capacity additions experienced a decline that lasted until the new Millennium. Annually installed wind energy capacities shrunk from 400 MW in 1996 to about 180 MW in 2001[64]. Two main reasons are mentioned for the slump of wind energy investment during this period of time: While one argument blames the Asian financial crisis for reduced economic activity, the other suggestion focuses on changes in national policy making regarding the incentive structures for the Indian wind industry. This dissertation proposes that negative wind power deployment rates are likely to be explained by characteristics and dynamics of stakeholder networks in Tamil Nadu and Kerala.

Even though not to the same intensity as compared to the rest of India, wind energy deployment in *Tamil Nadu* took place rather hesitantly during this second phase of development.

Firstly, as regards who governs, actors' characteristics had a low orientation to renewable energy deployment, particularly true for the TNEB and the private sector participants, especially investors. As for investors, interests in wind sector investment decreased in response to a changing taxation regime. While prior to this period huge tax benefits could be availed of due to accelerated depreciation benefits of 100% guaranteed by the Indian government, this option was curtailed by 1996. The Indian government noticed that a great deal of investments was either spent on faulty machinery, a lack of maintenance service or was hidden investment, i.e. not actually going into real projects. In order to ensure proper processes and to force private sector participants to take care that investments actually materialised in worthwhile, power-producing installations, taxation benefits were reduced as to separate "*scum from the cream*" of investors (INT 23). This in turn limited the appetite to invest in wind power by those parties mostly interested in availing of tax benefits. Furthermore, investors were disappointed by the insufficient outcome in terms of actual power generation for captive consumption due to the faulty installation, repair and maintenance serviced provided. Hence resources going into wind energy were reduced. As for the TNEB, wind power procedures all of sudden were procrastinated due to the imposition of greater obstacles and hurdles. Power purchase agreements were signed with great hesitance and grid management was neglected. TNEB's reputation during this period of time turned to the downside as negatively biased against wind energy. This is amounted to the change of director and chairman put in charge of the TNEB during this time. Being a personality not convinced of and rejecting wind energy, negative attitudes trickled down or influenced the organisation's outlook and behaviour to-

[64] In fact, at first sight one might be surprised to notice that installed capacities over this period of time actually doubled. However, looking at the statistics in a bit more detail and in comparison to the other phases of wind sector development, it becomes apparent that not only was growth a lot slower but also existed spans of years in which regressions occurred.

wards a certain hostility towards wind energy. State government and TEDA still retained the political will as well as the resources to promote wind energy deployment. As for private actors, manufacturers and service providers suffered from insufficient capabilities and human resources ensuring the proper process of wind energy deployment. In summary, declining actors' attributes contributed to the slump in wind energy deployment. Interesting to notice was that some critical actors constituting the stakeholder network had the power to influence sector developments.
Secondly, with regard to governance processes, some critical stakeholders referring to TNEB and private sector stakeholders were not embedded in the wind energy network which is why deployment activities stagnated. Due to the changed attitudes and interests of the TNEB, trust and expectation certainty were absent in relational arrangements particularly to the private sector. Some social capital in terms of mutual understanding and reciprocity as for the exchange of favours persisted in the relations to state government and TEDA – as otherwise the chairman would not have been put in this position which was approved of by the government. Similarly, disharmony entered into arrangements involving private sector stakeholders. Due to failures in machinery, supply of components, repair and maintenance service, expectation certainty, coordination and trust in each other were called into question. Service providers could not be counted upon in a similar manner due to the technical failures experienced. Manufacturers were not able to rely upon investors to retain previous customer relationships. Investors in turn lacked the confidence in manufacturers and service providers to guarantee high-quality services for their money invested. Public stakeholders were hesitant to extend relations to private actors that according to their perception were responsible for the slacking procedures and developments in the wind sector. On the other hand, private actors found it difficult to gain access to public stakeholders the same way as previously. Consequently, decreasing deployment activities were the response to dis-embedded relations characterising actors' arrangements in the wind energy network of Tamil Nadu.

During the second stage of wind sector development in India no additional wind power deployment took place in *Kerala*. This reflects several characteristics of the stakeholder network related to actors' attributes, stakeholder embeddedness and network structures.
Firstly, regarding governance actors' characteristics, attributes were by and large insufficient to get wind energy deployment going. This was attributed to the lack of political will that translated into unsupportive human and financial resources politics in the respective public institutions. Not only on the parts of the KSEB but also for state government, were attitudes towards wind energy somewhat hesitant if not hostile. This translated into the inability of state government to issue any kind of wind energy policy as well as the inapproachability of the KSEB to engage in any kind of wind sector activity. Indeed, up to 2006 no policies for wind energy as well or orders were stipulated. Organisational culture and routines did not embrace wind power deployment as an attractive option. Processes and practices in line with the organisational environment reflected this questioning of wind power as valuable and normatively attractive. Private actors displayed very little interest at this stage due to the lacking or/and unfavourable experiences in this context and the limited resources given the general macro-economic situation. Even though Suzlon operated an office in Kerala, this did not much affect actual wind deployment activities since processes were stuck in the bureaucracies.

Secondly, governance processes emerged as such that public stakeholders were trapped in actors' arrangements characterised by highly formal relations with great obligatory elements and dependencies. The main concern among public actors was to establish and secure their power position and local constituency. Even though outspoken private-sector hostility did not prevail, societal concerns clearly predominated. Interactions remained extremely closed within the public and private sector constellations and contributed to high entry barriers for any private sector activities in the wind energy sector.

In conclusion, decreasing deployment activities in the wind energy sector of Tamil Nadu and Kerala related to actors' attributes particular the limited political will. This implies that governance actor characteristics constitute an important explanatory factor but pertains to a critical mass of stakeholders. The diffusion of interests and levels of awareness and attention seem to create a domino effect which does not necessarily affect actual resources or capabilities but attitudes and mindsets. Actors' relations constitutive of stakeholder networks in both settings at that period of time were also characterised by lacking embeddedness. Particularly the Tamil Nadu case reflects that shortcomings on political will had repercussions with regard to the level of trust and reciprocity invested in relationships. This rendered interactions less intense and restricted to closed network constellations.

Phase III – Commercialisation Stage

From 2001 onwards until the point of investigation in 2007 the third phase of wind sector development took place. During the commercialisation stage installed wind power capacities in India soared from 200 MW in 1996 to over 1800 MW in 2006. Two distinct factors accompanied these developments: Firstly, changes in the Indian electricity sector occurred in 2003 when the Electricity Act came into force. At the macro level, the second important event was the establishment of the Clean Development Mechanism that promised new avenues for investment in the energy sector and particular the renewable energy business.

The turning point for an exponential growth in installed capacities in *Tamil Nadu* set in after 2004 and most and foremost reflected changes in the renewable energy stakeholder network. Firstly, as regards governance actor characteristics, the orientation of stakeholders towards wind energy deployment in this phase evolved to be quite favourable. Due to changing leadership and enhanced staff qualification the TNEB that had previously displayed a lack of political will turned into a "*good guy*" again (INT 24). State government continued to play a favourable role in the wind energy business and maintained its reputation as proactive leader that even reached model status across India. While TEDA experienced a slight reversal in actors' attributes due to the reduced role and hence resources attributed, a new actor, the TNERC, emerged equipped with well-qualified expert staff and an inherent interest to engage into neutral and objective arbitration. The private sector turned failed developments and negative experiences of the previous phase into lessons learnt and put more emphasis on the quality of capacities in terms of staff qualification.
Secondly, with concern to how was governed in this phase, stakeholders were embedded more intensely in relations within actors' arrangements. Bonds and contracts among private stakeholders were revisited on the basis of shared experiences and mutual leaning. This led to greater amounts of coordination. Slowly, trust and confidentiality started to be re-enter relations through repeated interactions. As for public

stakeholders, the degree of formality in relations was reduced on account of an increased mutual understanding about the common objectives of power generation and wind energy promotion. Social capital as regards trust and reliability in particular increased. Interaction intensity and quality especially between manufacturers and TNEB as well as state government reached another level. Expectation certainty was re-established and the coordination of activities was enhanced. Drawing on shared experiences and lessons learnt created a feeling of mutuality and contributed to increased trust. Going hand in hand with increased intensity and quality of relations between stakeholders, interaction constellations started to open up more and more during this stage of wind sector development in Tamil Nadu.

This third stage of wind sector development in India has only recently triggered de facto changes in terms of wind energy capacity installed in *Kerala*. During the commercialisation phase, the stakeholder network remained stuck to the characteristics and dynamics of renewable energy governance arrangements of the second phase which explains the lack of additional deployment activities.
Firstly, the characteristics of public actors towards wind energy promotion were unfavourable. This was due to the lacking political will of public sector stakeholders that translated into insufficient human resources policies. Adverse attitudes and interests in wind energy contributed to insignificant engagement in this sector. Private stakeholders also lacked crucial capacities especially as concerns experiences operating in this local context and in this sector.
Secondly, governance processes reflected in actors' arrangements were characterised by a lack of embedded relations due to the high formality constitutive of relations among public actors and between public and private actors. Private-public interactions were nearly absent due to the lack of trust, common understanding and reciprocity. This translated into relatively closed interaction constellations of dense ties among public actors.
Despite some favourable tendencies within the Kerala renewable energy stakeholder network during the third phase, no visible change in network characteristics and dynamics occurred. This might be explained by the famous "implementation gap" that firstly, reflected lacking political will, collective and individual leadership; and secondly, the absence of a favourable critical mass of actors.
Today, an increasing interest of private stakeholders in wind energy deployment as well as a slow transformation of public sector activities is observed. The stipulation of policy and regulations for the wind energy sector implies the first move towards greater openness to wind power. Applications for land and the commissioning of approvals for wind farm installations are under way. This reflects slowly changing attitudes of public stakeholders, emerging relations between public and private stakeholders and an evolving stakeholder network.
As concerns the characteristics of governance actors, some government interest is noticeable even though conflicts between the different departments exist. Additionally, the TNEB starts to show some interest in the business potential and benefits accruing for them. Private actors are very keen to start investment and deployment activities as given the mature stage of wind sector development in other Indian states ventures into yet unexplored areas such as Kerala are attractive.
Governance processes still take place in closed interaction constellations even though public stakeholders slowly move towards tentatively mutual understandings with private stakeholders. Stakeholders are embedded in scattered arrangements in

which a lack of trust, social capital and high formalism remain dominants characteristics.
In conclusion, this dissertation proposes that right now the tipping point for wind power deployment in Kerala is reached with the stakeholder network gearing up towards harnessing this natural resource potential.

In summary, tracing the process of wind energy development in Tamil Nadu and Kerala through the three different phases establishes the validity of the explanatory model for wind power deployment in India. Some general conclusions are drawn scaling up the perspective from the case of wind energy deployment in Tamil Nadu and Kerala to renewable energy governance in developing countries. Actors' attributes, relations between actors as well as network structures constitute factors that are decisive for explaining why and how wind energy deployment takes place in the stakeholder network of respective local settings. Regarding who governs, stakeholder characteristics with regard to the political will of key players, staff qualification and historic experiences matter. As concern how is governed, the structural openness of interactions between relationally-embedded stakeholders matter. Above all, a critical mass of stakeholders is crucial to trigger the emergence of networks that exert positive effects on renewable energy governance given locally-specific characteristics and dynamics. Characteristics and relational dynamics constitutive of local governance arrangements are decisive for explaining why and how wind energy deployment takes place in the stakeholder network of the respective local setting.
All in all, this analytical process tracing of wind power deployment delivers two main results. Firstly, the explanatory model of wind power deployment in India on the whole is considered confirmed and further specified. Secondly, this exercise stresses the importance to take history into account for understanding and explaining empirical realities such as wind power deployment. The following chapter will discuss the implications of these findings with regard to the effects of renewable energy networks on renewable energy governance in the light of respective theory debates.

CHAPTER VI

Turning Wind into Power – Effective Renewable Energy Governance in Developing Countries

CHAPTER VI
Turning Wind into Power – Effective Renewable Energy Governance in Developing Countries

Why do local stakeholder networks in some states manage to effectively govern renewable energy and harness the natural resource potential and fail to seize these opportunities for sustainable development in other states?
Spatial variations of wind energy deployment in Tamil Nadu and Kerala show that "*turning wind into power*" in order to reap sustainable development benefits in developing countries like India is not simply done by installing wind turbines. The central focus of this dissertation is on the preconditions for effectively governing renewable energy. In a nutshell, the characteristics and dynamics of locally-specific multilevel stakeholder networks were identified as conditions for successful wind energy deployment. Consequently, effective renewable energy governance depends on "who" governs and "how" in specific local contexts. The purpose of this chapter is to interpret the research findings in the light of research-guiding theories alongside two main questions: Firstly, how do renewable energy stakeholder networks with regard to their constitutive elements provide conducive conditions for successful wind power deployment and how might this be explained? Secondly, what are the impacts of stakeholder networks on external stimuli like the CDM in local contexts?
In the following, the first section will discuss empirical findings in the light of theory debates with the purpose to explain how and why stakeholder networks contribute to renewable energy governance. The objective is to specify underlying causal mechanisms, and to highlight why stakeholder networks evolve in a locally-specific manner. The second section of this chapter will discuss implications of the research findings for the practical reality of wind power deployment in India. The main purpose is to examine the role of stakeholder networks in renewable energy governance with regard to how opportunity structures can be tapped more efficiently. The CDM in the Indian wind sector provides an example alongside which causal mechanisms and effects of stakeholder networks are explored.

VI.1 Stakeholder Network Effects – Theory Implications

What are the theory-relevant implications of stakeholder network explanations of the Indian wind energy puzzle?
In a nutshell, stakeholder network analysis explored characteristic actors' attributes, and dynamic actors' relations constituting locally-specific multilevel renewable energy governance arrangements. These explain who governs wind energy how as to successfully deploy wind energy. Previous discussions left open questions about the mechanisms and reasons underlying the emergence of particular renewable energy stakeholder networks. This section will interpret these empirical results in the light of the governance framework in order to draw implications on the explanatory scope of the respective theories debated. In the following, firstly, questions about how governance takes place are discussed to specify the causal mechanisms relating governance actors and dynamics within stakeholder networks. Organisational sociology, neo-institutional approaches and new economic sociology theories are drawn upon to interpret research findings and draw implications. In a second step, the effects of re-

newable energy stakeholder networks on respective governance are examined taking a holistic approach on local governance arrangements[65].

VI.1.1 Who Governs? Governance Actors in Stakeholder Networks

In response to questions who governs wind power, actors' characteristics emerge as decisive for shaping the policy conditions as well as economic opportunity structures for renewable energy governance. Wind power is successfully deployed if actors involved in the process actively seek to create supportive framework conditions. In the following, potential causal mechanisms and underlying reasons linking actor's attributes to effective renewable energy governance are interpreted in the context of neo-institutionalist and organisational sociology debates.

The purpose of exploring the characteristics of who governs through stakeholder network analysis was to understand agency with regard to the role and effects of renewable energy stakeholder networks.
Summarising the empirical results, firstly, Tamil Nadu and Kerala differ particularly in relation to norms and values characterising governance actors. Secondly, the most striking differences in actors' attributes are noticed with regard to public authorities. As a matter of fact, most business actors engage in operations elsewhere in India or internationally integrated in the global economy. Characteristics in terms of resources, experience and norms do not depend on specific locations as the wind sector production and value chain is less dependent on local inputs than for example biomass energy undertakings. Local people or civil society actors reveal slightly varying attributes related to the different kinds of protest or popular culture as well as previous experience with regard to environmental projects.
At this stage, this dissertation will turn towards theory implications of these empirical findings to gain an understanding of causal mechanisms through which, and why actors' characteristics shape the governance of wind power. For this purpose, the four theory debates of neo-institutionalism and organisational sociology drawn upon in this dissertation will discuss the following questions:

(1) How exactly do stakeholder characteristics impact on wind energy deployment?
(2) What explains the emergence of the respective characteristics of governance actors in the two local settings?

(i) Transaction Cost Arguments – Rational Choice Institutionalism

This section will discuss how rational choice institutionalists (Peters 1999) firstly, view the causal mechanisms relating actors' attributes to wind power deployment and secondly, account for the emergence and origins of actors' attributes.
Rational choice institutionalism contributes two main approaches to explaining political outcomes and institutional effects. Firstly, transaction cost arguments relate institutional performance to the effect of information and trust on activities in and between stakeholder organisations (Koelble 1995). Secondly, principal-agent relations in and between stakeholders are considered decisive for influencing policy outcomes. Rules and contracts impact on decision costs, information, and transparency aligning individual and collective rationality (Hall 1996). The key causal mechanisms through

[65] These interpretative discussions refer again to the current wind power deployment in India as of 2007 even though the relevance of the historic dimension is acknowledged.

which governance effects are exerted concern changes of actors' preferences and goals as well as of power distribution:
How do governance conditions come to be shaped?

In Tamil Nadu, transaction costs arising in public institutions such as TEDA and state government energy department are less compared to Kerala due to fewer staff, a clear division of labour and regular information transfer. Transaction costs between public actors involved in wind energy procedures are also less in Tamil Nadu due to the single window clearance system established that involves fewer authorities. The principal agent approach provides explanations for why actors' attributes in Tamil Nadu are more favourable towards wind power. While decision centres and loci of authority are clearly delineated in Tamil Nadu, the hierarchy between public institutions and ultimate decision-making authorities remains ambivalent in the Kerala setting. Even though the same kind of institutions with similar basic functions are involved in renewable energy governance with regard to the wind sector in both states, the Tamil Nadu public institutional setting is characterised by a clear-cut responsibility and task division. As for Kerala, state government has not really taken the lead and leaves the responsibility for organising wind power deployment with both the KSEB and ANERT. In contrast to Tamil Nadu, state government does not act as the visible and official principal. Other public actors have hence an incentive to shirk, i.e. not to behave as to support effective renewable energy governance vis-à-vis wind power deployment. Public authorities do not benefit from efforts to render renewable energy governance effective that without doubt require additional upfront investment like acquiring new knowledge and engaging in new procedures. Yet, they also do not have to reckon with official sanctioning.
In summary, transaction costs within and between public authorities are higher in Kerala's renewable energy governance network. Hidden or unclear information and complexities in responsibilities, goals and preferences open windows for opportunistic behaviour. Since principal agent relations specifically between public authorities are not clearly defined, no real sense of collective or individual rationality arises. This means that neither formal nor informal contract mechanisms provide constraints or incentives for effective renewable energy governance.

What explains the emergence and origin of respectively characterised actors' attributes?
Rational choice institutionalism argues that the alignment of rules and distribution of resources is a question of implicit and explicit power. Underlying power relations account for the fact that rules and procedures are not clearly defined among the participating stakeholders in Kerala. This implies that within the set of public authorities, vested interests or preferences unfavourable of wind energy are entrenched in certain power loci. The fact that state government does not vociferously claim support for wind power deployment means that government's interests obviously reside with competing undertakings such as hydro power or the fossil fuel industry.

In summary, rational choice institutionalism provides insights into the causal mechanisms relating actors' characteristics to renewable energy governance. The main criticism is that rational choice institutionalism remains silent on the question of why power is distributed in this manner and why underlying preferences disfavour wind energy deployment in Kerala as opposed to Tamil Nadu. The two states initially emerged in a similar energy governance context in which preferences and interests became – for some reason – shaped differently. Assuming rational behaviour, the fact that public authorities in Kerala established ineffective renewable energy governance with regard to the wind sector is not logically explained. While cost benefit calculi with regard to wind power deployment obviously turn out differently in the two settings – the question of why this is the case still remains unsolved.

(ii) History Matters – Historical Institutionalism

How do historical institutionalists discuss the impact of actors' attributes on renewable energy governance?

Historical institutionalism looks at the paths alongside which actors' strategies and goals evolve and relations of conflict and cooperation emerge (Ebbinghausen 2005). The main contribution of historical institutionalism relates to the causal mechanisms while this approach remains silent on explanations about the emergence and origins of actors' attributes. Renewable energy governance evolved through a specific path that has become institutionalised in stakeholder networks. The design and emergence of this path is rendered visible through resource allocation, the incentive structures, mental modals and organisational routines established in respective stakeholder organisations (Stefes 2009). The evolution of a particular path takes place once a critical mass of stakeholders exists or a threshold is reached. Two processes that lead to such points are either conjuncture, i.e. the interaction of social processes, or a learning process culminating to a tipping point (March 1989). Consequently, the design and institutionalisation of certain support structures and processes for wind energy deployment characterising renewable energy governance in Tamil Nadu and in Kerala is explained by the specific paths both states have entered:

Wind power deployment in Tamil Nadu took of with the emergence of a certain mental model that considered wind power deployment as something good, societal beneficial and worthy of institutional back up. This viewpoint entered from highest level as the MNRE official involved in the conceptualisation of wind power guidelines in India was of Tamil Nadu origin. Due to the intimate relations held with the state officials in the Tamil Nadu government at that time, it was not difficult to convince key actors of the potentials and opportunities inherent in wind power. This mental model diffused the public organisations that were involved in wind power deployment. The TNEB constituted the crucial and the most difficult stakeholder to get on board. As a state-related monopoly with its own logics, power struggles, competition and ambivalence vis-à-vis state government existed. Previous energy trajectories locked into conventional fuels were hence difficult to disrupt due to potential threats to vested interests and authorities. State government, the MNRE and official development assistance aligned in the conscious undertaking to actively involve the TNEB. This occurred through the transfer of ownership of demonstration wind projects and later on assigning crucial responsibilities and functions. Involved as stakeholder in this learning process, the TNEB became less reluctant to wind power engagement. An evolving state-government and TNEB alliance remained the loci of ultimate authority. The diffusion and institutionalisation of this mental model was expressed in the design of incentive structures, resource allocation and the establishment of organisational routines. With regard to incentive structures, more informal systems of reward prevail. The TNEB receives official government support related to re-compensation of expenditures occurred through wind power support in infrastructure set-up. This does not only take place on remunerative level but rather in negotiating other issues, e.g. free power obligations. Resource allocation is more visible particularly as regards the task specification and staff qualification in public organisations. Among the public stakeholders, tasks involved in the wind power deployment process are clearly delineated and distinguished, and staff is assessed as qualified to meet required tasks (cf. Chapter V).

In contrast, Kerala has not experienced the emergence and institutionalisation of an equivalent path. Even though the overall energy governance opportunity structure, i.e. MNRE policy guidelines were similar to those in Tamil Nadu, this implicit mental model did not meet the *"right people at the right time"* (INT 1) to take the same effect. As regards state government, phases of wind power support existed, however, no alliance across party or interest groups and no individual leader promoting wind power per se emerged. The idea of wind power and the requirement of effective renewable energy governance prevailed and up and again appeared in policy notes, but did not manifest in concrete policy conditions. In contrast to the TNEB, the public utility could not be convinced of the advantages offered by wind power. Despite a similar design of demonstration activities, these undertakings were described as doomed to fail right from the beginning due to the lack of support from all parties involved (INT 3). Even though ANERT took a natural interest in renewable energy promotion, real enthusiasm for wind energy never emerged. Also the KSERC to date has not taken a clear stance and seems to engage in a laissez-faire policy oriented towards the current power situation. Hence no incentive structure was conceived of and institutionalised that could persuade the KSEB to support wind power deployment or render ANERT a vociferous proponent. KSEB resided with the preconception that wind energy implies costs only that have to be met out of its own restricted budget and for which the burden of risk and responsibility had to be carried. Resource allocation did not do much to improve this scenario. Despite sufficient staff strength and also finances, no attempt was made to create a dedicated unit for wind power activities in the KSEB. Likewise, ANERT's staff number is very impressive and above that of Tamil Nadu. However, its reputation and staff qualification is way below standards. Consequently, the routines established in and between the public stakeholders momentarily pose obstacles to rather than encourage wind energy deployment. Procedures are lengthy and are spread across institutional stakeholders that increase transaction costs rendering the entire wind deployment process vague and uncertain.

In summary, historical institutionalism provides a valuable entry point to understand how certain developments have come about, i.e. paths established, by recurring both to arguments of rationality as well as norms and mental concepts. It lacks insights into the specific triggers leading to the conjuncture, learning processes or critical mass that gives rise to the emergence and evolution of certain paths. Despite a similar macro setting at the point wind power entered the energy scenario in India, how come stakeholders took up opportunities in contexts like Tamil Nadu whereas the business-as-usual scenario prevailed in Kerala? One explanatory entry point, going beyond the scope of this dissertation at this stage, lies in the in-depth exploration of this conjuncture at the concrete local level. This would reflect on interactions between societal needs, business interests and government representation which also recur to actors' relations.

(iii) Norms Again – Sociological Institutionalism

How are governance conditions shaped the way they are, and what is the implicit role of actors' characteristics in this regard?

Sociological institutionalism takes up the argument that certain institutional templates influence governance outcomes (DiMaggio 1991). Within an external environment, a certain logic and perception of a policy issue develops. This translates into organisational routines (Ingebritsen 2002) which are institutionalised in terms of resources allocation, infrastructure and knowledge and define actions as appropriate (Finnemore 1998).

Applied to the empirical wind power puzzle, sociological institutionalists argue that a certain logic regarding the importance of supportive governance conditions for wind power deployment has emerged in the external polity environment of Tamil Nadu. This contributed to the establishment of routines and institutional templates, which defined appropriate behaviour in terms of wind power support in Tamil Nadu. For example, this translated into the set-up of a single-window clearance system that efficiently and effectively mainstreams wind power applications. Conversely, in the context of Kerala, appropriate organisational behaviour was not acknowledged as extending support to wind power investment. The self-perception of state government and the KSEB as people's institution reflects the emphasis on domestic consumers as primary constituency (INT 4). Wind power is not considered to meet the interests of broader society since additional costs of public investment are perceived to outweigh potential benefits. Appropriate behaviour does not integrate the promotion of an alternative energy source, i.e. wind energy that is perceived to benefit primarily industry as well as individual consumers only. Institutional and inter-organisational mechanisms in the wind sector are not streamlined as to minimise transaction costs and facilitate wind power deployment. The establishment of routines and institutional templates went alongside and reinforced an organisational culture characterised by the prevalence of certain norms that guide this type of appropriate behaviour. This explains the lack of vision admonished in ANERT or the wind-adverse reputation of the KSEB. Tamil Nadu is perceived as wind state in which public stakeholders tend to align to serve the interests of the wind business community. In contrast, Kerala's image is that of a wind power investment hostile environment in which public stakeholders are lethargic to promotional activities.

In summary, sociological institutionalism provides plausible explanations for the impact of mental models and institutional templates on respective governance outcomes and institutional behaviour. Meanwhile, sociological institutionalism remains misty on explanations for the evolution of such an institutional template. The influence of polity, i.e. regime types, on the emergence of organisational culture and routines is underspecified. As one entry point, sociological institutionalism implies that norms and values exert agency-related impacts on establishing organisational routines. Applied to the empirical puzzle, remaining questions concern the definition and the identification of norms and values, and beyond that the evolution of and triggers for the diffusion of norms. Following from that, the path or diffusion process of these norms with regard to the effects in and on institutions as well as the role of individual stakeholders could be traced. At this stage, such an analysis goes beyond the theory-related discussion of the empirical results and remains subject to future considerations.

(iv) Inside Out – Organisational Sociology

How does organisational sociology explain the emergence and impacts of organisational behaviour characterising governance actors on renewable energy governance? The main interest is to understand how and why public stakeholders, i.e. government authorities such as SEB, SERC and SNA, that are constituted similarly and are assigned with similar tasks and responsibilities act in support wind power in the Tamil

Nadu case and neglect wind power promotion in Kerala[66]. Organisational sociology takes a slightly different approach to governance conditions looking concretely at stakeholder agency in terms of organisational deviance (Vaughan 1999). Organisational sociology explains organisational behaviour in terms of environment, organisational characteristics and individual and cognitive practices (Kieser 2001).

Environment as the first organisational characteristic impacts on the organisational work culture, shared values and norms, and hence shapes respective institutions, instruments and experiences. Applied to the context of Tamil Nadu, the environment defined by wind investment orientation translates into a work culture reflecting similarly-oriented state government interests. In contrast, interest in the societal well-being characterises the environment in Kerala and reflects the lack of support to wind entrepreneurship.

The second set of organisational characteristics includes the three aspects of structure[67], processes and specific tasks. With regard to the Tamil Nadu context, organisational structures and structured relations between public stakeholders are characterised by the clear division of labour and an informal but internally known and respected hierarchy. Between public stakeholders in Kerala, the labour division pertaining to wind power deployment is dispersed between organisations and no clear hierarchy is perceptible. For example, responsibilities for policy making are ambiguously divided between two state government wings. Processes in and interactions between public organisations reflect power struggles about goals and objectives that entail certain feedback mechanisms. Without doubt, in both local contexts, public organisations are intimately linked with each other in the energy governance system and hence power struggles exist with regard to resources, staff and objectives. In Tamil Nadu, wind power promotion has emerged as a common public objective and burden sharing mechanisms are designed as to minimise issues of contestation. In Kerala, the power game within and between public utilities and state government is implicit in the context of wind power promotion. State government does not take a clear stance on wind power due to internal rivalries and the strong influence of the KSEB. The (until recently) rather wind-averse KSEB exerts influence on but also depends on government. Without common goals and mutually supportive feedback mechanisms effective policy making and implementation for wind energy is impeded. The completion of specific organisational tasks involved in wind power governance depends on occupational skills, risk distribution and technical uncertainty. As for Tamil Nadu, the overall situation emerges relatively straightforward as the ultimate risk and responsibility resides with the government and skills are distributed according to the specific roles and functions. In Kerala, the definition of risk and the delegation of potential losses are not clearly specified. Government shies away from taking the ultimate responsibility and the KSEB is not really willing to jump in given the perceived uncertainties and costs. Reflected in the bad staff qualification of ANERT, skills do not match the requirements to fulfil the assigned highly technical tasks.

The third organisational characteristic considers individual cultural beliefs, social expectations as well as organisational leadership which results in creating rules and a certain social fabric through practices such as staff hiring. The importance of leadership in particular becomes apparent looking at the Tamil Nadu case. The role of the

[66] The focus is on public stakeholders since they reflect the starkest differenced in characteristic attributes as compared to private actors and local people.

[67] Structures describe the ties between actors in terms of labour division, hierarchy, specialisation, information and language. This aspect is taken up and will be elaborated in the examination of actors' relations.

energy minister has been pre-eminent in initially pushing wind power deployment. Furthermore, the change of director in the TNEB to a so-called wind power sceptic at one point in time contributed to the stagnation of wind power deployment. The Kerala case emphasises that not only leadership but also rule following, i.e. trickle-down effects of leadership directions to the bottom of an organisation, and the acceptance of leadership is crucially important. Even though ANERT is headed by a skilful and experienced official supportive of wind power, organisational lethargy prevails and translates in highly transaction-cost intensive complex wind power procedures. Talking about leadership does not go without mentioning the issue of power. The opportunity and possibility to lead within and among public organisations is related to the extent to which power resources can be deployed, e.g. the freedom to hire qualified staff, and enforcement mechanisms can be implemented. In both cases, Tamil Nadu and Kerala, the autonomous power of individual public organisations is curtailed since public authorities are mutually dependent within the energy governance system[68]. In Tamil Nadu the relative autonomy and independent say of public institutions is characterised by greater leeway in governing renewable energy. This is reflected in the more autonomous role and the acceptance of the regulatory authority with regard to the implementation of contested regulations.

In summary, explanatory approaches inherent in neo institutional perspectives discuss transaction costs, relations, norms and values, and routines as plausible causes of how governance conditions take certain shapes. Historical institutionalism and the notions of power at this point suggest a potential entry point to understanding and explaining actors' attributes. Going beyond the scope of this dissertation, this would require recurring to meta-theories related to power in order to gain more systemic insights into why actors behave they way they do. Organisation sociology provides meaningful insights into causal mechanisms explaining the evolution of unsupportive or supportive governance conditions. Organisational environment, characteristics and practices shape actors' attributes and influence public stakeholder agency in terms of wind power support. This makes the case for organisational sociology as insightful theoretical perspective on how governance conditions are shaped by the agency of characteristic governance actors. However, organisational sociology remains silent on explanations of why organisations behave in one way or another and in terms of why actors' attributes adopt certain characteristic features.

In summary, new institutional and organisational sociology theories contribute to understanding plausible mechanisms and potential underlying conditions. What is still lacking from this interpretative analysis is a more explicit discussion of how stakeholder relations come into play and render this scenario more dynamic. The next section hence turns towards how governance takes place in the context of the relational dynamics in stakeholder networks.

VI.1.2 How is Governed? Governance Agency in Stakeholder Networks

In the following, the question how governance takes place with regard to relational dynamics between stakeholders is interpreted with focus on causal mechanisms and underlying reasons that explain effective renewable energy governance.

---

[68] This issue is taken up looking at the relations and interactions between stakeholders.

In conclusion of previous research, stakeholder networks emerged and exerted effects through relations and interactions with respect to wind energy deployment. Stakeholder network analysis explored the evolving functionally equivalent actors' arrangements going beyond the public/private binary divide in the deployment of wind power to better understand renewable energy governance. Stakeholder networks in the Tamil Nadu case emerged as arrangements with network relations of greater familiarity and less formalism. Stakeholder network analysis also considered interactions between the stakeholders involved in governing renewable energy with regard to openness and closure of emerging networks. The stakeholder network in the case of Tamil Nadu revealed more openness as interactions in evolving constellations went beyond pre-set public/private divides.
This section will tackle the following two issues:

(1) What effects and what causal implications do the dynamics of relational governance arrangements have on wind energy deployment?
(2) How is the emergence of the relational interaction arrangements explained in the respective local settings?

The following section will take up the new economic sociologist debates on how relational governance arrangements affect renewable energy governance. For this purpose two aspects are considered in turn to tackle the questions above. Firstly, the characteristics of stakeholder relations in evolving actors' arrangements are examined. Secondly, attention is turned towards the dynamics characterising interaction constellations.

(i) Dyadic Ties in Evolving Actors' Arrangements

Actors' arrangements shaped by different dyadic ties are examined in order to understand causal mechanisms explaining relational stakeholder network effects on renewable energy governance. New economic sociology suggests valuable interpretations with focus on relational characteristics of dyadic ties such as formality, social capital and reciprocity (Bathelt 2003):
One aspect characteristic of dyadic ties in actors' arrangements is the degree of formality. In the context of non-OECD countries and specifically India, this is considered as potentially powerful explanatory argument for why certain arrangements and interactions produce effective results in terms of wind energy deployment. Empirical research led to the conclusion that unofficial procedures rather than abiding to formal rules render wind energy governance effective. The art of making business is to know how to get things done and whom to associate with. Without doubt, formal rules and procedures guiding private-public actor as well as public-public actor interactions in specific business fields exist and are enforced. Beyond this layer, a web of tacit understandings prevails that serves to facilitate and speed up many processes. For example, applications for a government subsidy for wind power installations are available to any entrepreneur and openly announced as well as institutionalised and implemented. Following official procedures takes an enormous amount of time due to the administrative complexities involved in the institutional process. However, informal and unofficial mechanisms of speeding up this process exist, for example through a mechanism of paying "service fees" to institutional brokers. The extent to which these parallel shadow systems come into play depends on the overall stakeholder setting as well as the respective stakeholders and their interrelations. Based on rational theories, one might expect that higher degrees of formality contribute to a decrease in transaction costs since procedures clearly lined out and executed in-

crease transparency. Given the specificities of the highly complex and overstaffed Indian bureaucracy, this kind of formality in reality increases the administrative burden and hence impedes efficient governance agency such as wind power deployment. In the context of Tamil Nadu, relations between public and private as well as among these stakeholders are characterised by greater informality. In contrast, relations among public stakeholders in Kerala are described as more formal due to a higher perceived interdependence and less autonomy as regards staff allocation and decision making. This is partly amounted to higher overall transparency due to the pressure from a more attentive citizenry. For private parties, gaining a similar kind of informal access to public stakeholders as in Tamil Nadu is far more difficult.
The other relational features constituting actors' arrangements refer to social capital and reciprocity that to a large extent seem to compensate for formality. This implies that effective renewable energy governance is defined through a specific kind of proper process. This puts greater emphasis on mechanisms of trust, personal knowledge and repeated interaction. The case of wind project developers in Kerala shows that repeated interactions and frequent contacts as well as personal knowledge alone are not sufficient for establishing relations that work in support of wind power. Despite familiarity with the local setting and a sense of personal knowledge from previous interactions, access to the level of influence concerning procedures and framework conditions for wind power deployment is hard to establish (INT 5). Due to resistant and sticky individual attitudes as well an adverse public stakeholder alliance, time, personal knowledge and local expertise are required for stakeholder network evolution. What turned out most decisive in the Tamil Nadu case was the sense of common understanding that pervaded relations between private and public as well as among public and among private actors. The mechanisms leading to this sense of mutually-shared objectives of a wind power support as worthwhile, profitable and intrinsically "good" undertaking only peripherally relate to the frequency of interactions. Information and asset exchanges also seem less relevant. However, personal understanding combined with the creation of mutual trust and confidence in each other as well as renewable energy governance facilitated evolving embedded ties between stakeholders.

(ii) Interactions in Evolving Actors' Arrangements

Causal mechanisms are not only implicit in the characteristics of dyadic ties but also in the evolving interaction constellations between stakeholders in the network of relations. This section will explore relational characteristics of interaction constellations between public-public, private-private and public-private actors[69]. The objective is to understand plausible mechanisms through which functionally equivalent actors' arrangements trigger wind power deployment.
Firstly, *public actors' constellations* emerge not voluntarily but are institutionally embedded, are subject to several compliance and regulatory mechanisms and are assigned with certain functions and procedural codes. This implies that as a matter of fact some degrees of formality always characterise dyadic ties between public stakeholders. The level of autonomy and the extent to which more informal elements constitute the arrangements varies according to the local contexts and the types of public stakeholders. In the Kerala case, most actors' arrangements involving public stakeholders are characterised by a higher degree of formality. This implies that elements

[69]For the sake of reducing complexities, "private actors" includes all non-governmental stakeholder, i.e. also civil society.

of dependency with regard to resource allocation and decision making are perceived as predominating more informal means to attain common goals with a sense of mutual understanding. Even though public stakeholder constellations in Tamil Nadu particularly as concerns the regulator also do not reflect very intimate and familiar relations, a similar kind of obligation and restrictive dependency is not observed. The fact that informal elements based on mutual understanding are inherent in arrangements, might render public stakeholders in Tamil Nadu potentially more open-minded to the wind sector. In contrast, the perceived high formalistic interdependencies between public stakeholders in Kerala implies little space for interactions involving private actors beyond the confined boundaries of what is defined as appropriate. With regard to implications for wind power deployment, less focus on mutual interdependencies and greater inclusion of personal understanding in public actor arrangements allow for ideas, assets and exchange relations to pervade established procedures and arrangements. The characterisation of arrangements and interaction constellations between public stakeholders adds value to explaining renewable energy governance as regards the implications on access to other parties.

Secondly, *private actors' constellations* in the network of relations are explored with regard to the effects and mechanisms in renewable energy governance. Considering business actors only, one might assume that regardless the local context, private actor arrangements will reveal similar characteristics. This is related to the fact that Indian wind sector activities from a business perspective take place mostly at the national level. Turbine manufacturers operate as vertically integrated enterprises across India. Consequently, service providers are involved and integrated in arrangements in a similar manner across the entire territory. Issues such as business and investment finance are also matters of national concern as financial institutions do not tend to operate in a locally differentiated manner. Looking into the empirical reality indeed supports this argument. As finance decisions are based primarily on the client, i.e. recurring back to established business relations, relations to private actors do not vary significantly despite perceived higher risks in local contexts such as Kerala. Private stakeholder arrangements in Kerala as well as in Tamil Nadu reveal little formality and are characterised by varying degrees of trust, mutual understanding, personal knowledge and reputation. Of course, differences exist that are, however, mainly related to the actors' status and to how long private stakeholders have engaged in the local context. In the case of Tamil Nadu, newcomers to the wind power business have emerged. These are on the one hand independent power producers that offer new models of wind power procurement. On the other hand, entirely new types of actors entered the stage in emerging new business opportunities such as carbon finance. So-called traditional private stakeholders such as service providers and manufacturers to date hold more formal relations to those kinds of actors with less pronounced levels of social capital and reciprocity. This points towards the importance of time to establish embedded relations and implies that arrangements are dynamic and subject to change. All in all, private stakeholder arrangements do not add great explanatory value for effective renewable energy governance as they do not vary significantly between the local contexts. To some extent this is a slightly unanticipated finding as one might argue that the local context in terms of environmental uncertainty or power structures impact on the constitution and characteristics of private actor arrangements. This implies that private business activities in the wind sector per se seem to be applicable as 'one-size-fits-all' model at least in the Indian context.

The last point of consideration looks at *multi-stakeholder constellations* between public and private actors. New economic sociology assumes that economic activities such as wind power deployment are socially embedded in relational arrangements. Interactions between public and private stakeholders are thus fundamental for understanding and explaining wind power deployment (Granovtter 1992). Unfortunately, new economic sociology has so far provided neither theory background nor empirical studies that tackle the specific aspect of private/public relations and embeddedness. Political sciences and organisational sociology are also vague on the impact and power of relational arrangements and interactions that explicitly involve public and private stakeholders. In order to consider the effects of multi- stakeholder networks on governance, network analysis gave rise to several typologies of evolving actors' arrangements. This dissertation argues that the characteristics of multi-stakeholder arrangements either result in opening windows of opportunities for new emerging and alternative forms of economic agency such as wind power deployment or locking into pre-existing and dominant logic of exchanges, i.e. fossil fuel-based energy systems. One key finding of this dissertation is that public-private stakeholder relations provide significant explanatory accounts for the effects of renewable energy stakeholder networks in terms of successful wind power deployment.

With regard to Tamil Nadu, most of the public/private stakeholder constellations are constituted of dyadic relations characterised by so-called "hybrid-immediate" ties. Elements of formality or obligatory interdependence are mostly absent meanwhile a certain degree of mutual understanding prevails. Even though trust and reputation are not overly pronounced, a sense of common objectives, issue orientation and personal understanding exists. Initially, one might expect a greater extent of reciprocity and social capital in these stakeholder arrangements considering the amount of time, frequency of relations and informal channels. However, due to the rigidity of the energy governance system particularly since the institutionalisation of the regulator, too familiar relations would not be possible to establish and maintain. The level of informal interrelations characterising the majority of stakeholder arrangements is sufficient for creating support systems, mechanisms and procedures for successful wind energy deployment in Tamil Nadu. New economic sociology interprets that relational characteristics enable the transfer of intangible assets, information and tacit resources required as inputs for economic decision making and governance (Adler 2002).

In contrast, public-private stakeholder constellations in Kerala are dominated by "hybrid-distant disembedded ties" of formality and low reciprocity and social capital. Without doubt, interactions are also frequent, relations also evolve on an informal level and personal knowledge between public and private actors likewise exists to some extent. However, these kinds of arrangements have not reached the point at which a similar kind of underlying trust, common understanding and sense of mutual goals and action orientation prevails as in Tamil Nadu. An exchange of key resources required by public as well as private actors to mutually constitute supportive framework conditions as well as to enable effective renewable energy governance has not taken place. One explanation resides with the characteristics of public stakeholder arrangements since private stakeholders engage in comparable relational actors' arrangements as in Tamil Nadu. Another explanatory variable to be explored in the following refers to the attributes of respective stakeholders or specifically the impact of certain stakeholder characteristics on relational arrangements and interactions.

Empirical observations concluded that the Tamil Nadu stakeholder network reflects not only greater interaction extent but also more intensity among constellations of public-private actors than in the Kerala setting. This means that public and private stakeholder interactions go beyond the minimal business-as-usual required for wind energy deployment. Consequently, public-private stakeholder constellations appear more open to engage in frequent and intimate interactions than in the case of Kerala. This observation supports Burt's structural hole argument that openly structured interaction constellations between public and private stakeholders open a window of opportunity (Burt 1992). Private stakeholders gain access to crucial information and assets and have the chance to influence public agendas and processes arguing their case for business support to wind energy undertakings. Meanwhile, public stakeholders have the chance to shape private actors' decision-making on how and where wind power activities are executed. In the light of new economic sociology, evolving interaction constellations beyond formal procedural lines enable the exchange of intangible assets and the creation of mutual understandings and common concerns. Private as well as public stakeholders are less concerned containing their interests and status quo within closed constellations but are open to allow for flexible exchange and information flow.

In summary, private-public stakeholder constellations and interactions are significant for successful wind power deployment in as far as they constitute and characterise renewable energy stakeholder networks.

(iii) Stakeholder Network Constituents: Actors' and Relations

This section will take an interpretative perspective that considers arrangements and interactions constituting the renewable energy stakeholder networks. Firstly, causal mechanisms through which actors' relations exert effects on renewable energy governance are discussed in the light of new economic sociology. Afterwards, the focus is shifted towards interpreting why renewable energy stakeholder networks in respective local contexts have come to be shaped in these characteristic manners.

The first debate tackles what actors' relations mean for effective renewable energy governance with focus on the causal mechanisms on stakeholder network dynamics. New economic sociology allows for analysing and interpreting how stakeholders are embedded into renewable energy networks and how this impacts on wind energy deployment. The first causal mechanism that explains why wind energy deployment is supported and facilitated in the context of Tamil Nadu looks at the impacts of repeated interactions and the ongoing construction of common understanding. As discussed alongside case studies, this works mainly through instilling trust and confidence in the integrity of other stakeholders and the wind power process as such (Bathelt 2006). The second causal mechanism relates to the significance of brokers, i.e. stakeholders that manage to add quality to certain relations and interactions and that pioneer the establishment of common understandings and trust. Brokers are stakeholders that mediate between stakeholders and connect information and resources in the formation of new exchange relations. This requires the pre-existence of structural holes which means network structures in which interactions and relations between public and private stakeholders are encouraged (Coleman 1991).

In the context of Tamil Nadu, turbine manufacturers are typical examples of brokers. Due to the integrated nature of business operations, they hold relations to other private actors along the supply chain which they share with each other. Since turbine manufacturers belong to the most established stakeholders in this setting, interactions with public stakeholders such as the TNEB also look back at a history of common experiences. Consequently, new actors emerging in this setting as for example

CDM business lock on to these stakeholders in order to find hold in this new context. Service providers such as C-WET are another example of key stakeholders that broker several relations across public and private boundaries and at times even across geographical boundaries. They function as conveyors of novel information, and contribute to constituting trust and confidence in wind power deployment as well as pioneer new settings. Effective brokerage by predominantly private actors requires and depends on openly structured interactions of embedded relations in public and private stakeholder arrangements.

The second debate tackles the origins and reasons why actors' relations in renewable energy stakeholder networks develop in one direction or the other. How is the emergence of the specific actors' arrangements and interactions explained in the respective local settings? In the following, three approaches are discussed to shed light on these questions. While two perspectives consider literature of new economic sociology and neo institutionalism the third approach relates to previous empirical examinations.

In the theory context of new economic sociology, the first input to this debate is provided by Jones et al (1997) focussing on the impacts of environmental conditions that give rise to renewable energy stakeholder networks. Transaction cost reasoning underlies this study and suggests three triggers for the formation of specific network arrangements and interactions. Firstly, environmental uncertainty creates the need for adaptation and establishes the demand for networks. Secondly, asset-specific exchanges require coordination between stakeholders as these processes involve unique knowledge and/ or equipment. Thirdly, safeguarding stakeholders' specific needs calls for networks that balance the competing demands on exchange conditions. This is prerequisite to meet the requirement for overseeing and integrating different stakeholders' interests in wind energy undertakings (Jones et al 1997: 923). The main argument holds that the greater these environmental uncertainties, the more evolving stakeholder networks are shaped according to specific demands on asset and information exchange.

As second contribution, Biermann et al (2004: 193) add the debate on external contextual variables looking at the effectiveness of global environmental institutions. This approach derives at general points[70] applicable to the discussion of scope conditions for the evolution of stakeholder networks in the Indian setting. On the one hand, certain macro-economic or macro-societal triggers create a window of opportunity and provide incentives for the evolution of the appropriate opportunity structures that encourage renewable energy deployment. These might relate to global economic developments or crises, the emergence of an innovative technology, a sudden natural disaster that creates unanticipated effects or revolutionary events. Such external shocks contribute to the formation of an alliance, often located at multiple levels, in support of innovative and new developments such as wind energy deployment. On the other hand, general changes in belief systems, perceptions and discourses might support the formation of such an alliance to push developments in the wind energy sector. These processes are more difficult to anticipate, to perceive and to assess due to their longevity and potential unknown impact. In some cases, changes in belief systems go hand in hand with external triggers. In other cases, implicit or subtle

[70] This debate bears resemblance with historical institutionalist arguments.

processes of persuasion and insinuation led and carried on by various change agents at multiple levels might arise out of eventual ideas, coincidences or visions. Similarly to external shocks, the impacts and progression of these normative and ideational developments are difficult to predict. Ultimately, this can be boiled down to the question of power ranging from the power of discourses, power inherent in stakeholders as well as power to act and get things across. This argumentation concludes that external shocks or other kinds of implicit influences create a demand for specific renewable energy stakeholder networks in order to respond to ensuing changes and challenges.

The third approach calls for the integration of characteristics and dynamics constituting the stakeholder networks to explain the evolution of local governance arrangements. Characteristics of governance actors are expressed through the distribution of different soft and hard power resources and gain visibility in respective organisational processes. Evolving arrangements and constitutive relations that integrate these characteristics might work either in favour of or in lack of support to wind energy deployment.
Biermann et al (2004: 1993) amount explanatory power to internal contextual variables, i.e. actors' characteristics, which supports this argument. They argue that specific personal characteristics of certain leaders or organisational stakeholders might create windows of opportunities for stakeholder network formation oriented towards the support of wind energy. The aspects of leadership and work culture are also discussed in organisational sociology literature. Indeed, the power of individuals to get things moving sounds appealing. However, this still leaves open the question why some personalities manage to create change and others do not, or why work cultures find hold in some organisations but not in others. At this stage, one might only assume that feedback mechanisms and triggers exist between external and internal contextual conditions that contribute to stakeholder networks in which certain power processes and hierarchies arise in support of wind energy deployment. Ultimately, integrating attributes and relations between stakeholders to interpret network governance provides insights into causal mechanisms that impact on wind energy deployment as well as into stakeholder network constitution and evolution.

In summary, this section focused on two explanatory factors of the Indian wind puzzle – actors' attributes and relations constituting local stakeholder networks and answering to who governs how. The main purpose was to understand how and why these impact on successful wind energy governance and on what grounds they evolve. The following section will move beyond the constitutive network characteristics and dynamics to take a conceptual and holistic perspective on respective renewable energy stakeholder networks.

VI.1.3 Why Governance? Power of Renewable Energy Stakeholder Networks

This section aims to take a holistic perspective and to consider the locally-specific networks in/through which stakeholders interact for the purpose of renewable energy governance. The objective is to understand and explain how and why complex multi-level agency in evolving renewable energy stakeholder networks produces certain governance effects. This is distinct from the previous discussion where the explanatory focus was on who and how renewable energy is governed.

The points of debate in this section are how, why and with what effects stakeholder networks contribute to renewable energy governance taking the Indian wind sector case. In the following, network effects on renewable energy governance will be specified in relation to renewable energy stakeholder networks. On this basis, the relevance of stakeholder networks is explored alongside two different but integrative interpretive approaches – the power of networks on a conceptual, and the power in networks on an analytical level.

As point of departure, renewable energy stakeholder networks exert certain network effects. This means that in the course of renewable energy governance certain processes take place that impact on concrete activities such as wind power deployment. These are explained by locally-specific dynamic network configurations characterised by governance actors engaged in relational interactions. Table 32 summarises the network effects and mechanisms that determine renewable energy governance in the case of wind power deployment. Ultimately, the characteristics and dynamics of local stakeholder networks create effects that explain the research puzzle of spatially varying wind power deployment.

Table 32: Network Effects and Mechanisms

	Network Effects	Network Mechanisms
Output (stakeholder activities)	➢ Design framework conditions for wind deployment ➢ Getting policy & regulations right ➢ Bringing the key stakeholder constituency on board	- Organisational routines established - Evolution of norms - Establishment of leadership - Creating access – move beyond formal interactions
Outcome (stakeholder behaviour change)	➢ Implementation of and abidance to policies & regulations ➢ Deployment of wind projects	- Engagement in public-private interactions - Manifestation of common goals and understandings - Exchange of assets - Manifestation of reputation and trust
Impact (change in policy target)	➢ Access to additional support mechanisms ➢ Push for supportive policies & attitudes ➢ Extension of lobby activities	- Brokerage – technology diffusion - Norm entrepreneurship – policy and norm diffusion

Applied to the Indian wind puzzle, stakeholder network effects on renewable energy governance are distinguished with regard to the three levels of output, outcome and impact. Network effects on the output dimension regard local-level processes and activities within networks. These comprise the design of the right framework conditions for investment also related to policy and regulations as well as the process of aligning the constitutive private and public stakeholders in a common task of wind energy deployment. With regard to the outcome dimension, network effects refer to the de facto installation of wind projects and to de facto observable behaviour changes of stakeholders at local and national levels. This results from the implementation of policies and regulations as well as the abidance to this framework. The impact dimension reflects changes in the overall polity setting which in the wind case comprises the energy governance system, integrates multiple levels and extends the range of stakeholders. Concrete example are the creation of access to additional support mechanisms such as the CDM, the extension of existing support systems and processes as well as the up-scaling of influence and lobby activities towards a greater embracement of wind power.

The discussion so far put the focus on the outcome-related effects of networks in the context of renewable energy governance. This means that the deployment of wind power was explained in relation to actors' characteristics and exchange relations in the context of stakeholder networks. Stakeholder network analysis was initially conceptualised and applied as an analytical tool to tackle the questions who and how wind power is governed. This structured the exploration of actors' attributes and actors' relations as constitutive features and explanatory variables of networks. The previous empirical analyses and the theoretical interpretation give rise to two main concluding observations. Firstly, characteristics of governance actors prove valuable to understand the imminence of public actors but considered on their own are not sufficient to explain the research puzzle[71]. Secondly, actors' relations provide meaningful insights into the relational arrangements and interactions but do not sufficiently account for the actors' characterisation. While the previous interpretation shed some light on causal mechanisms, it fell short of explaining why networks exert effects on renewable governance and why they emerge in the first place. This calls for the integration of both network constituents, i.e. actors' attributes as well as relations into one holistic, explanatory approach that captures the multilevel complexities of renewable energy governance. For this purpose, the renewable energy stakeholder network as a whole is examined in the following two different stages:

The first stage breaks with the stakeholder network concept as analytical tool to consider these kinds of governance arrangements as theoretical concept (Häußling 2005). In this sense, locally-specific stakeholder network configurations are assigned with explanatory power due to the inherent mechanisms and processes emerging in the context of renewable energy governance. The second stage ventures further to explore stakeholder network effects with focus on the constitutive actors and interactions. In order to identify and explain network effects in renewable energy governance the dynamics and power emerging out of this locally-specific governance arrangements in the network context are of interest. Stakeholder network effects will be examined in the light of three issues related to trust, time and power in order to understand the power of stakeholder network configurations in relation to effective renewable energy governance.
In summary, the question to be examined at this stage asks how and why stakeholder networks lead to effective renewable energy governance. In contrast to the previous interpretation, this implies a changed conceptual understanding of networks as specific governance arrangements. In the following, two approaches are explored

[71] The main reason is that actors' attributes appear too static and only imply but do not make explicit the interactive relations to other stakeholders and the local context dynamics required in order to grasp the richness of processes in the empirical reality.

to understand and conceptualise network effects in a broad sense that differentiate between (a) the power of networks and (b) the power in networks[72].

(a) The Power of Networks – Governance in Stakeholder Networks
With what means and why do local stakeholder networks contribute to effective renewable energy governance?
This section will turn away from stakeholder networks as analytical tool to understand stakeholder networks in the light of governance debates as governance arrangements. The added value is that this concept integrates network constituents, i.e. actors' attributes and relations, into a holistic framework. The network concept has so far been understood as analytical tool in order to analyse and explore individual stakeholders and their characteristics as well as the relational arrangements and interactions. This also allowed integrating the two explanatory sociological theories. The main conclusion is that not network properties alone but the interplay and linkages are relevant for determining effective renewable energy governance. This justifies a different approach to networks as theory concept that assigns some explanatory power to stakeholder networks as governance mechanism. In this sense, stakeholder networks are regarded as functional equivalents to governing renewable energy. This conceptual turn provides the basis to explore in-depth the dynamic interplay between network properties and their effects on renewable energy governance in a second stage. Yet, in the following first stage, the interpretation of several theory perspectives will give rise to conceptualising renewable energy stakeholder networks as explanatory concept.

Stakeholder networks have been studied from different perspectives in various academic disciplines. This dissertation adopts an interdisciplinary approach drawing on neo-institutional perspectives of political sciences as well as organisational and new economic sociology. The manifold of ways in which network concepts are applied range from analytical tools, metaphors, methods of social structure analyses and theoretical power concepts. This dissertation has so far referred to the network terminology as analytical concept void of theoretical implications. The purpose was to explore constitutive elements such as actors and relations and to integrate interdisciplinary theory approaches. The interplay and interactions within these stakeholder networks as well as the multi-dimensional effects calls for assigning more than just descriptive value to stakeholder networks. Understanding networks as theory concept assigns power to networks and enables specific explanations of network effects on renewable energy governance.

Two broad theory debates support scaling up stakeholder networks from an analytical tool to a theory concept in order to add explanatory value to the wind puzzle:
Firstly, new economic sociology (Jones et al 1997: 925) argues that the greater environmental uncertainties that create problems of adapting, coordinating and safeguarding exchanges, the greater the propensity for network governance to emerge. Network governance provides the potential of mitigating these challenges to renewable energy deployment through social mechanisms instead of authority, bureaucratic rules, standardisation or legal recourse. Social mechanisms characterising actors' arrangements and interactions assign power to stakeholder networks.

---

[72] The concept of power this dissertation draws upon comprises notions of hard and soft power and is confined to the definition as "capacity and intentional agency towards the purpose of effective wind energy governance".

Secondly, political sciences literature on policy networks grasps the multi-dimensional patterns of interaction between different actors (Soerensen & Torfing 2005: 201). This approach fits neatly into the previous stakeholder network analysis as it regards negotiation rationality as central mechanism driving and characterising governance networks. It departs from the assumption that governance networks are constituted by individual actors with rules and resources of their own. Trust and obligation are considered the main causal mechanisms through which agreements are realised and implemented, i.e. governance takes place.

Based on these considerations, political science-based governance perspectives provide an insightful conceptual approach into network governance to understand stakeholder network effects on renewable energy governance in India. Network governance in this sense is the regulation of a certain policy field and the production of public purpose (Soerensen & Torfing 2005: 204) in the shadow of hierarchy within a certain issue area, i.e. wind energy. Governance networks exert several effects such as the identification and the negotiated response to problems as well as the establishment of a framework and joint responsibility through which information, arguments and resources are exchanged. The objective of discussing governance networks is to understand the mechanisms that drive the intended outcome, i.e. wind energy deployment and impacts. This goes beyond the idea of networks as an analytical tool and to conceptualise a new form of governance through which policy issues are tackled (Soerensen & Torfin 2005: 208). Core issues that ultimately determine the power of renewable energy networks relate to legitimacy and effectiveness. The effectiveness or the power of networks depends on the extent to which problems as of wind energy deployment are solved. Governance debates argue that governance through stakeholder networks intended to pursue common goals explains and determines the effectiveness and legitimacy of wind power.

Going into the depth of policy network literature, two ways of conceptually scaling up governance networks appear insightful. These are briefly discussed in order to assess the extent to which they respond to the question how renewable energy stakeholder networks exert multidimensional effects on renewable energy governance.

A first approach based in the interest intermediation school argues that network structures impact on interactions within networks that hence influence policy processes and outcomes (Börzel 1998: 255). In this sense, network types which in the Indian wind puzzle case constitute renewable energy stakeholder networks are assigned with some explanatory power. This interpretation acknowledges the importance of the entire network with its respective characteristics as well as stakeholder relations. It considers the emergence as well as change of specific configurations characterised by interactions and interplays between the mutually constitutive elements. The explanatory power with regard to the issue area of wind energy lies in the network as such, which means the interactions evolving between stakeholders. Since it departs from the assumption that interests and agency towards certain objectives are intentional, the multi-dimensionality of effects is considered. The notion of renewable energy stakeholder networks appears to be more locally-sensitive and useful in studying dynamics, effects and causes of renewable energy governance in developing countries. In summary, interest-intermediation perspectives argue that stakeholder networks are powerful since they exert effects on multi-level dimensions of outputs, outcomes and impacts of renewable energy governance.

The second approach emerges in the context of governance debates and regards networks as specific forms of public-private interaction. Networks as so-called new modes of governance constitute informal, functionally interdependent, interorganisational relationships. The principal mechanism through which renewable energy stakeholder networks exert effects on renewable energy governance is through joining common interests to provide solutions to coordination problems (Börzel 1998). Networks serve to intentionally produce collective outcomes despite diverging interests. Applied to the Indian wind puzzle the interpretation of networks as "renewable energy governance arrangements" appears valuable. It acknowledges modes of governance beyond the state and the market and explains how critical issues such as wind energy deployment are solved differently. Renewable energy stakeholder networks are considered as functionally equivalent modes of delivering renewable energy governance services. This goes beyond Western conceptualisations of both politics as well as economics and adds value to an interdisciplinary analytical understanding of empirical challenges. In this sense, the power of stakeholder networks is the capacity to get things done which is inherent in the attributes, relations and interactions of stakeholders constituting renewable energy stakeholder networks. In summary, the governance approach suggests the most insightful conceptual take on stakeholder networks. Consequently, this conceptual perspective is now applied to discuss governance arrangements in relation to the empirical reality of wind power deployment in India:

As point of departure, renewable energy governance is expressed through the multi-dimensional effects of locally-specific governance arrangements characterised by certain attributes and engaged in relational interactions. The power of renewable energy networks related to the wind sector differentiates firstly the output dimension. In this sense, renewable energy networks contribute to the emergence of certain stakeholder constellations and the provision of framework conditions for respective deployment processes. Secondly, with regard to the outcome dimension, network governance relates to mechanisms of interactions taking place in the course of different wind energy governance processes. Lastly, network impacts explained by renewable energy governance relate to the diffusion of supportive practices.
Despite the explanatory power, this concept is wrought with some problems. The first point is that the governance concept seems to focus uni-dimensionally either on actors or on interactions but lacks the richness of conceptualising interplays. Secondly, this way of conceptualising renewable energy governance networks implicitly introduces certain theory assumptions such as actor-centred institutionalism (Scharpf 2000b). This restricts the scope for integrating alternative accounts on actors' attributes, relational arrangements and interactions. The last point refers to the implicit assumption of intentionality. The stakeholder network configurations grasped in the Indian context are not constructed with a certain purpose to deliver energy services as functional equivalents to government and/or market but evolve situational. Wind energy deployment constitutes not only a governance issue but also a business activity. Without doubt, public and private actors are both involved in either process – creation of governance conditions for wind energy deployment and economic agency. While a common orientation towards the objective of wind power deployment pervades, the case of Kerala illustrates that the mere intentions to scale up power generation do not get renewable energy governance very far.

In summary, the governance network concept proves useful in providing explanations for the emergence and origins of stakeholder networks. Ultimately, it has less explanatory significance when it comes to understanding the dynamics that lead to network effects on wind energy deployment. One way to circumvent the problem inherent in the governance concept is suggested by the equivalence functionalist approach (Draude 2007). This considers the specific stakeholder network configurations of actors, actors' arrangements and interactions as functional equivalents engaged in the overall undertaking of wind energy governance. The equivalence functionalist approach to renewable energy governance hence allows for explaining the power of networks. In addition, this perspective enables discussions about the legitimacy and effectiveness compared to other governance/network arrangements. This dissertation argues that the power of networks on multidimensional wind energy governance is inherent in these stakeholder constellations. This implies firstly that renewable energy stakeholder networks themselves explain governance effects though the way they are constituted. Secondly, in order to understand how and why stakeholder networks exert effects on renewable energy governance, the integration of explicit theory perspectives is required. As will be demonstrated in the following section, the "power of networks" approach provides the conceptual basis upon which network effects and dynamics are explored. Organisational and new economic sociology perspectives will contribute to understand and explain these network mechanisms and effects involved in renewable energy governance.

(b) The Power in Networks – Governance in the Context of Stakeholder Networks
This section will discuss the power of stakeholder networks to create network effects in terms of how these networks are constituted and work. Network effects arise out of the linkages between the attributes characterising individual stakeholders and the kinds of relational arrangements and interactions evolving between stakeholders. Point of departure is the previous conceptualisation of stakeholder networks and network governance in the light of governance concepts. Integrating organisational and new economic sociology theories, stakeholder networks are empowered through the causal mechanisms the locally-specific constellations of actors' attributes and relations exert. In order to understand the power emerging in networks, governance actors as loci of power and relational processes as power channels are analysed in turn. Following from that, the interplay between actors' attributes and relations is examined with regard to the network effects on output, outcome and impact of wind energy governance. The added value of this integrative approach is to understand network effects, mechanisms and origination based in the renewable energy stakeholder networks.

The first approach to explain the power in networks is to focus on governance actors as one constitutive element of renewable energy stakeholder networks. Stakeholders exert power in these networks and influence renewable energy governance due to their positionality or hierarchy. This is expressed in the quantity and quality of knowledge inherent in public and private stakeholder organisations that is most and foremost reflected in staff qualification and experience. However, knowledge, qualification and material resources translating into higher positions in networks of relations are not sufficient to explain the positive effect on renewable energy governance. Capacity to employ these so-called power resources in a manner that contributes towards the common objective of successful wind power deployment is required to exert network effects. Power of individual stakeholders to influence renewable energy governance not only resides in characteristics such as resource equipment but lies firstly in the willingness and ability to employ these resources. Secondly, power depends on the perception and acknowledgement of stakeholders as power loci by

other constitutive network actors. Only if leaders are recognised as leaders, rules and regulations will be implemented. Only if individuals and organisational stakeholders are recognised as norm holders, trust and certain routines will become established. Positionality or hierarchy as actors' characteristics says a lot about the potential to get things done. Yet, as a static concept it does not actually reveal much about the de facto realisation of these potentials for successful wind power deployment. In this sense, power as the capacity to act appears the more suitable explanatory concept when it comes to understanding stakeholder network effects. This goes beyond the focus on stakeholder and actors' attributes as it implies the inter-linkages between stakeholder characteristics and relations reflected through various channels and expressed in the course of interactions in terms of network effects.
The second approach to explain the power in networks is to examine relational arrangements and interactions between stakeholders, i.e. actors' relations. The effects of differently constituted relational arrangements on wind power deployment were discussed alongside the typologies identified in the previous chapter. Qualitative characteristics of actors' arrangements in relation to social capital, mutual trust and reciprocity are important as they determine the extent to which actors lead and tap on the capacities to act towards the objective of effective renewable energy governance.
In summary, the interlinkage between stakeholder characteristics and relations in a network emerges as a mutually constitutive one. On the one hand, the attributes characterising certain stakeholders influence the kind of relational arrangements evolving. Stakeholder characteristics define the possibilities and potential effects of actions and interactions. On the other hand, emerging relational arrangements and interactions shape actors' characteristics. They render stakeholders visible and enable the application of power resources.

The third approach to explain the power in networks goes beyond the distinction between actors and relations which is often emphasised in social network analysis. The identification of elements that render networks dynamic and hence consider both stakeholders characteristics and relational arrangements is required to understand causal mechanisms. The previous brief discussion already emphasised that certain actors' attributes with regard to norms, reputation and leadership find expression through processes such as organisational routines in the context of relational arrangements, i.e. through interactions in network systems. The ways in which, how, and with what impact processes take place are influenced by the level of trust and informality inherent in relations. These dynamics are shaped by actors' attributes, reflect in actors' relations and ultimately exert effects on wind energy governance. The power in networks as networked governance is now analysed alongside the three renewable energy governance dimensions:
Firstly, the output dimension refers to the stakeholder network formation in terms of setting the right framework conditions and getting together the constituency. Key mechanisms through which stakeholders' characteristics and arrangements become mutually constituted in renewable energy networks relate to the establishment of organisational routines. This refers to the emergence and manifestation of certain norms that regard wind power as worthy of support. Furthermore, leadership of and leading in organisations in favour of the wind energy undertaking becomes established. This goes hand in hand with greater openness and accessibility of public authorities to different kinds of private stakeholders and interactions. Organisational routines are a crucial point since these integrate the above-mentioned features and enable access to and interactions between stakeholders. At this stage, public stake-

holders and their characteristic attributes predominate in stakeholder network formation as they are required for interactions between other public stakeholders as well as business actors. In summary, the power of stakeholder networks on output-dimensional effects is explained by the way in which certain organisational routines are established.

Secondly, with regard to the outcome dimension, the focus shifts towards the implementation of policies and regulations as well as the deployment of wind projects, i.e. the operation of networks. Causal mechanisms that lead to renewable energy governance outcomes are the actual engagement of stakeholders in multiple-actor interactions in multilevel energy governance systems. This argues for the relevance of evolving actors' arrangements and interactions as regards network effects on the outcome dimension. Exchange of assets such as information, knowledge, expertise and material resources between private and public stakeholders is one important effect as well as condition to stimulate wind energy deployment. This partly requires and also contributes to agreements on and the manifestation of common goals and understandings. Touching upon the intangible aspect of resource exchange points towards the exchange of good-will, faith and belief in the overall project of wind power deployment. This leads to the emergence of certain reputation, trust and a deeper sense of mutual and personal knowledge. This aspect gains importance only in the context of repeated interactions, i.e. relations between stakeholders. Actors' attributes are important properties in concrete interactions through the previous experiences of engagement in related issues, observation of previous action or word-of-mouth. In this sense, interaction quantity in terms of frequency, directionality and repetition of exchange might facilitate the emergence of qualitative aspects in relations. However, stakeholder attributes come into play as regards the determination of the kinds of ties and characteristics of arrangements evolving. For a start, in the context of network agency the manifestation social capital through mechanisms of trust is crucial. This refers specifically to the relation between public and private stakeholders for the effectiveness of renewable energy governance. Both are mutually involved in and affected by the constitution and implementation of framework conditions as well as concrete business undertakings in the wind sector. However, dissimilar attributes reflect asymmetrical power distribution and often constitute obstacles to the evolution of trustful and familiar interrelations that render complex undertakings such as wind energy deployment difficult. In summary, network effects on the outcome dimension in terms of successful wind energy deployment depend on the stakeholders' attributes, on the arrangements evolving and on the dynamic interactions.

Thirdly, with regard to the impact dimension, networks exert effects that move beyond local and national levels to include other stakeholders as well as the international arena. In this sense, access to additional support mechanisms for wind energy deployment is established and support activities through lobby work are enhanced. Two multi-level causal mechanisms that explain network impacts on effective renewable energy governance are technology transfer and the diffusion of policies and norms. Technology diffusion concerns the opening of stakeholder networks to seize opportunities provided by international mechanisms and instruments such as FDI, CDM or other sources of international technology collaboration, development assistance or finance. Networks play a crucial role as they constitute the link between the multiple levels and the respective stakeholders involved in these activities, i.e. renewable energy stakeholder networks broker relations and interactions. Private stakeholders take up interactions with other private or public actors at international level and manage, facilitate or mitigate interactions to stakeholders in the local context. The establishment of relations particularly to key public stakeholders when accessing the locally-specific renewable energy stakeholder network from outside is confronted by many implicit and or explicit obstacles and or perceptions. This is reflected in the information flow or exchange of assets. Only public or private stakeholders that are part of the pre-existing network facilitate the opening of windows of opportunity and access. Policy and norm diffusion relates to the adaptation of certain internationalised ideas or concrete policy instruments. Intuitively, this falls into the domain of public actors and hence requires norm entrepreneurs or leaders within pre-existing stakeholder networks to enable such processes. In summary, network impacts on effective renewable energy governance relate to the interplay between actors' relations and attributes within the respective network.

Why do renewable energy stakeholder networks emerge and evolve in locally-specific manners?

The previous debate focused on network governance as the ways in which renewable energy stakeholder networks configured by actors' attributes and relations exert multidimensional effects on renewable energy governance. Social constructivism appears a suitable meta-theory for elaborating on the wider reasons for stakeholder network rationales and evolution. Historical institutionalism as well as organisational sociology suggests concrete concepts such as 'timing' in terms of conjunctures and 'power' in terms of capacity to get things done. With regard to 'conjuncture', a critical mass of certain events that are time- and context-specific explain the emergence and increasing (perceived) importance of networks for renewable energy governance.

In the case of Tamil Nadu, a combination of macro-economic energy scenarios, societal needs, business requirements, inspired leaders, external influence and chance triggered the successful developments described above. As regards the concept of power, not only capacities but the capabilities to get things done appear essential and depend on the recognition of these power conceptions as legitimate and appropriate. In the case of Kerala, power capacities in terms of staff strength and resources of public organisations such as ANERT and KSEB were not sufficient for promoting the wind undertaking.

As a word of caution, the application of power as meta-theoretical concept to add explanatory value to renewable energy stakeholder networks is wrought with difficulties. The notion of power required by this dissertation encompasses not only intentional exercise of influence but also capacities and willingness to act. One option to circumvent an entanglement in complex debates but nevertheless capture similar issues is to draw back on theories of leadership. Even though disregarded as theoretical concept particularly in the political science domain, such concepts tackle similar issues without normative implications. Leadership as theoretical concept appears valuable contribution to answer the why-questions of renewable energy stakeholder networks since it considers mechanisms that make things happen in collaboration (Vangen & Huxham 2003: 62). The leadership approach considers three levels of internal functioning, capacity and types of leadership and differentiates certain media through which leadership is exerted as well as purposes of leadership.

Applied to the Indian wind puzzle, leadership approaches discuss specific questions of why public organisations manage to produce effective results, why some heads of organisations or crucial units exert different kinds of influence, why some organisations adopt certain practices and why relational arrangements evolve and are structured in specific ways. Examples are the case of ANERT that despite abundant resources and a pro-active director produces inefficient outcomes as regards the timely

and accurate issuance of clearances. In Tamil Nadu, changes in the unit head of the TNEB resulted in the wind farm clearances system suddenly halted and grid management not being responsive to the requirements. In contrast, TEDA with an insignificant director provides a reliable and continuous support base and services to wind entrepreneurs. Despite repetitive state government announcements to support wind energy deployment in Kerala, neither KSEB nor KSERC have taken significant steps towards implementing these policy requirements. Lastly, arrangements between manufacturers and ANERT as well as KSERC fail to establish greater relational quality and work less in support of private interests in wind power promotion in spite of personal knowledge, shared experiences and common background.
This discussion provided a glimpse into the dynamics of renewable energy stakeholder networks even though a wholehearted analysis of leadership goes beyond the scope of this dissertation. Organisational outcomes appear to be driven by the charismatic leadership of the director that is dependent on the recognition and acknowledgement as legitimate, credible and purposeful by the rest of the organisation. This implies the importance of value-based or normative leadership that also comes into play looking at why certain organisations such as KSEB and KSERC do not take up government directions. Lastly, capacity or resource-based leadership depend on support base, voice, reputation and actual resources. This explains why certain relational arrangements are productive in the sense of furthering certain interests.

In summary, these integrative conceptual discussions focused on the power in renewable energy stakeholder networks configured by stakeholder characteristics and relational interactions with regard to multidimensional effects on renewable energy governance. The added value of conceptualising stakeholder networks as governance arrangements is to understand the dynamics, origins and effects on renewable energy governance. Particularly the focus on mechanisms through which stakeholder network constituents affect renewable energy governance benefits from the integration of various theory perspectives. This kind of analysis provides scope for discussing normative questions about legitimacy and effectiveness which goes beyond this dissertation.

In conclusion, this section widened the perspective to explore network effects on renewable energy governance in two ways. On the one hand, networks were examined in relation to their multidimensional effects on the wind sector and conceptualised as governance arrangements. On the other hand, network governance in the context of governance networks was interpreted on the basis of this conceptual turn. Scaling up network conceptions in the light of governance debates provided interesting insights into the power of networks in the context of renewable energy governance. Theory interpretations left one main interpretative gap that relates to the question why governance arrangements evolved as locally-specific configurations.
The next section turns towards the practical implications of empirical research to tackle the impact-related governance effects of stakeholder networks in the context of new opportunities arising through the CDM. This discussion is taken up at this stage as Chapter III concluded that the CDM somehow relates to successful wind power deployment but does not constitute an explanatory factor. In the course of this dissertation, the explanatory power of stakeholder networks was explored and interpreted which suggested new perspectives on how the CDM fits into local wind energy governance to be discussed as follows.

VI.2 Empirical Implications of Stakeholder Networks

This dissertation has derived at the conclusion that local stakeholder networks play a powerful role in renewable energy governance in developing countries as they exert certain effects on multiple dimensions. Stakeholder network effects have so far been examined with focus on the outcome dimension regarding concrete deployment activities. This also touched upon the output dimension concerning the design of framework conditions. At this stage, the perspective is shifted to the impact dimension including the international level and respective stakeholders. Spatial variations in local wind power deployment challenge renewable energy options for sustainable development dearly required to integrate global climate change and developmental concerns. Due to the global nature of issues such as renewable energy deployment and the interconnectedness of the global system, local issues such as wind energy deployment become a matter of global concern. This means that national and international opportunity structures impact on local developments. This section will explore the concrete example of the CDM in the Indian wind sector. The CDM provides new potentials to meet sustainable development requirements as a new mode of governance. As will be argued in the following, governance arrangements emerge as precondition for external stimuli such as the CDM to impact on local settings. Stakeholder networks matter and suggest how external assistance and opportunities might be targeted to stimulate local developments such as wind power deployment. The purpose of this section is to illustrate the interplay between wind energy deployment, stakeholder networks and external stimuli. The added value of is justified on three grounds. Firstly, the CDM as external stimulus and a new mode of governance provides unique opportunities to respond to sustainable development challenges. Secondly, with regard to theory discussion, the CDM contributes to the debate on new modes of governance in non-OECD contexts. Thirdly, the implications of stakeholder network research for the practical realities of wind power deployment as regards the impact-dimension of network effects are discussed. The discussion of network impacts in relation to external stimuli will proceed firstly, by exploring the role and effects of the CDM in the Indian wind sector. Secondly, the interplay between effects of the CDM and renewable energy stakeholder networks is discussed.

VI.2.1 External Stimuli for Renewable Energy Governance

This section explores the role and effects of the CDM in the Indian wind power case and departs from the general discussions about external stimuli for developmental activities (Chapter II). Summarising literature debates in global environmental and governance traditions, the main arguments emphasise the potential of new modes of governance to impact positively on renewable energy governance in developing countries in as far as they contribute to technology transfer and policy diffusion.

Developing countries like India nowadays are confronted with multiple challenges relating to development and growth imperatives and climate change threats. In the quest for alternative and sustainable development paths, renewable energies have turned into a promising solution. At the international level, several options and opportunity structures are provided in order to support, assist and promote the deployment of renewable energy sources such as wind power in developing countries like India. Traditionally, international development assistance in terms of financial and technical support as well as foreign direct investments (FDI) are the main mechanisms to promote renewable energy deployment. Since many of these attempts failed (Foley 1993), questions arise about the appropriate framework conditions and requirements. The most important lesson learnt out of past experiences is that attention to local context conditions is more than a prerequisite to any external intervention (Altomonte 2000). During the last couple of years, changes in global governance led to the emergence of new forms of external assistance institutionalised and/or channelled via the international level and involving various stakeholders in e.g. PPPs or issue networks. One example of such a new mode of governance is the Clean Development Mechanism (CDM). The CDM fundamentally differs from previous external stimuli in as far as it constitutes a new mode of governance (Benecke et al 2008). This means that the CDM is a policy device using market mechanisms and is characterised by new kinds of actors, actor constellations, interactions and exchanges across multiple levels (cf. Chapter II). Renewable energy projects are one example of how the CDM extends support to clean development through various channels, e.g. transferring funds or hard and soft technologies. Even though the Indian wind sector is characterised as a relatively mature renewable energy market, empirical studies point to many barriers that impede the dissemination of wind power in India (Basu 2007). Furthermore, looking at the spatial variations in wind power deployment makes the case that external stimuli such as the CDM are required. The CDM might provide material financial or policy incentives for local stakeholders to engage in renewable energy deployment[73]. Contrary to alternative explanations, this dissertation will argue that the CDM cannot be considered an explanatory factor driving renewable energy deployment. As will be demonstrated, the potential of the CDM to be effectively tapped depends on the configuration of the renewable energy stakeholder network in the specific local context.

In order to explore these arguments in the following, the concrete demand for the CDM is illustrated as an option for technology transfer and policy diffusion in the light of environmental governance perspectives. This aims to shed light on the questions how and why the CDM as external stimulus taking the form of a new mode of governance plays a role in promoting wind power in India.

As point of departure, wind technology constitutes an advanced renewable energy sector in India, and the Indian wind energy market is characterised as mature despite policy and financial insecurities. As shown previously, India has emerged as a significant player at the global level in terms of wind capacities since Indian companies such as Suzlon increasingly expand activities beyond national boundaries. For these reasons one might wonder why any kind of assistance or support is required at all to scale up booming activities. Indeed, the eligibility of the Indian wind sector for international opportunities emerging in the light of new modes of governance such as the CDM has often been questioned. One main point of critique concerns the issue of financial additionality. This relates to the fact that CDM revenues are merely an icing on the cake and do not constitute the tipping point that makes a project viable. This implies that wind projects take place anyway.

But many Indian states have not managed to harness their existing wind resources potential and hence the CDM as external factor promoting wind energy deployment has a significant role to play. Furthermore, the geographical distribution of CDM pro-

[73] Even though most CDM wind power projects in India are unilateral, this dissertation assumes that the CDM nevertheless provides incentives such as global participation than render wind power more attractive for investment.

jects[74] logically follows similar lines of distribution of wind energy activities and repeats the research puzzle (Map G).

Map G: CDM Wind Projects

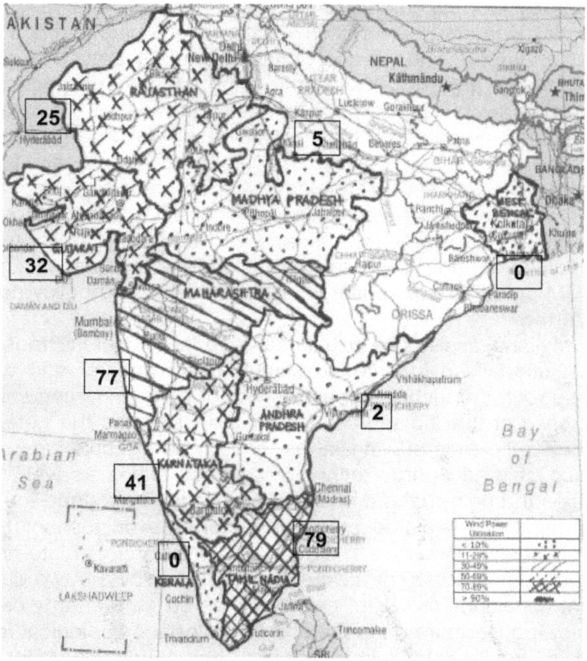

Source: UNEP Risoe 2007 and 2008 (b) (No. of CDM wind projects in rectangular boxes)

States that anyway managed to harness existing resource potential like Tamil Nadu and Maharashtra also benefit from carbon credits under the CDM. States lagging behind the installation of wind power projects also see little CDM activity. Surprisingly, for states like Karnataka and Rajasthan, CDM projects in the wind energy sector take a substantial share of carbon market activities despite the to-date hesitant installation of wind power projects. This scenario of spatially varying wind deployment and respective CDM activities gives rise to three implications. Firstly, the demand for the CDM arises in order to level out spatially uneven wind power deployment patters. Secondly, questions about the preconditions for wind power activities gain anew significance as the CDM distribution follows a similar pattern as wind power deployment. Thirdly, the cases of Kerala and Rajasthan give rise to wondering about the potential impact of the CDM in local contexts, e.g. in terms of policy diffusion.

This section briefly demonstrates the demand for the CDM in order to provide common ground on the need and the CDM potential to scale up wind power dissemination. The dismal success rate of wind plants calls for technical technology transfer

[74] This analysis considers CDM projects in the wind sector only.

and opens the scope for international support mechanisms such as the CDM. The Indian wind energy sector is on the verge of commercialisation. Most wind power projects in the country in fact use equipment produced by local manufacturers (Dechezlepetre et al 2008). Yet, future prospects and the transition towards a mature market are hampered by the implementation deficits reflected in the large untapped wind resource potential. Barriers to the dissemination of wind energy relate to the two aspects of technical underperformance and of institutional lethargy. The one aspect that hampers future growth and technical leapfrogging concerns the poor plant performance and installation practices (Lewis 2007: 217). In most of the cases this relates either to inadequate equipment or inappropriate operation and maintenance. This contributes to the gap in local deployment rates and hence calls for external stimuli to contribute to (hard and soft) technology transfer. The other aspect relates to policy and regulatory barriers as well as lacking political will that impede wind power deployment (Amin 1999). This raises the demand for the transfer of ideas and wind energy-supportive values to trigger mental and ultimately governance changes. In this respect, the CDM presents a potential vehicle for norms and policy diffusion via the hardware transferred.
In the context of providing assistance to climate-friendly economic transitions, the CDM has been institutionalised in order to stimulate the transfer and diffusion of technologies as to support the technical leapfrogging of emerging economies like India. In the specific case of the Indian wind sector the CDM has the potential to increase the operational efficiency of wind farms through hard technology transfer, to catalyse the diffusion of wind energy technology across states as well as to instil changes in mentalities through norms diffusion. In order to understand the dynamics and mechanisms behind CDM-related processes, literature on technology transfer and policy diffusion is briefly discussed that deals with the impact of international opportunities on local developments. The discussion is guided by the two questions of how and under what conditions the CDM exerts effects on wind energy deployment. The purpose of reviewing literature on technology transfer and ecological modernisation at this stage is to specify potential mechanisms and conditions under which the CDM might contribute to wind power dissemination.

The following debate will derive at the conclusion that these technological shortcomings can be met via hard and soft technology transfer. The theoretical background to these debates embraces concepts like ecological modernisation that assign great importance to external stimuli such as the CDM. As elaborated in Chapter II, ecological modernisation integrates the idea that transiting to greener production and consumption is feasible but requires leapfrogging from traditional to cleaner and modern industrial processes (Dryzek 1997). Leapfrogging in general implies the existence of mechanisms and strategies for transiting to greener development paths. Leapfrogging encompasses the entire process of shifting away from the reliance on traditional energy sources towards incorporating the broad utilisation of advanced renewable energy technologies (Lewis 2007). Ecological modernisation debates argue that technology transfer plays an important role for leapfrogging to alternative development paths. Wind turbines provide one concrete example of past experiences and learning curves with regard to North-South technology transfer. Particularly the case of India shows how local manufacturers obtained technology expertise through global learning networks (Lewis 2001: 221). Drawing on global experiences, technical expertise was transferred back and applied in the local context. Supported by national policy development, the emergence of a domestic industry was facilitated. Nevertheless, the practices that characterise the Indian wind industry today are that turbine production is geared towards global export while low quality technology is installed at home (INT 2). International technology transfer in the Indian context hence requires to go beyond the rudimentary vertical equipment transfer from industrialised to developing countries and to consider how to foster the development and diffusion of domestic manufacturing. Given the need for proper technologies and processes to tackle issues of spatial variation in wind energy deployment creates the demand for hard and soft technology transfer. In this context, there might be a role for the CDM to stimulate such transfers across multiple levels. Literature debates in Chapter II provided evidence that the international mechanisms institutionalised under the Kyoto Protocol facilitate international technology transfer. Looking at the CDM specifically, Haites et al (2006) derive at the conclusion that technology transfer takes place in about one third of the CDM projects even though this is no obligatory element mandated by this market-based mechanism. Contributions of the CDM towards technology transfer are the reduction of technological barriers (Schneider et al 2008). Ecological modernisation literature remains rather silent on concrete local requirements for such modes of governance to find hold in specific contexts. Empirical literature (Haites et al 2006) points towards the importance of local conditions as CDM contributions to technology transfer are independent from host country income or general macro-economic features. In summary, this section identified technology transfer as one mechanism through which the CDM works towards tackling technology-related barriers to wind power dissemination.

Policy diffusion literature based in environmental governance debates discusses issues of political and attitudinal obstacles and argues that changes in attitudes and practices require certain triggers and new ideas (Arts 2006). For a start, the debate about (multi-level) impacts of policy instruments acknowledges policy diffusion as a rather extendable concept applied to studies on the diffusion of environmental policy innovations (Tews 2002: 4). This comes closest to understanding how the CDM is conceptualised as a new mode of governance and how it is conceived to work in international climate politics. As a matter of fact, the CDM constitutes a policy option that is based on the application of market instruments. Essentially, governance in the light of the CDM is not a domestic issue and the way in which the CDM operates and exerts effects is dependent on the characteristics of stakeholders that make use of it. This is due to the different responsibilities assigned to parties under the Kyoto Protocol. For developing countries like India the immediate effect of this policy instrument is to channel finances and hard as well as soft technologies in order to assist the realisation of climate-friendly activities such as wind power deployment.

In a nutshell, environmental governance theories examine general mechanisms and conditions for external stimuli to exert positive effects on clean development (Esty 2002).

The first debate concerns how policies are diffused looking at communication channels and actors. On the one hand, relations between interaction partners are considered as one explanatory variable why policies are diffused. The different causal mechanisms explaining policy diffusion relate to imitation through competitive effects, the desire for reputation and recognition or due to uncertainty, dependency and power asymmetries. Another important channel is the creation of a critical mass of certain actors in distinct positions that intentionally produce and disseminate information and form supportive alliances. On the other hand, vertical communication relations are examined to explain policy diffusion. Particularly the participation and impact of external actors such as international organisations and transnational actors is relevant for the diffusion of policies. The causal mechanisms identified in this process are either hard power or alternatively instilling values and norms. Applied to the CDM, this allows for drawing implications on how this policy device contributes to technology transfer and norms diffusion hence exerting effects on renewable energy governance. In this sense, the CDM is diffused through actors or brokers not only as project-based mechanism for finance and technology but also as vehicle implying norms and governance.

The second debate concerns reasons for why innovations are diffused and draws upon interstate factors such as socio-economic, political-institutional, social structures of relations as well as the responsiveness to innovations inherent in regulatory styles, structures and practices. One key argument holds capacities inherent in actors' constellations and structural framework conditions determinative for the success of policy diffusion. The other argument scales up the perspective to focus on administrative traditions, regulatory structures and policy styles and looks at power distribution, sensibility to public opinion, decision making, and commensurability with existing policy practices. Applied to the CDM this debate tackles the conditions required for the diffusion of practices and ideas and emphasises the important of the local context. Yet, the concrete characteristics or constitution of local preconditions and actors' constellation are left as a black box.

In summary, empirical studies and environmental governance theory-based literature recognise the multi-dimensionality of policy diffusion and integrate multiple explanatory entry points on the role, effects, mechanisms and conditions under which the CDM might impact on the Indian wind sector. On the one hand, external factors and the dynamics that trigger certain effects on domestic actors through technology transfer and norm diffusion are considered. According to that, the CDM functions as vehicle for technology transfer as well as policy diffusion in order to meet the obstacles that contribute to the spatial variations in wind power deployment. On the other hand, the importance of national and local levels in terms of stakeholders, framework conditions and interactions is acknowledged as prerequisite for effective policy/norm diffusion. This dissertation argues that questions how and why policy innovations such as the CDM take certain effects in some local contexts are most and foremost related to the configuration of local stakeholder networks. This argument based on research findings confronts alternative explanations proposed by environmental governance and historical institutionalist literature. These consider the CDM as trigger or external shock that creates a critical mass of stakeholders and stimulates the emergence of respective renewable energy stakeholder networks.

In the following, this argument is empirically examined in the case of the CDM in the Indian wind sector.

VI.2.2 New Modes of Governance and Stakeholder Networks

This section will explore the extent to which the CDM triggers policy diffusion and technology transfer under the conditions that supportive local stakeholder networks allow for the access and acceptance of external stimuli.

Due to the lack of empirical analyses on CDM effects on technology transfer and policy diffusion in the Indian wind sector, this section proceeds as follows. First of all, the contributions of the CDM to technology transfer will be demonstrated. Secondly, the role of the CDM in terms of policy diffusion related to wind energy deployment is discussed. Finally, attention is turned towards the question why the CDM triggers certain effects. Exploring renewable energy stakeholder networks as local preconditions for external stimuli gives rise to wider implications on the role of the CDM for wind power dissemination.

The first central question to be tackled is to what extent and how the CDM provides the scope and the potential to trigger wind technology dissemination.
As point of departure, to date, Indian wind energy projects are characterised by a low performance. This is partly due to the less favourable wind regime as compared to industrialised countries and difficulties in resource assessments and localisation of wind farms. However, the greater problem is inherent in the low technical performance in terms of the efficiency of wind electricity generators. This is due to the utilisation of wind technology that is not geared primarily to Indian conditions and demands. This leads to plant load factors that deliver significantly less output than optimally feasible and achieved under different conditions. In addition, domestic Indian and foreign turbine manufacturers have come under criticism (INT 2) for orienting local manufacturing towards global markets and producing for export purposes only. The challenge is hence how to provide incentives for wind project developers, i.e. equipment sellers, to adapt wind energy technology as to better fit Indian circumstances. This is prerequisite to diffuse appropriate wind energy technology across the country accordingly-designed to locally-specific wind regimes. The other obstacle to the efficient utilisation of wind power is related to plant failure due to operation and maintenance practices. Lack of skills, knowledge and expertise contribute to sloppy procedures that result in component failure and power generation disruption. This creates the demand for soft technology transfer in order to improve staff qualification as well as awareness.
As new policy device the CDM holds the potential to supply a mechanism or set incentives to transfer efficient turbine technology as well as to apply appropriate operation and maintenance practices. In the practical reality of wind power deployment, this would translate into higher plant load factors (PLF). Nowadays, not much illusion exists as to whether the CDM in the Indian wind sector might cover the high upfront costs of wind projects, might trigger new investments in wind energy projects or might contribute to de facto vertical technology transfer (Purohit 2007). Given the maturity of the Indian wind sector, no outstanding contributions are expected. However, due to the need to prove the additionality of project activities to business-as-usual scenarios as eligibility criterion for CDM activities, chances are that project developers feel incentivised to increase the efficiency of wind power projects. This might occur through improvements in the "proper process" in operation and maintenance and/or the manufacturing and application of locally-appropriate technology components. Efficiency increases would translate into higher capacity utilisation rates, i.e. plant load factors. In this sense, the CDM is expected to contribute to technology transfer either through the development and dissemination of more efficient hardware equipment or through diffusing skills and capacities required for more effective project management[75]. As proposed at this stage, the CDM contributes to the diffusion of wind energy technology in India either related to hardware or software components since a significant share of CDM projects reveals higher efficiency than the baseline conditions and projects. In the following, the empirical analysis will demonstrate indications

[75] Either way, the plant load factor serves as an indicator to measure efficiency rates.

of technology transfer alongside the analysis of CDM activities in the Indian wind energy sector.

According to the UNEP Risoe CDM/JI Pipeline (December 2008)[76], CDM wind projects are the second largest category of CDM activities in India (after biomass energy projects) constituting 22.6 % (275 out of 1219 projects). For analytical purposes, the project design documents (PDD) of all CDM projects registered and at validation stage are analysed with regard to the PLFs and the technology transfer mentioned. The sample size taken in this paper comprises 233 CDM wind projects selected according to the status 'registered' and 'at validation'. All of the associated project documents were searched for indications of the plant load factor as well as any mention of technology transfer. The results of the PDD analysis (UNEP Risoe CDM Pipeline December 2008) are compared firstly to the plant load factors cited by critical studies on the Indian wind business (Narain et al 2008) and secondly to the plant load factors stipulated by the respective state electricity regulatory commissions (SERC)[77]. The proposition to be explored holds that the higher the PLF in CDM projects compared to business-as-usual scenarios (given the CSE scenario and the SERC stipulation), and the more mention of technology transfer in the PDDs, the greater the contribution of the CDM to technology transfer.

Table 33: Plant Load Factors

State	CSE Scenario	SERC Stipulation
West Bengal	/	18.6 – 20 %
Kerala	/	22 %
Madhya Pradesh	13.9 %	22.5 %
Andhra Pradesh	9.4 %	17 – 19 %
Rajasthan	14.4 %	20 – 22 %
Karnataka	20.8 %	23 %
Gujarat	7.8 %	23 %
Maharashtra	11.7 %	18 & 20 %
Tamil Nadu	17.9 %	25.29 & 26.7 %

Sources: Narain et el 2008; most recent tariff orders of respective states

The first analytical step contrasted the results of the PDD analysis, i.e. PLFs stated in CDM project documents, with the PLFs referred to in the CSE study (Narain et al 2008). This makes a clear case for technology transfer in the context of the CDM. All of the CDM wind projects in all the Indian wind states claim significantly higher plant load factors than those referenced in the CSE report. This dissertation argues that this interpretation should be taken with a pinch of salt as the sources used by the CSE authors to make their case are rather ambiguous[78]. For this reason, the analysis

[76] The research puzzle of spatially varying wind energy deployment as well as CDM distribution refers to 2007 data. For the purpose of this analysis I had to draw on data of December 2008 as otherwise insufficient data (i.e. in terms of registered CDM wind projects) would not have allowed for precise assessments. As the data source is the same, this does not change anything on the quality of the argument but only substantiates the validity of the claims by enhancing the number of observations.

[77] An inherent difficulty analysing the Indian wind sector is the lack of proper baseline data as regards PLFs. The CSE reference as one source is deemed unreliable since no data source is provided. However, SERC stipulations as second source are only second best options since baselines refer to new installations and scenarios and are not plant-specific.

[78] Sources are not referenced and indicated.

of technology transfer in the context of the CDM takes as baseline the plant load factors stipulated in the latest tariff orders by the respective state regulatory commissions[79] (Table 33).

Table 34: Technology Transfer in CDM Wind Projects

State	CDM (total)	Higher PLF	Similar PLF	Lower PLF	N.N.
West Bengal	/	/	/	/	/
Kerala	1	1	/	/	/
Madhya Pradesh	4	1	3	/	/
Andhra Pradesh	2	/	1	/	1
Rajasthan	21	7	7	2	5
Karnataka	31	23	1	/	7
Gujarat	31	7	11	5	8
Maharashtra	71	21	28	1	23
Tamil Nadu	72	19	6	9	37

State	CDM (registered)	Higher PLF	Similar PLF	Lower PLF	N.N.
West Bengal	/	/	/	/	/
Kerala	/	/	/	/	/
Madhya Pradesh	1	1	/	/	/
Andhra Pradesh	2	/	1	/	1
Rajasthan	9	3	2	2	2
Karnataka	14	12	/	/	2
Gujarat	2	2	2	/	/
Maharashtra	18	9	3	2	2
Tamil Nadu	17	5	5	3	4

Sources: UNEP Risoe CDM Pipeline December 2008

The second analytical step hence contrasts the PLFs mentioned in the PDDs of CDM projects with the PLFs stipulated by the respective state regulatory commissions. Presumably, higher or at least similar PLFs claimed in the wind energy CDM projects than the stipulated baseline indicate some degree of technology transfer[80].

Table 34 summarises the general observation that the majority of CDM projects claim a higher or similar PLF than stipulated by the regulatory commissions. This tendency is even more pronounced when looking only at the registered projects. Hence a decent share of CDM projects states higher degrees of efficiency in wind power generation than implied by previous practices. However, results vary according to specific local contexts. In states like Rajasthan, Gujarat and Maharashtra no great variation in plant load factors from the business-as-usual scenario is observed. Consequently, no significant technology diffusion takes place even though a couple of projects claim

[79] SERCS use different methodologies and procedures to fix baselines and tariffs. This renders the comparability and the extents to which these match the local realities extremely difficult contrasting the different Indian wind states.

[80] As an initial word of caution, the predominant share of CDM wind projects is still at validation stage.

higher efficiency rates. In states that are newcomers to the wind business such as Madhya Pradesh, the CDM has some impact on increasing the efficiency baseline of wind projects. The most striking case for technology diffusion appears with regard to the state of Karnataka where almost all projects state higher efficiency rates, i.e. PLFs. Karnataka constitutes one of the states with a surprisingly high number of wind CDM projects despite previously low wind energy utilisation rate. This allows for tentatively positive assessments on technology diffusion taking place in the context of the CDM. This conclusion needs to be taken with a pinch of salt as in some cases like in Tamil Nadu and Rajasthan some projects claim an even lower than baseline plant load factor. This might relate to specific local wind regimes and indicates the demand for more careful and rigorous data monitoring and accounting. Another explanation refers to the fact that good windy sites resulting in higher wind farm outputs (i.e. PLF) that served as baseline for PLF stipulation by regulators are nearly gone (particularly for Tamil Nadu). Lower PLFs hence indicate a de facto less favourable wind regime despite which the CDM still provides incentives to install wind projects. Surprisingly, some PDDs particularly for projects at validation stage recurrently either not mention the PLF or make general statements only. This is puzzling as the PLF constitutes one factor used for calculating the project income (internal rate of return) on which grounds financial additionality is argued. This again cautions against overstating the role of the CDM for stimulating wind power dissemination.

In the third analytical step, the CDM project documents were screened for written arguments about technology transfer that might provide more clarity on the extent and direction technical leapfrogging takes, i.e. the role of the CDM in meeting barriers to wind power deployment. Almost all PDDs make some remark on technology transfer. Most of them state that no (vertical) technology transfer took place or that at least no foreign technology was used. However, the majority of cases argue that technical well-being is enhanced through demonstration and promotion of previously less common technologies and practices. Additionally, many projects, particularly those executed by Suzlon and Enercon, state that R&D has taken place abroad with focus on Indian circumstances and that domestic manufacturing plants implementing this knowledge have been installed. Consequently, even though no country-to-country technology transfer takes place, the case of the Indian wind sector implies some horizontal technology diffusion in India. This makes the case for the role of the CDM and clarifies the mechanisms that respond to wind power challenges.

In summary, as regards the role of the CDM in wind energy deployment, empirical analyses confirm a tendency towards contributing to the higher efficiency of wind farm installations. This indicates some level of technology diffusion and regional diversification of wind energy activities in the context of CDM projects. The observations provide scope for two main interpretations with regard to the de facto role of the CDM in the Indian wind sector. Firstly, a negative interpretation of the CDM with regard to technology transfer holds that the previous underperformance of wind resource assessments and the lack of proper data management account for the positive changes in PLFs claimed in the PDDs. PLF increases do not reflect more suitable and effective equipment or operational practices but are the result of the demand for proper accounting. Furthermore, the sole effect of the CDM on the diffusion of wind energy technologies in India needs to be qualified for the following reasons. The PDDs give projections on the PLF expected to be achieved in the wind projects. The extent to which these figures will stick to their promises in many cases still remains to be seen. Additionally, the PDD analysis does not clearly indicate whether efficiency increases are due to improved technologies or changes in management and operational practices. Unfortunately, due to the lack of data no inferences can be made with regard to the kinds of technologies utilised previously. However, the PDDs usually argue that R&D has been taken place in advance of project activities that then contributed to the application of more appropriate technologies. Throwing additional doubt on the PLF as an indicator for higher efficiency and technology transfer is the fact that project developers have an incentive to cheat on these figures. In order to make the case for the need of CDM in the respective project, stating lower PLFs justifies the financial additionality. The nevertheless rigorous verification by the Executive Board might explain why many PDDs do not to mention PLFs in the first place.

The second interpretation disagrees and argues that the CDM has in fact catalysed wind power dissemination and the installation of more efficient equipment or operational practices in a range of states. Due to lacking information on previous practices, this can be proven neither right nor wrong. Ultimately, it might be argued that the CDM has a role to play in the Indian wind sector in the broader sense. This relates to the fact that project developers are forced to employ proper processes either installing efficient hardware or operating more effectively as well as to account for current practices.

In summary, this brief empirical excursion demonstrates that the CDM as new mode of governance plays a role in responding to the demands of the Indian wind energy sector for external stimuli to counter sustainable development obstacles. The key mechanisms through which the CDM increases the efficiency of wind farms relate to technology transfer in terms of technical components diffused and the improved operational performance of wind projects.

At this stage, the attention is turned towards the potential role of the CDM and implicit mechanisms for contributing to policy/norms diffusion in response to political and attitudinal barriers[81]: What are the causal mechanisms explaining the adoption of CDM policy innovations in mature or emerging wind industries? How come local contexts such as Kerala that have a high theoretical demand for policy innovations to trigger and incentivise wind energy deployment have not benefited from diffusion effects?

In the context of environmental governance literature (Benz 1995) the role and importance of key actors or brokers to realise opportunities provided through the CDM are emphasised. This is the precondition for norms and meanings such as wind power being beneficial for energy sustainability to be disseminated through the CDM. Since norms and values are observed only through qualitative in-depth studies, the scope of this dissertation restricts preliminary assessments to the two case studies of Tamil Nadu and Kerala. In the context of these two case studies, the role of brokers for norms and policy diffusion was explored in terms of how linkages between local business community and international-level CDM actors were established and maintained. In the case of Tamil Nadu, brokers such as external consultancies or manufacturers taking up service provision in local contexts facilitated reaping the benefits of the CDM in terms of technology and finance provided. This went alongside the diffusion of norms and practices within private business as well as the public policy community on the favourable contributions of the CDM towards wind power dissemi-

[81] Empirical observations and qualitative interviews are taken as basis on which to assess the role of the CDM in meeting political and attitudinal obstacles through mechanisms of policy/norms diffusion.

nation. As demonstrated in the case of Kerala, despite the eagerness of the business community to tap CDM opportunities, technology transfer and policy diffusion has so far hardly been observed. This is due to the negative take of public actors on wind energy and the at this time rather obstructive demands to tax CDM revenues even though activities have barely taken off. In summary, the CDM has a role to play in policy/norms diffusion as to meet attitudinal and political obstacles to wind power deployment. Interpretations of the causal mechanisms involved in policy/norm diffusion processes imply that renewable energy stakeholder networks matter as preconditions for dissemination processes. This empirical venture indicates that renewable energy stakeholder networks provide the conditions for such brokerage to take place through enhancing the receptiveness and political will or openness of public stakeholders.

On this basis, the following interpretation will discuss under which conditions and why the CDM plays a role in wind power promotion through technology transfer and policy diffusion.

This dissertation emphasises the importance of local governance arrangements for disseminating wind energy technology in India. As lesson learnt from numerous attempts of transferring technologies and diffusing Western norms to developing countries, no long-term sustainability is achieved unless technologies and changes are accepted and integrated into local circumstances. This requires the existence of stakeholder networks in which transactions involved in the diffusion of technical hardware and software as well as norms are embedded. Consequently, characteristics and dynamics of stakeholder networks constitute the condition for renewable energy governance to be effective as concerns impacts like tapping emerging opportunities, e.g. CDM, to materialise.

The first argument in support of this hypothesis is the previous empirical analysis recognising that CDM distribution reflects the research puzzle of spatially varying wind power deployment. This leads to the preliminary conclusion that supportive stakeholder networks constitute a precondition for wind business activities to take place at all. In Tamil Nadu, stakeholder networks of favourable characteristics and dynamics facilitate access for new actors and allow new information to be disseminated. In contrast, Kerala's less favourably-constituted local governance arrangement exerts negative network effects in as far as impeding technology transfer and policy diffusion through the CDM. As long as supportive renewable energy governance networks in the wind power sector are not established, the CDM per se is not effective as new mode of governance to assist technological leapfrogging. The observation that lacking wind power deployment matches the absence of CDM projects is not trivial given the objective of the CDM and the widely-held expectation to trigger activities in this sector.

The second argument makes the case for external stimuli such as the CDM to contribute to some degree to technology transfer and policy/norms diffusion through catalysing the formation of supportive renewable energy stakeholder networks. Indeed, empirical observations on diffusion processes allows for qualifying the role of the CDM in technology and policy diffusion. As a matter of fact, potential trigger effects for norms and mentality change are not immediately translated into deployment rates, PLFs or supportive practices. Yet, for states that have to date not benefited from wind energy projects such as Kerala, West Bengal and Madhya Pradesh, the CDM might constitute a necessary catalyst for supportive stakeholder networks to emerge. As empirical research concluded, interest from the wind power and CDM community in henceforth untapped potentials exist which might likely lead to a diffusion and adaptation of respective ideas and activities in these local settings. In this sense, the CDM might provide the means to restructure relations and interactions between public and private stakeholders given that all sides consider mutual engagement profitable. However, this depends on the levels of trust already established, the history of relations and experiences of former collaboration as well as the extent to which mutual commitments and expectation certainty are established. The extent to which the CDM meaningfully contributes to stimulating wind power depends on the embeddedness of previous project activities. In any case, the CDM might play a role in as far as it creates an additional demand for the formation of supportive renewable energy stakeholder networks that pay respect to local context conditions. Particularly the case of Karnataka and of date Kerala reflect a high number of CDM activities compared to previously little if any wind power deployment. This gives empirical justice to the potential role of the CDM as external shock or incentive instilling the transformation of energy governance paths.

In summary, this section explored potential impacts of the CDM as new mode of governance to respond to sustainable development challenges inherent in concrete shortcomings of the Indian wind sector. The main findings are that the CDM plays some role in meeting technical, political and attitudinal barriers through technology transfer, policy and norms diffusion. Renewable energy stakeholder networks, in the end, explain the extent to which technology transfer takes place and how policy/norms are diffused in the concrete local context. Stakeholder network effects emerge due to certain dynamics and characteristics of local governance arrangements and determine how CDM potentials are taken up in the specific local context. However, external stimuli such as the CDM over time play a role in opening windows of opportunity to instil an awareness of or interest in wind energy deployment. This means that stakeholder networks are not only the condition for effective renewable energy governance on the impact-level, but might also be the consequence of new modes of governance encountering local realities. The final and concluding implication is that for external stimuli like the CDM to exert supportive effects in local contexts, a critical mass of a supportive stakeholder network is required for enabling access and seizing opportunities.

This chapter moved beyond the analytical exploration of stakeholder networks in renewable energy governance in the case of wind energy deployment in India. The conceptual turn to consider stakeholder networks as functionally equivalent governance arrangements in the light of governance debates provided a holistic perspective. Implications of stakeholder networks with regard to theory debates as well as empirical practices were interpretatively discussed. As concerns theory implications, questions remain about causal mechanisms and the reasons why stakeholder networks exert certain effects on renewable energy governance. With regard to empirical implications, the analysis of the CDM as external stimulus to the Indian wind sector confirmed the explanatory power of stakeholder networks for renewable energy governance.

The next and last chapter will summarise the main results of research on renewable energy governance in developing countries. The research findings will be integrated into an explanatory model of renewable energy governance in developing countries that is suggested as conceptual theory input for further research. Finally, shortcomings and limitations are discussed with respect to further research implications.

CHAPTER VII

Conclusions - Beyond Stakeholder Networks for Renewable Energy Governance

CHAPTER VII
Conclusions - Beyond Stakeholder Networks for Renewable Energy Governance

In the light of Lincoln's introductory quote, this dissertation argues that in order to turn into a "*power to move things*" wind energy deployment is conditional upon how stakeholder networks emerge as "power to get things moving". In other words, the characteristics and dynamics of local stakeholder networks determine effective renewable energy governance. The research puzzle of spatially varying wind deployment in India is explained by the differences in local governance arrangements.

This concluding chapter will firstly draw together the main research findings. The purpose is to derive at a general explanatory model of renewable energy governance in developing countries that remains to be validated. Secondly, relevance, contributions, shortcomings and implications for future research are discussed.

(a) Summary

Global climate change threatens developmental prospects particularly of developing countries like India. Renewable energy emerges as central option for sustainable development that embraces developmental imperatives and environmental concerns. The research puzzle of spatially varying wind power deployment in India challenges these prospects of sustainable development. The research interest hence focused on identifying contextual conditions for successful wind power deployment alongside the concrete examples of Tamil Nadu and Kerala. In the light of wider theory debates about effective governance in developing countries a two-stage research approach was adopted to give justice to novel, empirical complexities of renewable energy deployment in a multilevel system.

The purpose of the first inductive research stage was to identify explanatory factors through theory-guided explorations. A literature review responding to previous empirical research on renewable energy deployment gave rise to an analytical framework that guided the inductive exploration of the Indian wind puzzle.

The first approach to explain wind power deployment in India looked at developmental conditions. The role of geography in terms of natural resource potential as well as socio-economic characteristics was taken into consideration. Development studies and economics elaborated on the impact of these factors on developmental activities and economic growth. In a nutshell, any kind of rationally-decided economic activity such as wind power deployment would primarily examine if the environment suited specific investment requirements. Case study analysis, however, led to the conclusion that none of these factors significantly accounted for the variations in wind power deployment between Tamil Nadu and Kerala. Specificities regarding the differences in terrain in Kerala were considered negligible when compared to other Indian states.

The second approach focused on policy conditions. Empirical studies identified obstacles to wind power such as untargeted and inappropriate policies and regulations, and respective practices in addition to unfavourable attitudes and perceptions. Neo-institutional theories and public policy literature suggested several explanatory mechanisms: Public policy debates focused on the policy regime, stating that the type of government, the stability and continuity of the political system influence investment decisions in renewable energy. In comparison to other Indian states differences in the Kerala and Tamil Nadu policy regimes did not significantly impact on policy conditions. Neo-institutional perspectives discussed policy conditions in terms

of institutions, regulations and implementation involved in wind energy deployment. Three causal mechanisms were suggested for explaining governance processes in relation to wind power deployment. These considered the role of transaction costs, policy diffusion and organisational routines. Case study analyses concluded that indicators such as policy effectiveness and efficiency, financial status and corruption revealed little about organisational practices and behaviour in terms of renewable energy support. In conclusion, differences as concerns the characteristics of organisational stakeholders in the two local settings of Tamil Nadu and Kerala emerged as important explanatory issue deserving closer attention.
The third explanatory approach considered economic conditions for renewable energy deployment responding to deployment obstacles identified like inappropriate technology, infrastructure and labour. Economic geography theories recurring to spatial economics discussed how the assessment of location factors drives rational investment activities. Behind this background, a range of factors such as industrialisation, infrastructure and the general opportunity structure inherent in the respective power sector context were considered. Case study analyses derived at the conclusion that not location conditions per se mattered for wind energy deployment. Moreover, the availability and accessibility to opportunity structures emerged as potential entry to understand what stakeholders make of local conditions. This led to the assumption that the ways in which stakeholders dealt with the given settings exchanged resources and related to each other were more conclusive on how and why wind power deployment takes place. This called for the theoretical and empirical exploration of actors' relations and interactions in the context of renewable energy stakeholder networks.
The last explanatory approach responded to barriers to renewable energy engagement such as insufficient finances, lacking capacities and capabilities discussed in empirical literature. Global environmental governance discussed the potential of external factors to promote renewable energy deployment creating diffusion effects and positive local stimuli. Foreign assistance through financial mechanisms, e.g. direct investment, or development activities with a long tradition particularly in the renewable energy sector of developing countries were considered as explanatory factors. Even though case study analyses at first sight supported these arguments on FDI, this did not hold to comparisons with other Indian states. The impact and role of new modes of governance such as the CDM on wind energy deployment emerged as point for further examination.

As a result of the first inductive research phase, the theory-guided empirical analyses of the Indian wind puzzle derived at the conclusion that locally-specific governance agency and processes mattered. Particularly two factors appeared of greater potential explanatory value. These referred to "actors' attributes" in terms of stakeholder characteristics and "actors' relations" as concerns interactions and respective constellations between stakeholders involved in renewable energy governance.
The purpose of the second deductive research phase was hence to embark upon an in-depth exploration of these variables. A literature review of organisational and new economic sociology theories was conducted in order to illuminate these factors and causal mechanisms. The result of this discussion was an explanatory model of wind energy deployment in India that served to guide the ensuing empirical analysis. Case study explorations explicitly focused on Kerala and Tamil Nadu that as demonstrated previously, constituted the most puzzling instances of the Indian wind sector. In addition, governance debates suggested a framework to examine functional equivalents

to electricity service provision in developing countries. This guided the explorations and interpretations of the research puzzle alongside the question "who" and "how" wind energy was governed. Stakeholder network analysis was applied as analytical and conceptual approach to venture into the examination of the respective governance arrangements. Stakeholder networks were examined according to actors and their characteristic attributes as well as actors' relations in emerging interaction arrangements in the locally-specific multilevel context of renewable energy governance. Deeper understandings of who governs required elaborations on the constitution and effects of actors' attributes on renewable energy governance as first explanatory variable. Organisational sociology provided valuable theory guidance to the analysis of actors according to certain properties and characteristics. Point of departure was the hypothesis that actors' characteristics influenced renewable energy governance through constitutive elements like capacities, interests and capabilities. Particularly attitudes and qualitative resource characteristics of public stakeholders turned out relevant as they translated into organisational routines and cultures.
"How" and with what means wind energy was governed was explored in the context of the second explanatory variable "actors' relations". Stakeholder network analysis was applied to map network relations and interactions between stakeholders. This departed from the assumption that the greater the quality of ties characterising actors' arrangements, the more open interactions in networks constellations. This led to ensuing positive network effects on the exchange of tangible and intangible assets and contributed to effective renewable energy governance. Differences between these relational interaction arrangements of emerging stakeholder networks in Tamil Nadu and Kerala were considered of explanatory value. Successful wind power deployment in Tamil Nadu was explained by the openness of interactions in the relationally-embedded actors' arrangements. The second deductive research stage derived at the conclusion that stakeholder networks in terms of specific dynamics and characteristics of local governance arrangements hold the power to render renewable energy governance effective. Responding to the research puzzle, this dissertation demonstrated that successful wind power deployment is conditional upon the effects of local stakeholder networks characterised by favourable attributes of stakeholders embedded in relational, public-private interaction arrangements.

In a last step, research findings were interpreted with regard to theory and empirical implications. The purpose was to discuss the mechanisms how governance occurred and why stakeholder networks emerged in locally-specific manners.
With regard to theory implications, networks effects could not be fully explored with focus on the individual theories only. Consequently, stakeholder networks were scaled up as theory concept which in the light of governance debates provided for more holistic and complex understandings. Given this conceptual turn, stakeholder networks were assumed to exert supportive effects on the output, outcome and impact dimensions of renewable energy governance.
In this sense, local governance arrangements were interpreted as functional equivalents to channel public and private efforts and provide renewable energy services. As regards practical implications, stakeholder network effects were explored with regard to newly emerging potentials like the CDM. Stakeholder network characteristics turned out to determine the extent to which external stimuli were locally adapted in a way as to encourage wind power deployment. This dissertation argues that the stimuli for renewable energy deployment are combined effects of the international geopolitics of energy and renewable energy as well as the national energy and renew-

able energy scenario. These are reflected in the characteristics and dynamics of local renewable energy stakeholder networks (Table 35).

Table 35: Network Effects of Renewable Energy Governance

Network Effects/ Renewable Energy Governance	**Output Dimension** (Framework; Policies & Regulations; Key Stakeholders)	**Outcome Dimension** (Deployment activities)	**Impact Dimension** (Expanding options, seek & realise new opportunities)
Actors' Attributes	Political will and attitudes; Resource availability	Leadership of and in organisations; Effective resource deployment	Vision; Risk capital
Actors' Relations	Common ground; Tacit exchange	Reduced formalism; Emphasis on trust and reciprocity	Embedded at multiple levels
Stakeholder Networks	Internally closed interactions to establish confidence base	Opening structures and dynamics; Sharing bounds	Open networks; temporally closed alliances to tap niches

The above table summarises the overall research results related to stakeholder network effects on renewable energy governance. In a nutshell, the constitutive characteristics and dynamics of multilevel governance arrangements, i.e. local stakeholder networks, exert effects related to the output, outcome and impact dimensions of renewable energy governance.

The Tamil Nadu case showed supportive network effects on all dimensions translating into successful wind power deployment. Not only governance outputs such as policy frameworks and outcomes in terms of successfully installed wind capacities but also impacts are noticed with respect to venturing into new wind power options, e.g. the CDM.

In contrast, the Kerala case demonstrated some network effects on the output dimension in as far as pre-conditions, e.g. supportive policies and regulations, for successful wind power deployment are evolving. However, dysfunctional stakeholder network aspects relating to closed interaction constellations and adverse public stakeholder attributes restricted the scope for effective outcome- and impact-related renewable energy governance. The Kerala case showed that the stakeholder network exerting supportive effects on renewable energy governance outputs was insufficient to support wind power deployment as effective governance outcome.

In summary, renewable energy governance emerged as a dynamic multilevel and -staged process that requires locally-specific conditions to be in place in order to turn out effectively. Local stakeholder networks constitute these basic preconditions. Local stakeholder networks shape respective governance mechanisms as functionally equivalent governance arrangements to supply alternative energy services. Not only the characteristics and dynamics but also the fit of stakeholder networks to the specific requirements on renewable energy governance in a certain setting determines the governance effectiveness.

(b) Towards an Explanatory Model

On the basis of these research findings and interpretive discussions, an explanatory model of renewable energy governance in developing countries will be elaborated in the following. The purpose is to contribute theory input to the study of renewable energy governance in developing countries on the basis of the research findings.

As point of departure, several assumptions constitute the basis of the explanatory model. Stakeholder networks are analytical and conceptual devices that from an overall governance perspective grasp who, how, why and with what effects renewable energy governance takes place in non-OECD settings. Local stakeholder networks constitute the necessary condition for effective renewable energy governance. Organisational and new economic sociology provide theoretical approaches to analyse and explain renewable energy stakeholder networks in the governance framework. Renewable energy networks mutually constitute of stakeholders characterised by specific attributes as well as relational interactions. Networks are dynamic and evolutionary. Distinctive forms of renewable energy networks emerge out of the interactions between these constitutive elements in the context of emerging, locally-specific demands on renewable energy governance. Stakeholders behave both rationally and in a social-constructivist sense inspired by norms, values and the logic of appropriateness. Interactions and relations in networks produce facilitating and impeding effects in an intentional and unintentional manner on renewable energy governance. These effects involve multi-level agency and dimensions and are exerted through several causal mechanisms. External stimuli in terms of new modes of governance might trigger renewable energy governance in specific local contexts. Given a critical mass of stakeholders or a conjuncture of internal and external circumstances, implicit processes or changes might be promoted in stakeholder networks.

This dissertation focused on the Indian wind sector as an example of renewable energy deployment in developing countries. In order to render these findings applicable and comparative, theory-guided and empirically-derived observations are scaled up into a general explanatory model. The main argument is that local stakeholder networks exert multi-dimensional effects on renewable energy governance through the characteristics and dynamics of stakeholder engagement in distinct multi-level interrelations.

The following hypotheses specify the role and effects of stakeholder networks on renewable energy governance in developing countries:

(a) Given that a critical mass of public stakeholders dispose of strategic assets in term of (recognised) leadership and are characterised by established, purpose-oriented organisational routines, renewable energy deployment is encouraged

(b) Renewable energy governance takes affect if relational arrangements between public and private stakeholders are characterised by structurally open interactions and the embedded ties reflect social capital, reciprocity and little formality.

(c) Stakeholder networks display locally-differentiated characteristics and dynamics exerting effects and hence respond differently in the specific renewable energy governance context. Renewable energy stakeholder networks emerge, change and take shape in a path-dependent manner in response to internally and externally-stimulated conjunctures and given a critical mass of stakeholders.

Causal mechanisms that specify how renewable energy networks exert effects on respective governance are considered on the output, outcome and impact dimensions. Embraced by the overall governance framework, questions of how and why certain stakeholder networks exert differential effects on renewable energy governance are discussed in the underlying theory framework of organisational and new economic sociology. In this regard, network effects are related to and depend on the access, brokerage, information and asset exchange and involve mechanisms of trust as well as transaction cost considerations. Conditions under which stakeholder networks are effective in terms of renewable energy governance refer to the environment and the entire multilevel energy governance system. This provides specific windows of opportunities and channels for the diffusion of ideas and practices.

In summary, this model acknowledges that renewable energy stakeholder networks are subject to change and involve an important time dimension. This is also of explanatory value for the origins and characterisation of network configurations. Intervening variables such as internal network factors or external stimuli exert different impacts depending on the respective stage of network formation, operation or expansion. The validity and explanatory power of this model remains to be tested in similar or different issue areas and/or contexts.

(c) Research Relevance and Contributions
The following five points will summarise the main contributions of this dissertation and hence prove its relevance to the research on renewable energy governance in developing countries.
The first point regards this dissertation's thematic relevance: Global climate and developmental challenges that affect particularly developing countries like India render relevant the examination of renewable energy alternatives to realise sustainable development. Vast literature on potentials, barriers and challenges of renewable energy deployment for power generation exists. Yet, research gaps prevail concerning triggers, conditions and mechanisms that transform institutional, political and economic systems into opportunity structures. The contribution of this dissertation to empirical debates on renewable energy in developing countries is the assessment of the requirements for and conditions under which renewable energy deployment takes place.
Secondly, this dissertation's contribution to qualitative research arises out of the combination of inductive and deductive procedures. This was reflected in the conduct as well as the structure of research in the course of this dissertation. A mix of methods and methodologies including ethnographic, participant-based research were applied in order to approximate local realities. This enabled a non-Eurocentric approach to examining local dynamics based on and involving the relevant actors. Concerning theory contributions, with regard to new economic sociology, unlike most studies in this discipline, this dissertation considered the importance of time and public agency in order to gain an understanding of the origins, causal mechanisms and changes in the relations between stakeholders. Further contributions relate to the integration and operationalisation of embeddedness concepts in the development of relational typologies to add to the understanding of mechanisms and dynamics inherent in stakeholder networks. With regard to organisational sociology, this dissertation moved be-

yond the static positional and hierarchical public stakeholder characterisation to consider mechanisms and dynamics inherent in relations and interactions. As concerns stakeholder network analysis, applications as analytical tool and at later stages as conceptual instrument allowed for integrating both theory debates fruitfully in exploring actors and ties in emerging governance arrangements. This dissertation made relevant contributions to global environmental governance debate with the determination of an explanatory model that emphasises local stakeholder networks as condition for effectively countering governance challenges. Taking into consideration the importance of local stakeholder networks as condition for effectively providing governance services while acknowledging the specificities of non-OECD settings, provides relevant input to governance debates.

Thirdly, this research bears implications for respective deployment activities and energy governance. This dissertation focused on the options and the potential provided by renewable energy alternatives in terms of technological leapfrogging and ecological modernisation. Even though incentives for renewable energy deployment exist at national and international levels, many countries and regions such as in the Indian case do not make use of the opportunities available. In the context of climate negotiations today, this is of relevance as new windows of opportunity are opening for renewable energy in developing countries. This dissertation contributed to assessing conditions and determinants for tapping new modes of governance such as the CDM that require knowledge about local governance arrangements. Understanding the pre-requisites for embarking upon such development pathways allows suggesting options for the advancement of alternative energy paths through assistance and support efforts. The findings of stakeholder network research in this dissertation might prove insightful for private actors to assess entry points to business engagement on the basis of situational analysis of certain sectors and stakeholders. Public administration might draw on several implications as regards how to scale up effective and efficient, internal and coordinated working as well as collaboration with private actors towards meeting common ends. International assistance might more effectively encourage and externally support the preconditions for activities, i.e. stakeholder network formation.

(e) Research Shortcomings

In the following, shortcomings are discussed with regard to three aspects related to theory, empirical data and research procedure in order to draw lessons-learnt for future research.

Firstly, in the course of this dissertation adaptations of the research design were unavoidable as further explorations of previously unstudied empirical realities suggested new entry points for examination. Since an inductive-deductive research design was adopted, going back and forth between empirics and theory was required. Consequently, in order to provide for conceptual clarity further research could consciously subscribe to one research methodology only.

Secondly, with regard to the theory and methodology, one might argue that this dissertation departs from an over-determined research design. In fact, this dissertation consciously integrates several theory literature debates. Research limitations with respect to theory frameworks emerged because this dissertation did neither embark upon testing theory propositions nor engaged in theory development due to the em-

pirical novelty and lack of previous research on the Indian wind puzzle[82]. The richness of theories is also the main point of criticism related to parsimony. The predictive value of this dissertation was restricted by its general applicability since no specifically defined causal relations was examined. These produced at times contradictory hypotheses that were based on different assumptions regarding actors' behaviour or causal mechanism. This had been taken into account as trade off against the richness of analysis and non-bias against alternative explanations. Yet, a more general claim on the extent to which hypotheses could indeed be falsified was rendered more difficult. Apart from contrasting different theory accounts in the context of empirical results, no satisfactory conclusion on the transferability and explanatory value of specific theories was possible. An additional criticism is the inherent bias of theory debates to the OECD context.
One suggestive remedy might be to confine the research design to a selected and distinct set of theories and ensuing hypotheses on the grounds of the newly-explored empirical reality. Suggestions for further research are a purposeful restriction to either new economic sociology or organisational sociology that turned out to deliver the most valuable insights into renewable energy governance in developing countries.

A third shortcoming was that this dissertation suffered from the lack of sufficient empirical case studies despite the attempt to increase the number of observation by integrating a temporal component in the sense of process tracing. This was due to the lack of case study analysis on the governance of the Indian wind sector prior to own research. This means that replicating the research by enhancing the number of cases would add more theoretical knowledge to the debate and understanding of empirical realities.

As fourth point, difficulties encountered researching wind energy deployment in India were due to intrinsic obstacles to empirical research in developing country contexts. This dissertation had to be limited to selecting just one renewable energy technology, i.e. wind power, in the context of only two Indian states, i.e. Tamil Nadu and Kerala. Reasons for the confined selection were related to lacking access, feasibility considerations, and time and resource constraints. This limits the application to other technology and cases and renders questionable the general validity of research results and implications. Further research could thus seek to expand the number of states as well as to examine other renewable energy technologies.
Finally, in the course of the research process several challenges were encountered due to the respective research methods applied. The greatest obstacle to research was access to the field, institutions and persons considered of relevance for information gathering. In addition, data was not only difficult to obtain but also proved to be inconsistent among and within the cases or contradictory with regard to content. This restricted not only the analytical options as no proper process tracing could be undertaken and sound comparison could be made but ultimately also casts doubt on the validity of research findings[83]. Potential biases were inherent in the research strategy as data sampling was subject to access opportunities and limited due to time and resource constraints.

---

[82] The main purpose of theories was consciously considered to guide the exploration of the empirical cases and to provide for different and also competitive explanatory accounts (Flick 2006).
[83] This was the major reason why this dissertation consciously introduced subjectivity and questioned the pretence of an objective science (Flick 2006).

Suggestions for future research are to involve local people and include indirect participant observations in order to achieve greater validity and richness of information.

(f) Research Implications

Discussions of research contributions as well as challenges encountered in give rise to some general implications as regards entry points to further research.

With regard to empirical research, one option to test the explanatory validity of the explanatory model developed on the basis of this dissertation's research findings is to enhance the number of case study states in the wind sector in India. Another option would be to examine the case of a different renewable energy technology or an entirely different sector in developing countries in order to focus on the aspect of stakeholder networks and their explanatory power. Yet, staying with the empirical reality of the Indian wind sector already examined, another alternative approach would be to contrast the explanatory model elaborated with competing hypotheses.

Another indication for future research refers to the elaboration of the explanatory model with the purpose of theory development. This would integrate the temporal dimension and focus on "how come" questions. In this sense, questions about conditions and reasons for the emergence of stakeholder networks and the constitution of their characteristics and dynamics are tackled.

Additionally, a comparison of stakeholder networks in different countries is conceivable even though the transferability and applicability of models developed in a certain context would be a crucial issue. In this sense, renewable energy governance explored in the context of this explanatory model in developing countries might shift the focus to the OECD world and contrast findings guided by a similar research design on the light of governance debates.

Lastly, one interesting option arises in applying research findings of stakeholder network analysis to action-oriented research. This implies the implementation of pilot projects in regions that do not experience effective renewable energy governance in terms of e.g. renewable energy deployment, focussing on the support of stakeholder network formation through e.g. capacity development.

To put this dissertation in a nutshell, renewable energy stakeholder networks emerge as powerful explanation for spatially varying wind energy deployment in India. This provides suggestions about how to scale up renewable energy alternatives in response to pressing global climate and developmental dilemmas. This dissertation contributes empirically-grounded, analytical as well as theoretical input to existing debates on the determinants of renewable energy governance in developing countries. *"Turning wind into power"* requires effective renewable energy governance which is conditional upon the dynamics, characteristics and effects of multilevel, locally-specific stakeholder networks.

APPENDIX

ANNEX I

List of Interview Partners

Name	Organisation	Title	Date
Ms Pamposh Bhat	GTZ India	Head Indo-German Energy Programme	10.01.2008 21.09.2007
Mr Rakesh Kacker	InWEA	CEO & Secretary General	09.01.2008 21.09.2007
Mr Satapathy	TNSG	Principal Secretary	30.10.2007
Mr Gopalakrishnan	TNEB	Chief Engineer, NCES	29.10.2007
Mr Raja Sukumar	IndoWind	Vice President	29.10.2007
Shri Ajit Gupta	MNRE	Advisor	18.09.2007
R. Ramesh	KSEB	Assistant Executive Engineer	26.10.2007
Mr Suresh Kumar	SUZLON	AGM – Corporate Affairs	24.10./25.10.2007
Mr Unnithan	ANERT/EMC	Dep. Director	25.10.2007
Mr Harikumar	ANERT		26.10.2007
Shri C. Balakrishnan	KSERC	Member	25.10.2007
C.P Bramod	Kerala State Government	Minister for Power and Electricity	25.10.2007
Mr Vijayakumar	Kerala State Government	Minister of Law	25.10.2007
Ramesh Kymal	NEG Micon	Managing Director	02.11.2007
Mr Saibaba	Suzlon	Business Head South	29.10.2007
Mr Vijayarajan	ABI Energy Consultancy	Director	29.10.2007
Thiru Elangovan	TEDA	Director	30.10.2007
N. Parasuraman	TEDA	Consultant	30.10.2007
B. Jeyaraman	TERC	Member	01.11.2007
DV Giri	Indian Wind Turbine Manufacturing Association	President	03.11.2007
R.K. Radhakrishnan	The Hindu	Deputy Chief	04.11.2007
A. Balraj	TNSERC	Ret. Chairman	04.11.2007
Dr E. Sreevalsan	C-WET	Unit Chief	02.11.2007
	Suzlon		27.10.2007
Harish Mehta	Suzlon		28.10.2007
A. Mohan	TNERC	Assistant General Manager	01.11.2007
K.S. Sridharan	IT Power	Consultant	07.11.2007
Kalipada Chatter-	Winrock Interna-	Sr Advisor	19.09.2007

jee Mr P Mukherjee	tional India		
Mr Padmanaban	USAID	Dep. Director & Senior Energy Advisor	20.11.2007 08.01.2008
Dr Navroz Dubash	JNU, New Delhi	Professor	29.11.2007
Dr Amit Kumar	TERI	Associate Director; Energy-Environment Technology Division	10.01.2008
Dr PC Maithani	MNRE	Director	10.01.2008 12.09.2007
V.S. Verma	Government of India, Planning Commission	Member	
Rajiv Arora	MNRE	Director	28.11.2007
Sarita Chauhan	Confederation of Indian Industry	Executive	26.11.2007
Mario D'Souza	CSE	Research Associate	09.01.2007
Vijay Chaturvedi	Development Alternatives	Programme Director	20.11.2007
Dr TC Tripathi	MNRE	Advisor	08.10.2007
S. Metha	VESTAS RRB	Vice President	17.10.2007
A Mohan Reddy	Zenith Energy	Director	03.10.2007
Debashish Majumdar	IREDA	Managing Director	15.04.2007 15.09.2007
Ajay Mathur	Bureau of Energy Efficiency, Ministry of Power	Director General	11.01.2008
Rajendar Kharul	WISE	Wind Energy Specialist	26.09.2007
GM Pillai	WISE	CEO & General Manager	26.09.2007
Mr Chintan Shah	Senergi Global	Director	13.09.2007

ANNEX II

Sample Interview Guideline

Interview Guideline – Project Stakeholders (Business – Manufacturers; Consultants; Service Providers)

I Generic Questions

(1) Stakeholder's Role and Position
- Describe organisational set up & development
- Describe mandate and main activities & development and trends
- Ratio of wind-specific activities to other REN support
- Main intentions and rationales – what strategies for promotion of wind
- How activities and objectives for wind development derived at (who involved)
- Main customers – how relationship/activities evolved

(2) Determinants of REN (wind) Deployment

a) Policy Framework
- Main characteristics and elements for wind (fiscal, financial incentives; soft loans etc.)
- How derived at and how changed
- To what extent influenced by other policy frameworks, other stakeholders
- Assess impacts on wind development
- To what extent other policies (energy/power; infrastructure development; irrigation; agriculture; private sector-relevant; FDI) influence wind investment

b) Regulatory Framework
- Main characteristics and elements - How derived at and how changed - Assess impacts on wind development

c) Business Environment
- Development & trends of power supply/energy situation – impacts on REN deployment
- Development & trends of grid connectivity and status (T&D losses) – impacts on REN deployment
- To what extent do issues on land acquisition and other clearances required for REN deployment play a role
- Access, availability and easy to obtain finance (credit) for REN deployment – main sources? Constraints?

d) Business Community
- Characterisation of wind entrepreneurs - competition
- Situation with suppliers, consultants, service providers – origins/location; availability
- R&D and technical support development
- Attitude local people, civil society – involvement in project

(3) Impacts of 'Risks' on Wind Energy Deployment

a) Operational Risks
- Characterisation of wind resource availability – how relate to wind turbine productivity – issue?
- Availability of grid and ease of grid integration
- Existence and responsibility for evacuation facilities

b) Financial Risks
- Assess danger of SEB contract breach – past cases?

- Potential of PPA contract negotiation (who and why)
- Existence of uncertainty on power sector restructuring

c) Regulatory Risks

- Extent to which uncertainty on tariff (rate; negotiation process)

(4) Potential of the CDM

- Ration CDM Wind project to non-CDM wind projects
- Main rationales
- Changes to wind deployment: plant deployment, O&M processes / negotiation with different, other stakeholders / increase in (?) transactions
- Affect of economic and financial risks of project
- CDM as driver for wind deployment?

II Specific Questions

(1) All Actors

- Why so much private sector participation in wind (REN deployment) while public sector dominates conventional power generation?
- What are the main features of the unique Indian business model ('one-stop-shop') in the wind sector? Why has it developed as such? What are central issues?
- To what extent does a lack in design capabilities (adapting wind technology to Indian circumstances) exist and why? – What status of R&D; technology and professional capabilities in India?
- To what extent are missing networks (research – industry) still an issue/impediment to wind deployment?
- To what extent does community support play a role for wind farm development?
- What kinds of limitations restrict the financial sector/Indian banks from financing wind power?
- To what extent does FDI in wind take place? – Which areas? – What main barriers?
- To what extent has a drop in SEB/SNA support for wind power dropped in 1997/98 and why?
- To what extent has CDM triggered private sector participation (increase in) renewable energy deployment by end-1990s?

(2) PS specific

- To what extent have investments in wind turbines yet let to a decrease in cost/kWh wind?
- To what extent are manufacturing/supply/entrepreneurial activities relocated to other (Indian) states? What are prospects/plans on strategic diversification/relocation?
- Regarding wind farm installation – what parts are still imported/to what extent/ why?
- What is phenomenon of 'gold plating'? Why/to what extent does it take place? (increase in capital costs of wind plant capital while reduction in international wind turbine prices (1996/97)?

(3) State Specific

- How has the state of power sector restructuring/unbundling progressed – What status now?

Tamil Nadu

- Why has Tamil Nadu become a leader in wind farm installations in the first place (1st demonstration plant in Gujarat)?
- When and why have wind turbine manufacturers/suppliers have settled in TN?

- How can the following paradox be explained: despite worst SEB status still highest installed wind capacity?

Kerala

- To what extent do regional service stations and R&D on wind exist? Why not?
- Why is the growth of the energy sector described as lagging? What techno-economic and environmental barriers are responsible for this?
- How does energy planning work and how has it changed in Kerala?
- To what extent does deforestation have an impact on wind?
- How come wind sector development has not yet taken off in Kerala? What are main policy characteristics for private investment – matter?

ANNEX III

Sample Stakeholder Network Map

Actors' Mapping
Expert: Mr Saibaba; SUZLON, Chennai
Date: 30.10.2007

What determines the (successful) deployment of a renewable energy project? How important are the respective stakeholders?

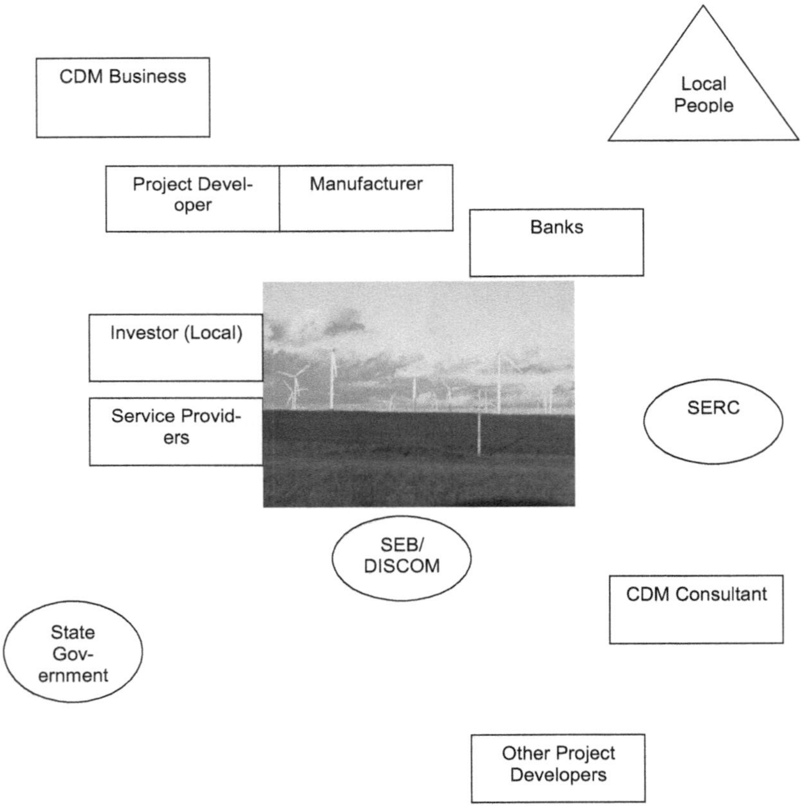

Bibliography

ABBOTT, M. (2001) Is the Security of Electricity Supply a Public Good? *The Electricity Journal,* 31-33.

ACKERMANN, T. (Ed.) (2005) *Wind Power in Power Systems,* Chichester, Wiley& Sons.

ADLER, P. K., SEOK-WOO (2002) Social Capital: Prospects for a New Concept. *Academy of Management Review,* 27, 17-40.

AECKERBERG, K.; JOAS, M. (2004) Multilevel Environmental Governance: A Concept Under Stress? *Local Environment,* 9, 405-412.

AHLUWALIA, M. S. (2000) State Level Performance under Economic Reforms in India. *Indian Economic Prospects: Advancing Policy Reform.* Stanford, USA.

AILAWADI, V.; BHATTACHARYYA, S.; SUBHES, C. (2006) Access to energy services by the poor in India: Current situation and need for alternative strategies. *Natural Resources Forum,* 30, 2-14.

ALTOMONTE, C. (2000) Economic Development and Institutional Frameworks: FDI in economies in transition. *Transnational Corporations,* 9, 75-106.

AMIN, A.-L. (1999) Institutional barriers to commercialisation of wind power in India. The case of Gujarat. Oslo, CICERO.

ANDRIESSE, E. (2008) *Institutions and regulatory development in South-East Asia,* Utrecht.

ARTS, B. (2006) Non-state actors in global environmental governance: New arrangements beyond the nation state. In: KOENIG-ARCHIBUGI, M.; ZÜRN, M (Ed.) *New modes of governance in the global system.* Basingstoke.

ASIF, M.; MUNEER, T. (2007) Energy supply, its demand and security issues for developed and emerging economies. *Renewable and Sustainable Energy Reviews,* 11, 1388-1413.

AUER, R. (2008) The Colonial and Geographic Origins of Comparative Development. *Swiss National Bank Working Paper No 08.* Swiss National Bank.

BACHE, I.; FLINDERS, M. (Eds.) (2003) *Multi-level Governance,* Oxford, Oxford University Press.

BAJPAI, N.; SACHS, J. (1999) The progress of policy reform and variations in performance at the sub-national level in India. *HIID Development Discussion Paper 730.* Harvard Institute for International Development, Harvard University.

BAKSHI, R. (2006) Development of Renewable Energy in India: An Industry Perspective. IN MALLON, K. (Ed.) *Energy Policy and Politics. A Handbook for Decision-making.* London, Earthscan.

BANERJEE, B. P. (2005) *Handbook of Energy and the Environment in India,* New Delhi, Oxford University Press.

BANERJEE, R. (2006) Comparison of Options for distributed generation in India. *Energy Policy,* 34, 101-111.

BANIK, A. (2000) Technology Transfer in Asia - Case of latecomer countries. Conference Paper presented at: *INSEAD Conference 20.07.2000, Fontainebleau.*

BARBIER, E. (2003) The Role of Natural Resources in Economic Development. *Australian Economic Papers*, 253-272.

BARNARD, CH. (1938) *The Functions of the Executive.* Cambridge, MA, Harvard University Press

BARNES, D.; FLOOR, W. (1996) Rural Energy in Developing Countries: A Challenge for Economic Development. *Annual Review of Energy and Environment*, 497-530.

BARNETT, A. (1990) The Diffusion of Energy Technologies in the Rural Areas of Developing Countries: A Synthesis of Recent Experiences. *World Development,* 18, 539-553.

BARNETT, A. (2000) Energy and the Fight against Poverty. Brighton, Sussex University Press.

BARNETT, M.; FINNEMORE, M. (2004) *Rules for the World. International Organisations and Global Politics,* Ithaca and London, Cornell University Press.

BARRO, R. SALA-I-MARTIN, X. (1995) *Wirtschaftswachstum,* Oldenbourg.

BARTELS, D. (1968) Zur wissenschaftstheoretischen Grundlegung einer Geographie des Menschen. *Erdkundliches Wissen,* 19.

BASU, S. (2007) Risks to Wind Power Projects in India: Assessing the role of the CDM. New Delhi, TERI.

BATHELT, H. (1994) Die Bedeutung der Regulationstheorie in der Wirtschaftsgeographischen Forschung. *Geographische Zeitschrift,* 82, 63-90.

BATHELT, H. (2003) Geographies of production: growth regimes in spatial perspective 1 - innovation, institutions and social systems. *Progress in Human Geography,* 27, 763-778.

BATHELT, H. (2005) Geographies of production: growth regimes in spatial perspective 2 - knowledge creation and growth in clusters. *Progress in Human Geography,* 29, 204-216.

BATHELT, H. (2006) Geographies of production: growth regimes in spatial perspective 3 - toward a relational view of economic action and policy. *Progress in Human Geography,* 30, 223-236.

BATHELT, H.; GLÜCKLER, J. (2003a) Towards a relational economic geography. *Journal of Economic Geography,* 3, 117-144.

BATHELT, H.; GLÜCKLER, J. (2003b) *Wirtschaftsgeographie. Ökonomische Beziehungen in räumlicher Perspektive,* Stuttgart, Verlag Eugen Ulmer.

BAYLIS, J.; SMITH, S. (Eds.) (2001) *The Globalisation of World Politics. An introduction to international relations,* Oxford, Oxford University Press.

BENECKE, G.; FRIBERG, L.; SCHROEDER, M. (2008) From PPP to Market. The Clean Development Mechanism as a New Mode of Governance in Climate Protection. *SFB Working Paper Series No.10.* Berlin, SFB 700

BENECKE, G. (2008) Der Clean Development Mechanism (CDM) als Instrument zur Bewältigung globaler Klimaprobleme? Der Mehrwert einer Analyse als "Neue Form von Governance". In: DE LA ROSA, S.; HÖPPNER, U.; KÖTTER, M. (Eds.) *Transdisziplinäre Governanceforschung. Gemeinsam hinter den Staat blicken.* Baden-Baden, Nomos.

BENECKE, G.; BRANOVIC, Z.; DRAUDE, A. (2008) Von Governance zu Raum. *SFB Working Paper Series No. 13.* Berlin, SFB 700.

BENZ, A (Ed.) (2004): *Governance. Regieren in komplexen Regelsystemen. Eine Einführung.* Wiesbaden, Verlag für Sozialwissenschaften

BIERMANN, F.; BAUER, S. (2004) Assessing the effectiveness of intergovernmental organisations in international environmental politics. *Global Environmental Change,* 14, 189-193.

BIERMANN, F.; BAUER, S. (2005) *World Environment Organisation: Solution or threat for effective international environmental governance?* Ashgate, Aldershot.

BIERMANN, F.; JACOB, K.; BUSCH, P.O.; FEINDT, P (2001) Politik und Umwelt. *Politische Vierteljahreszeitschrift,* Sonderheft 39, 94-114.

BIERMANN, F.; SIEBENHÜNER, B.; SCHREYÖGG, A. (Eds.) (2007) *International Organisations in Global Environmental Governance.* London, Routledge

BIERMANN, F.; PATTBERG, P.; CHAN, S.; MERT, A. (Eds.) (2007) Partnerships for Sustainable Development. An Appraisal Framework. *Global Governance Working Paper.* Amsterdam, Vrije Universiteit Amsterdam.

BIERSTEKER, T. (Ed.) (1992) *International Finance Negotiations,* Boulder, Westview Press.

BÖRZEL, T. (1997) What's so special about policy networks? - An exploration of the concept and its usefulness in studying European governance. *European Integration online Papers,* 1, 1-28.

BÖRZEL, T. (1998) Organising Babylon: On the Different Conceptions of Policy Networks. *Public Administration,* 76, 253-273.

BÖRZEL, T. (2006) What is Governance? Unpublished *SFB Working Paper. SFB 700,* Berlin.

BOUTELIGIER, S. (2009) Does networked globalisation need networked governance? An inquiry into the applicability of the network metaphor to global environmental governance. Conference Paper presented at the *International Studies Association Annual Convention 13th-15th February, 2009.* New York.

BRÜHL, T.; FELDT, H.; HAMM, B (Ed.) (2004) *Unternehmen in der Weltpolitik. Politiknetzwerke, Unternehmensregeln und die Zukunft des Multilateralismus,* Bonn.

BULKELEY, H. (2005) Reconfiguring environmental governance: Towards a politics of scales and networks. *Political Geography,* 24, 875-902.

BURT, R. (1992) *Structural holes: the social structure of competition,* Cambridge, MA, Harvard University Press.

CAMPBELL, J. (2002) Ideas, Politics and Public Policy. *Annual Review of Sociology,* 28, 21-38.

CANAN, P.; REICHMAN, N. (2002) *Ozone connections: expert networks in global environmental governance*, Greenleaf Publishing.

CASTELLS, M. (2000) Materials for an exploratory theory of network society. *British Journal of Sociology,* 51, 5-24.

CLEGG, S.; DUNKERLEY, D. (1980) *Organisation, class and control,* London, Routledge.

COLEMAN, J. (1991) Social Theory, Social Research and a Theory of Action. In: ABELL, P. (Eds.) *Rational Choice Theory.* Cambridge, Cambridge University Press.

COLEMAN, J (1988) Social Capital in the Creation of Human Capital. *The American Journal of Sociology, 94,S95-120*

COLLIER, D. (1991) *Shaping the Political Arena: Critical junctures in the labour movement and regime dynamics in Latin America*, Princeton University Press.

COLLIER, D.; MAHONEY, J. (1996) Insights and Pitfalls: Selection Bias in Qualitative Research. *World Politics,* 49, 56-91.

COOLEY, A.; RON, J. (2002) The NGO Scramble: Organisational insecurity and the political economy of transnational action. *International Security,* 27, 5-39.

DANG, H. (2007) Wind Energy: Challenges and Prospects in India. *INWIND Chronicle.*New Delhi, InWEA

DASGUPTA, P. (1997) *Environment and Resource Economics in the World of the Poor.* Washington, RFF Press

DECHEZLEPETRE, A.; GLACHANT, M.; HASCIC, I.; JOHNSTONE, N.; MÉNIÈRE, Y. (2008a) Invention and Transfer of Climate Change Mitigation Technologies on a Global Scale: A Study Drawing on Patent Data. Paris, CERNA

DECHEZLEPETRE, A.; GLACHANT, M.; MÉNIÈRE, Y. (2007) The Clean Development Mechanism and the International Diffusion of Technologies: An Empirical Study. *CCMP - Climate Change Modelling and Policy.* Paris, Fondazione Eni Enrico Mattei.

DECHEZLEPETRE, A. G., MATTHIEU & MÉNIÈRE, YANN (2008b) Technology transfer by CDM projects: A comparison of Brazil, China, India and Mexico. *Energy Policy,* forthcoming.

DIMAGGIO, P.; POWELL, W (1983) The iron cage revisited: Institutional isomorphism and collective rationality in organisational fields. *American Sociological Review,* 48, 147-160.

DINCER, I. (2000) Renewable energy and sustainable development: A crucial review. *Renewable and Sustainable Energy Reviews,* 4, 157-175.

DIVAN, S.; ROSENCRANZ, A. (2001) *Environmental Law and Policy in India: Cases, Materials and Statutes,* New Delhi, Oxford University Press.

DRAUDE, A. (2007) Wer regiert wie? Eine funktional-äquivalente Betrachtung von Governance in Räumen begrenzter Staatlichkeit. *SFB 700 Working Paper Series, No.2.* Berlin, SFB 700.

DRYZEK, J. S. (1997) *The Politics of the Earth: Environmental Discourses,* Oxford, Oxford University Press.

DUBASH, N.; RAJAN, S. (2002) India: Electricity reform under political constraints. World Resources Institute (Ed.) *Power Politics.* Washington, *WRI* Available at: http://pdf.wri.org/powerpolitics_chap4.pdf

DUNNING, J. H. (2006) Towards a new paradigm of development: implications for the determinants of international business. *Transnational Corporations,* 15, 173-192.

EBBINGHAUSEN, B. (2005) Can path dependence explain institutional change? Two approaches to welfare state reform. *MPIfG Discussion Paper,* 05, 1-28.

ECCLES, R. (1982) The quasi-firm in the construction industry. *Journal of Economic Behaviour and Organisation,* 2, 335-357.

ELLIOTT, D. (1997) *Energy, Society and Environment: Technology for a sustainable future,* London, New York, Routledge.

EMIRBAYER, M.; GOODWIN, J. (1994) Network analysis, culture and the problem of agency. *The American Journal of Sociology,* 99, 1411-1454.

ESTY, D.; IVANOVA, M.H. (2002) *Global environmental governance: options and opportunities,* Yale School of Forestry and Environmental Studies.

EXWORTHY, M.; POWELL, M. (2004) Big Windows and Little Windows: Implementation in the 'Congested State'. *Public Administration,* 82, 263-281.

FINNEMORE, M.; SIKKINK, K. (1998) International Norm Dynamics. *International Organization,* 52, 887-917.

FIORINO, D. (1995) *Making Environmental Policy,* Berkeley, University of California Press.

FISHLOW, A. (1990) The Latin American State. *Journal of Economic Perspectives,* 4, 61-74.

FLICK, U. (2006) *Qualitative Sozialforschung. Eine Einführung,* Hamburg, Rowohlt Taschenbuch Verlag.

FOLEY, G. (1993) Renewable energy in third world development assistance. Learning from experiences. *Energy Policy,* 20, 355-364.

FUHR, H. (1987) Economic Restructuring in Latin America: Towards the Promotion of Small-Scale Industry. *IDS Bulletin,* 18, 3.

FUHR, H.; LEDERER, M. (2009) Convergence and Divergence of Carbon Governance. *Journal for Environment and Development,* forthcoming

FUHR, H.; LEDERER, M.; SCHRÖDER, M. (2007) Klimaschutz und Entwicklungspolitik: Der Beitrag privater Unternehmen. In: RISSE, T.; LEHMKUHL, U. (Eds.) *Regieren ohne Staat? Governance in Räumen begrenzter Staatlichkeit.* Baden-Baden, Nomos.

GALLUP, J. L.; SACHS, J.; MELLINGER, A. (1998) Geography and Economic Development. *Harvard Institute of Economic Research Working Papers.* Harvard Institute of Economic Research, Harvard University.

GANGHOFF, S. (2005) Kausale Perspektiven in der vergleichenden Politikwissenschaft: X-zentrierte und Y-zentrierte Forschungsdesigns. In: KROPP, S.; MINKENBERG, M. (Eds.) *Vergleichen in der Politikwissenschaft.* Wiesbaden, Westdeutscher Verlag.

GEORGE, A. L.; BENNETT, A. (2004) *Case Studies and Theory Development in the Social Sciences,* Cambridge, MA, The MIT Press.

GERRING, J. (2004) What is a Case Study and What is it Good for? *American Political Science Review,* 98, 341-354.

GHOSH, D.; SHUKLA, P.R.; GARG, A.; RAMANA, V. P. (2002) Renewable energy technologies for the Indian power sector: mitigation potential and operational strategies. *Renewable and Sustainable Energy Reviews,* 6, 481-512.

GLOBERMAN, S; SHAPIRO, D. (2002) Global FDI Flows: The role of governance institutions. *World Development,* 30, 1899-1919.

GLÜCKLER, J. (2006) A relational assessment of international market entry in management consulting. *Journal of Economic Geography,* 6, 369-393.

GLÜCKLER, J.; ARMBRÜSTER, T. (2003) Bridging uncertainty in management consulting: The mechanisms of trust and networked reputation. *Organization Studies,* 24, 269-297.

GÖHLER, J. (2007) Weiche Steuerung. Regieren ohne Staate aus machttheoretischer Perspektive. In: RISSE, T.; LEHMKUHL, U. (Eds.) *Regieren ohne Staat? Governance in Räumen begrenzter Staatlichkeit.* Baden-Baden, Nomos.

GOLAIT, N.; MOHARIL., R.M.; KULKARNI, P.S. (2007) Wind electric power in the world and perspectives of its development in India. *Renewable and Sustainable Energy Reviews.*

GOLDEMBERG, J. (1996) *Energy, environment and development.* Paris, Académie international.

GOLDEMBERG, J. (2006) The promise of clean energy. *Energy Policy,* 34, 2185-2190.

GRABHER, G. (Ed.) (1993) *The Embedded Firm. On the Socio-economics of Industrial Networks,* London, Routledge.

GRANOVETTER, M. (1982) The strength of weak ties. A network theory revisited. MARSDEN, P. ; LIN, N. (Eds.) *Social Structure and Network Analysis.* Beverly Hillds, Sage.

GRANOVETTER, M. (1985) Economic Action and Social Structure: The Problem of Embeddedness. *American Journal of Sociology,* 91, 481-510.

GRANOVETTER, M. (1990) The Old and the New Economic Sociology: A History and an Agenda. In: FRIEDLAND, R.; ROBERTSON, A.F. (Eds.) *Beyond the Market Place: Rethinking Economy and Society.* New York, DeGruyter.

GRANOVETTER, M. (2005) The impact of social structure on economic outcomes. *Journal of Economic Perspectives,* 19.

GREENWOOD, C; HOHLER, A.; HUNT, G. (2007): Global Trends in Sustainable Energy Investment 2007. UNEP & NEF, New Delhi. Available at: http://sefi.unep.org/fileadmin/media/sefi/docs/publications/SEFI_Investment_Report_2007.pdf

GREENWOOD, R (1988): Organisational Design Types, Tracks and the Dynamics of Strategic Change In: *Organisation Studies* 9.3, 293-316

GRUBB, M. (1993) Politics and economics of climate change after Rio. *International Journal of Environmental and Pollution,* 3, 179-186.

GULATI, R. (1998) Alliances and Networks. *Strategic Management Journal,* 19, 293-317.

GUPTA, C. L. (2003) Role of renewable energy technologies in generating sustainable livelihoods. *Renewable and Sustainable Energy Reviews,* 7, 155-174.

HAANYIKA, C. M. (2006) Rural electrification policy and institutional linkages. *Energy Policy,* 34, 2977-2993.

HAFNER-BURTON, E. (2009) Globalisation and the Social Power Poltics of International Economic Networks. Conference Paper presented at the *International Studies Association Annual Convention* 13th*-*15th *February, 2009.* New York.

HAITES, E.; MAOSHENG, D.; SERES, S. (2006): Technology Transfer by CDM Projects. *Climate Policy 6.3,* 327-344

HAJER, M.; VERSTEEG, W. (2005a) A Decade of Discourse Analysis of Environmental Politics: Achievements, Challenges and Perspectives. *Journal of Environmental Policy & Planning,* 7, 175-184.

HAJER, M.; VERSTEEG, W. (2005b) Performing governance through networks. *European Political Science,* 4, 340-347.

HALL, P.; TAYLOR, R. (1996b) Political Science and the Three New Institutionalisms. *Political Studies*, 936-957.

HARDY, M (1995): *Power and Politics in Organisations.* Dartmouth Publishing

HÄUßLING, R. (2005) Netzwerke und Organisationen - konträre oder komplementäre gesellschaftliche Mechanismen? In: JÄGER, W.; SCHIMANK, U. (Eds.) *Organisationsgesellschaft. Facetten und Perspektiven.* Wiesbaden, VS Verlag für Sozialwissenschaften.

HAVNEVIK, K.; NEGASH, T; BEYENE, A. (2006) Of Global Concern: Rural livelihood dynamics and natural resource governance. Stockhom, SIDA.

HELLER, T.; SHUKLA, P.R. (2003) Development and Climate: Engaging developing countries. In: ALDY, J. E. (Ed.) *Beyond Kyoto: Advancing in International Efforts against Climate Change.* Arlington, VA, Pew Centre.

HESTER, R.; HARRISON, R. (2003) *Sustainability and environmental impact of renewable energy sources*, Royal Society of Chemistry.

HITE, J. M. (2003) Patterns of multidimensionality among embedded network ties: a typology of relational embeddedness in emerging entrepreneurial firms. *Strategic Organization,* 1, 9-49.

HITE, JM; HESTERLEY, WS (2001): The evolution of firm networks: From emergence to early growth of the firm. *Strategic Management Journal* 22.3, 275-286

HO, P. (2005) Greening industries in newly industrialised countries: Asian-style leapfrogging? *International Journal of Environment and Sustainable Development,* 4, 209-226.

HOEKMAN, B. M.; MASKUS, K.; SAGGI, KAMAL (2004) Transfer of Technology to Developing Countries: Unilateral and Multilateral Policy Options. *World Bank Working Paper Series*. Washington, World Bank.

HOLLSTEIN, B. (2006) Qualitative Methoden und Netzwerkanalyse - ein Widerspruch? In: HOLLSTEIN, B.; STRAUS, F. (Eds.) *Qualitative Netzwerkanalyse. Konzepte, Methoden und Anwendungen.* VS Verlag für Sozialwissenschaften.

HOPWOOD, B.; MELLOW, M.; O'BRIEN, G. (2005) Sustainable Development: Mapping Different Approaches. *Sustainable Development,* 13, 38-52.

INGEBRITSEN, C. (2002) Norm Entrepreneurs: Scandinavia's Role in World Politics. *Cooperation and Conflict,* 37, 11-23.

JACOBSON, S.; LAUBER, V. (2006) The politics and diffusion of energy systems transformation - explaining the German diffusion of renewable energy technology. *Energy Policy,* 34, 256-276.

JAEGER, A.; KANUGO, R. (1990) *Management in Developing Countries,* London, Routledge.

JAGADEESH, A. (2000) Wind energy development in Tamil Nadu and Andhra Pradesh, India. Institutional dynamics and barriers - A case study. *Energy Policy,* 28, 157-168.

JAHN, D. (1998) Environmental performance and policy regimes: Explaining variations in 18 OECD-countries. *Policy Sciences,* 31, 107-131.

JÄNICKE, M. (2000) Ecological Modernisation: Innovation and Diffusion of Policy and Technology. Berlin, FFU.

JÄNICKE, M (1991) Institutions and other framework conditions for environmental policy success. A tentative comparative approach. Berlin, FFU

JÄNICKE, M.; BAUER, M.; BLAZEJCZEK, J (2003) *The emergence of lead markets for environmental innovations,* München, oekonom.

JANN, W. (1998) Politik und Verwaltung im funktionalen Staat. In: JANN, W.; KÖNIG, K.; LANDFRIED, C.; WORDELMANN, P. (Eds.) *Politik und Verwaltung auf dem Weg in die transindustrielle Gesellschaft.* Baden-Baden, Nomos.

JANSEN, D. (2002) Netzwerkansätze in der Organisationsforschung. In: ALLMENDINGER, J.; H, T. (Eds.) *Organisationssoziologie.* Wiesbaden, Westdeutscher Verlag.

JASANOFF, S.; MARTELLO, M. L .(2004) *Earthly politics: Local and global in environmental politics,* Cambridge, MA, USA, MIT Press.

JESSOP, B. (1996) The rise of governance and the risk of failure: the case of economic development. UNESCO *Enjeux des debats sur la governance.* UNESCO, University of Lausanne.

JESSOP, B. (1999) The Dynamics of Partnership and Governance Failure. In: STOKER, G. (Ed.) *The New Politics of Local Government in Britain.* Oxford, Oxford University Press.

JESSOP, B. (2007) *The social embeddedness of the economy and its implications for economic governance.* Lancaster University. Available at: http/..eprints.cddc.vt.edu/digitalfordism_materials/jessop2.htm.

JOHANNSON, B.; Q, J. M. (2006) Agglomeration and networks in spatial economics. *Papers in Regional Science,* 83, 165-176.

JONES, C.; HESTERLEY, W.; BORGATTI, S. (1997) A general theory of network governance: Exchange Conditions and Social Mechanisms. *Academy of Management Review,* 22, 911-945.

JUMA, C.; FANG, K.; HONCA, D.; HUETE-PEREZ, J.; KONDE, V.; LEE, S. H. (2001) Global governance of technology: meeting the needs of developing countries. *International Journal of Technology Management,* 22, 629-655.

KACKER, R. (2007) The wind industry in India: policy, regulatory and attitudinal issues. *AEI Newsletter.* New Delhi, Asian Economics Institute

KHAN, J. (2004): *Local Politics of Renewable Energy. Project Planning, Siting Conflicts and Citizen Participation*, Lund University, Doctoral Dissertation

KHATOR, R. (1991) *Environment, Development and Politics in India,* Lanham, Md, University Press of America.

KIESER, A. (2001) *Organisationstheorien,* Stuttgart, Kohlhammer.

KINGDON, J. (1995) *Agendas, Alternatives and Public Policies,* New York, Longman.

KIRKPATRICK, C.; PARKER, D.; ZHANG, Y.-F. (2006) Foreign direct investment in infrastructure in developing countries: does regulation make a difference? *Transnational Corporations,* 15, 143-171.

KJELLSTRÖM, B.; ARVIDSON, A.; FORSLUND, H.; MARTINAC, I. (2005) Renewable Energy Technologies for Decentralised Rural Electricity Services. Stockholm, SIDA.

KOCHHAR, K.; KUMAR, U.; RAGHURAM, R.; SUBRAMANIAN, A.; TOKATLIDIS, I. (2006) India's pattern of development: What happened, what follows? *Journal of Monetary Economics,* 53, 981-1019.

KOELBLE, T. (1995) The New Institutionalism in Political Science and Sociology. *Comparative Politics,* 27, 231-243.

KRISHNA, K. (2004) Patterns and determinants of economic growth in Indian states. *Indian Council on International Economic Relations Working Paper.* New Delhi, Indian Council on International Economic Relations.

LALL, S. V.; CHAKRAVORTY, S. (2005) Industrial location and spatial inequality: Theory and evidence from India. *Review of Development Economics,* 9, 47-68.

LANGHELLE, O. (2000) Why Ecological Modernisation and Sustainable Development Should not be Conflated. *Journal of Environmental Policy & Planning,* 2, 303-322.

LEVINE, L. (Ed.) (1994) *Transfer and implementation of information technology,* Holland, IFPI.

LEVY, D.; NEWELL, P. (2005) *The business of global environmental governance,* Cambridge, MA, USA, MIT Press.

LEWIS, J. (2007) Technology acquisition and innovation in the developing world: Wind turbine development in China and India. *Studies of Comparative International Development,* 49, 208-232.

LINCOLN, A. (1860) Lecture "Discoveries and Inventions". LINCOLN, A. (Ed.) *Discoveries and Inventions.* Available at: http://www.todayinsci.com/L/Lincoln_Abraham/LincolnAbraham-Quotations.htm

LINTON, J. (2002) Implementation Research: state of the art and future directions. *Technovation,* 22, 65-79.

LIPSCHUTZ, R.; MAYER, J (1996) *Global civil society and global environmental governance: the politics of nature,* New York, State University of New York Press.

MABEL, C. M.; FERNANDEZ, E. (2008) Growth and future trends of wind energy in India. *Renewable and Sustainable Energy Reviews,* 12, 1745-1757.

MABUZA, L. O. K.; BRENT, A.; MAPAKO, M. (2007) The transfer of energy technologies in a developing country conext - Towards improved practice from past successes and failures. *Proceedings of the World Academy of Science, Engineering and Technology,* 22, 237-241.

MALLAMPALLY, P.; SAUVANT, K. P. (1999) Foreign Direct Investment in Developing Countries. *Finance & Development,* 36.

MARCH, J.; OLSON, J.P. (1989) *Rediscovering Institutions,* New York, The Free Press.

MARCH, J.; OLSON, J.P. (1995) *Democratic Governance,* New York, Free Press.

MARCH, J.; OLSON, J.P. (1999) The institutional dynamics of international political orders. In: KATZENSTEIN, R.; KATZNELSON, R.; KRASNER, S.D. (Eds.) *Exploration and contestation in the study of world politics.* Cambridge, MA, MIT Press.

MARSHALL, G. (1998) Organisation Theory. *A Dictionary of Sociology.* A Dictionary of Sociology.

MARTINOT, E. (2004) Global renewable energy markets and policies. *New Academy Review,* 3, 56-67.

MARTINOT, E.; CHAUREY, A.; LEW, D.; MOREIRA, J. R.; WAMUKONYA, N. (2002) Renewable Enery Markets in Developing Countries. *Annual Review of Energy and Environment,* 27, 309-348.

MASSEY, D.; ALLAN, J.; SARRE, P (1999) *Human Geography Today,* Cambridge, Polity Press.

MATHUR, K. (2003) Battling for Clean Environment: Supreme Court, Technocrats & Populist Politics in Delhi. *Working Paper.* New Delhi, Nehru University.

MAYNTZ, R. (2004) Governance im modernen Staat. In: BENZ, A (Ed.) *Governance. Regieren in komplexen Regelsystemen. Eine Einführung.* Wiesbaden, Verlag für Sozialwissenschaften, 65-76

MAYNTZ, R. (1993) Policy Netzwerke und die Logik von Verhandlungssystemen. *Politische Vierteljahreszeitschrift,* Sonderheft Policy Analyse, 39-56.

Mc DADE, S. (2005): *Energy Services for the Millennium Development Goals,* Washington, International Bank for Reconstruction and Development & UNDP, Available at: hpttp://www.unmillenniumproject.org/documents/MP_Energy_Low_Res.pdf

MEHTA, P.; KAPUR, D (Eds.) (2005) *Public Institutions in India. Performance and Design,* New Delhi, Oxford University Press.

MERTON, R. (1968) *Social Theory and Social Structure,* New York, The Free Press.

MEYER, J.; ROWAN, B. (1977) Institutional Organisations. Formal Structure as Myth and Ceremony. *American Journal of Sociology,* 83, 340-363.

MICHAELOWA, A. (2007) Efficient DNA Operation: Lessons from different DNA settings in non-Annex-B countries. Hamburg, HWWI.

MILLER, M. (1995) *The Third World in global environmental politics,* London, Boulder.

MINK, S. (1993) Poverty, Population and the Environment. *Environment Working Paper.* Washington, World Bank.

MINOGUE, M.; CARINO, L. (2006) Introduction: regulatory governance in developing countries. In: MINOGUE, M.; CARINO, L. (Eds.) *Regulatory Governance in Developing Countries.* Cheltenham, Edward Elgar.

MORAN, T. (1999) FDI and Development. A reassessment of the evidence on policy implications. In: OECD (Ed.) *FDI. Development and Corporate Social Responsibility.* Paris, OECD.

NAHAPIET, J.; GHOSHAL, S. (1998) Social capital, intellectual capital and the organisational advantage. *Academy of Management Review,* 24, 242-266.

NARAIN, S.; BHUSHAN, C.; D'SOUZA, M.; SINGH Y.; KUSHAL, P. (2008) Fanning an Alternative. *Down To Earth.* Centre for Science and Environment, New Delhi

NORTH, D. (1990) *Institutions, Institutional Change and Economic Performance,* Cambridge, Cambridge University Press.

NORTH, D.; THOMAS, R. (1973) *The Rise of the Western World,* Cambridge, Cambridge University Press.

NUNNENKAMP, P. (2002) Determinants of FDI in Developing Countries: Has Globalisation Changed the Rules of the Game? *Kiel Working Paper.* Kiel, Kieler Institute for World Economics.

OSANG, T. (2006) External and Internal Determinants of Development. Dallas, Texas, Department of Economics, Southern Methodist University.

PAINULY, J. P. (2001) Barriers to renewable energy penetration; a framework for analysis. *Renewable Energy,* 24, 73-89.

PARIKH, I. J.; GARG, P. K. (1990): *Indian organizations: value dilemmas in managerial roles.* In: JAEGER, A.; KANUGO, R. (Eds.): *Management in Developing Countries.* London & New York, Routledge

PERKINS, R. (2003) Environmental leapfrogging in developing countries: A critical assessment and reconstruction. *Natural Resources Forum,* 27, 177-188.

PETERS, G. (1997) The policy process: an institutionalist perspective. *Canadian Public Administration,* 35, 160-180.

PETERS, G. (1999) *Institutional Theory in Political Science. The New Institutionalism,* New York and London, Pinter.

PICOT, A.; DIETL, H.; FRANCK, E. (Eds.) (1999a) *Organisation. Eine ökonomische Perspektive,* Stuttgart.

PICOT, A.; FREUDENBERG, H.; GASSNER, W. (1999b) Die neue Organisation - ganz nach Maß geschneidert. *Harvard Business Manager,* 5, 46-58.

PIERSON, P.; SKOCPOL, T. (2002) Historical Institutionalism in Contemporary Political Sciences. In: KATZNELSON, I.; MILNER, H. (Eds.) *Political Science: State of the Discipline.* New York, Norton.

PILLAI, G. M. (Ed.) (2006) *Wind Power Development in India,* Pune, India, WISE.

PODOLNY, J.; PAGE, K (1993) Network forms of organisation. *Annual Review of Sociology,* 24, 57-76.

POWELL, W. (1990a) Neither Market Nor Hierarchy: Network Forms of Organisation. *Research in Organisational Behaviour,* 12, 295-336.

POWELL, W.; DI MAGGIO, P. (Eds.) (1991) *The New Institutionalism in Organisational Analysis,* Chicago, University of Chicago Press.

PRENGEL, R. (2004) Risk assessment in renewable energy projects under the CDM: Wind energy projects in China. TU Berlin.

PRZEWORSKI, A.; LIMONGI, F. (1993) Political Regimes and Economic Growth. *Journal of Economic Perspectives,* 7, 51-69.

PUROHIT, P.; MICHAELOWA, A: (2007) CDM potential of wind power projects in India. *HWWI Research Paper 1-5,* Hamburg ISSN 1861-504x

RAJSEKHAR, B.; VAN HULLE, F. & JANSEN, J.C. (1999) Indian wind energy programme: Performance and future directions. *Energy Policy,* 27, 669-678.

RAMACHANDRA, T. V.; JOSHI, N.V.; SUBRAMANIAN, D.K. (2000) Present and prospective role of bioenergy in the regional energy system. *Renewable and Sustainable Energy Reviews,* 4, 375-430.

RAMANA, PV; SINHA, CS; SHUKLA, PR (2001): Renewable energy technologies and climate change politics in India. *International Journal for Global Energy Issues* 15.1-2, 97-116

RANDERIA, S. (2003): Between Cunning States and Unaccountable Institutions: Social Movements and Rights of Local Communities to Common Property Resources, *WZB Discussion Paper Nr. SP IV 2003-502*, WZB, Berlin

RAO, K.; CHALAPATY; MURTHY; M. (2006) Towards understanding the state-wise distribution of foreign direct investmens in the post-liberalisation period. *ISID Working Paper.* New Delhi, Institute for Studies in Industrial Development.

RAO, S. L. (2004) *Governing Power: A new institution of governance: the experience with independent regulation of electricity,* New Delhi, TERI.

RAVINDRANATH, N. H. (Ed.) (2000) *Renewable Energy and Environment: A Policy Analysis for India,* New Delhi, Tata McGraw-Hill Publishing Company Ltd.

REINEBERGER, H. (2006) India's Electricity Sector in Transition: Can its giant goals be met? *The Electricity Journal,* 19, 77-84.

RENO (1998) *Warlord Policy and African States,* Boulder, Lynne Rienner.

RICHARDS, P. (2006) Renewable development: New strategis in rural electrification. *Renewable Energy World.* Tulsa, Penn Well Coporation.

RICHTER, R. (2001) New economic sociology and new institutional economics. *MPRA Paper presented at* Annual Conference of the International Society for New Institutional Economics September 13th-15th 2001. California, USA, University of Berkeley

RICHTER, R. (2006) On the social structure of markets: A review and assessment in the perspectives of the New Institutional Economics. *Marketsociology,* 13, 1-19.

RIEGER, E.; LIEBFRIED, S. (2003) *Limits to Globalisation: Welfare State and World Economy,* Cambridge, Cambridge Polity Press.

RIGGS, F. (1964) *Administration in Developing Countries: The Theory of Prismatic Society.* Boston

RISSE, T. (2002) Constructivism and International Institutions: Towards conversation across paradigms. In: KATZNELSON, I.; MILNER, H. (Eds.) *Political Science: The State of the Discipline.* New York.

RISSE, T. (2002) Transnational Actors and World Politics. In: SIMMONS, C. R. (Ed.) *Handbook of International Relations.* London, Sage.

RISSE, T. & LEHMKUHL, U. (Ed.) (2007) *Regieren ohne Staat? Governance in Räumen begrenzter Staatlichkeit,* Baden-Baden, Nomos.

RISSE, T.; LEHMKUHL, U (2006): Governance in Räumen begrenzter Staatlichkeit: Neue Formen des Regierens?, Das Forschungsprogramm des Sonderforschungsbereich 700. *SFB 700 Governance Working Paper Series No. 1*, SFB 700, Berlin

RITTBERGER, H. M. (1997) *Theories of International Regimes,* Cambridge, Cambridge University Press.

RODRIK, D.; SUBRAMANIAN, A.; TREBBI, F. (2004) Institutions Rule: The Primacy of Institutions over Geography and Integration in Economic Development. *Journal of Economic Growth,* 9, 131-165.

ROWLEY, T.; BEHRENS, D.; KRACKHARDT, D. (2000) Redundant Governance Structures: An Analysis of Structural and Relational Embeddedness in the Steel and Semiconducter Industries. *Strategic Management Journals,* 21, 369-386.

RUGGIE, J. (1998) What makes the world hand together? Neo-utalitarianism and social constructivist challenges. *International Organization,* 52, 855-885.

SACHS, J. (2003) Institutions Don't Rule: Direct Effects of Geography on Per Capita Income. *National Bureau of Economic Research,* Working Paper 5276.

SACHS, J.; BAJPAI, N.; RAMIAH, A. (2002) Understanding Regional Economic Growth in India. *CID Working Paper, 88.* Harvard University, MA USA, Centre for International Development

SANTISO, C. (2003) The Paradox of Governance: Objective or Condition of Multilateral Development Finance? *SAIS Working Paper Series,* 3.

SCHAMP, EW (2000): Vernetzte Produktion: Industriegeographie aus institutioneller Perspektive. Darmstadt, Wissenschaftliche Buchgesellschaft

SCHÄTZL, L. (2003) *Wirtschaftsgeographie 1 - Theorie,* Paderborn, UTB Schöning

SCHÄTZL, L (1998) *Wirtschaftsgeographie*, Paderborn, UTB Schöning

SCHARPF, F. (2000a) Institutions in comparative policy research. *Comparative Political Studies,* 33, 762-790.

SCHARPF, F. (2000b) *Interaktionsformen. Akteurszentrierter Institutionalismus in der Politikforschung,* München, UTB Leske & Budrich.

SCHMIDT, M. (1996) When parties matter: A review of the possibilities and limits of partisan influence on public policy. *European Journal of POlicy Research,* 30, 155-183.

SCHNEIDER, M.; HOLZER, A:; HOFFMANN, V. (2008): Understanding the CDM's Contribution to Technology Transfer. *Energy Policy 36.8, 2930-2938*

SCOTT, W. R. (1987) The Adolescence of Institutional Theory. *Administrative Science Quarterly,* 32, 493-511.

SHEPSLE, K.; BONCHEK, M. (1997) *Analysing politics: Rationality, behaviour and institutions,* New York, WW Norton.

SHERMAN, R. (2005) Energy: Energy Governance, Poverty and Sustainable Development. In: AYRE, G.; CALLAWAY, R. (Eds.) *Governance for Sustainable Development.* London, Earthscan.

SILAYAN, A. (2005): Equitable Distribution of CDM Projects among Developing Countries, *HWWI-Report 255,* Hamburg, HWWI

SIMONIS, U. E. (2002): Global Environmental Governance – Speeding up the Debate on a World Environment Organization, *WZB Papers FS II 02-404,* WZB Berlin

SIMSEK, Z.; LUBATKIN, M. H.; FLOYD, S. W. (2003) Inter-firm networks and entrepreneurial behaviour: A structural embeddedness perspective. *Journal of Management,* 29, 427-442.

SINGH, A. (2006) Power sector reform in India: current issues and prospects. *Energy Policy,* 34, 2480-2490.

SINGH, K. (1999) *Rural Development. Principles, Policies and Management,* New Delhi, Sage.

SKELCHER, C. (2000) Changing Images of the State: Overloaded, Hollowed-Out and Congested. *Public Policy and Administration,* 15, 3-19.

SKOCPOL, T. (1992) *Protecting Soldiers and Mothers: The Political Origin of Social Policy in the US,* Cambridge, MA, Belnak Press of Harvard University Press.

SMITH, C.; REES, G. (1998). *Economic Development, 2nd edition*. Basingstoke: Macmillan

SOERENSEN, E.; TORFING, J. (2005) Network Governance and Post-Liberal Democracy. *Administrative Theory & Praxis,* 27, 197-237.

SRIVASTAVA, L.; REHMANN, I. H. (2006) Energy for sustainable development in India: Linkages and strategic directions. *Energy Policy,* 34, 643-654.

SOUTHGATE, D. (1988) The Economics of Land Degradation in the Third World. *Environment Working Paper.* Washington, World Bank.

STEFES, C. (2009) Path dependence and renewable energy development in the United States and Germany. Conference Paper presented at the *International Studies Association Annual Convention 13th-15th February, 2009.* New York.

STEINMO, S.; THELEN, K; LONGSTRETH (1992) *Structuring Politics: Historical Institutionalism in Comparative Analysis,* Cambridge, Cambridge University Press.

STRECK, C. (2004) New partnerships in global environmental politics: The Clean Development Mechanism. *Journal of Environment and Development,* 13, 295-322

STURGEON, T. J. (2003) What really goes on in Silicon Valley? Spatial clustering and dispersal in modular production networks. *Journal of Economic Geography,* 3, 199-225.

SUBRAHAMANIAN, K. (2000) Industrial growth in Kerala: Trends and Explanations. Trivandrum, Centre for Development Studies.

SWEDBERG, R. (1997) New Economic Sociology: What Has Been Accomplished, What Is Ahead? *Acta Sociologica,* 40. 161-182.

TAYLOR, M (1987) *The Possibility of Cooperation.* Cambridge, Cambridge University Press

TEWS, K. (2002) Der Diffusionsansatz für die vergleichende Policy-Analyse. Wurzeln und Potenziale eines Konzepts. Berlin, Forschungsstelle für Umweltpolitik, Berlin.

TEWS, K.; BUSCH, PO; JÖRGENS, H. (2003) The diffusion of new environmental policy instruments. *European Journal of Policy Resources,* 42.

THATCHER, M. (1998) The Development of Policy Network Analysis: From Modest Origins to Overarching Frameworks. *Journal of Theoretical Politics,* 10, 389-416.

THELEN, K.; STEINMO, S (Eds.) (1992) *Structuring Politics,* Cambridge, Cambridge University Press.

THOMPSON, G. (2003) *Between hierarchies and markets: The logic of network forms of organisation,* Oxford, Oxford University Press.

TOOLE, L. (2000) Research on policy implementation: assessment and prospects. *Journl of Public Administration Research and Theory,* 10, 263-288.

TUKKER, A. (2005) Leapfrogging into the future: developing for sustainability. *International Journal of Innovation and Sustainable Development,* 1, 65-84.

UZZI, B. (1996) The Sources and Consequences of Embeddedness for the Economic Performance of Organizations: The Network Effect. *American Sociological Review,* 61, 674-698.

UZZI, B.; LANCASTER, R. (2003) Relational embeddedness and learning: the case of bank loan managers and their clients. *Management Science,* 49, 383-399.

VANGEN, S.; HUXHAM, C. (2003) Enacting leadership for collaborative advantage: Dilemmas of ideology and pragmatism in the activities of partnership managers. *British Journal of Management,* 14, 61-76.

VARUN; SINGAL, S. K. (2005) Review of augmentation of energy needs using renewable energy sources in India. *Renewable and Sustainable Energy Reviews,* 1-9.

VAUGHAN, D. (1999) The dark side of organisations: Mistake, misconduct and disaster. *Annual Review of Sociology,* 25, 271-305.

VENNESSON, P. (2008) Case studies and process tracing: theories and practices. DELLA PORTA, D.; KEATING, M. (Eds.) *Approaches and Methodologies in the Social Sciences. A Pluralist Perspective.* Cambridge, Cambridge University Press.

WASSERMANN, S.; FAUST, K. (1994) *Social network analysis: Methods and applications,* Cambridge, Cambridge University Press.

WEIR, M.; SCOKPOL, T. (1985) State Structure and Possibilities for Keynesian Responses to the Great Depression in Sweden, Britain and the US. EVANS, P.; RUESCHEMEYER, D.; SKOCPOL, T. (Eds.) *Bringing the State Back In.* Cambridge, Cambridge University Press.

WENDT, A. (1999) *Social Theory of International Politics,* Cambridge, Cambridge Univerity Press.

WIENGES, S. (2008) New Strategies for Global Governance: Sustainable Energy-Related Global Policy Networks. *Berlin Conference on the Human Dimensions of Global Environmental Change.* Berlin.

WILKINS, G. (2003) Renewables and Development. *Renewable Energy World,* 5, 202-215.

WILKINS, G. (2002) *Technology Transfer for Renewable Energy. Overcoming Barriers in Developing Countries,* London, Earthscan.

WILLIAMSON, O. (1975) *Markets and hierarchies, analysis and antitrust implications: A study in the econonmics of internal organisations,* New York, The Free Press.

WILLIAMSON, O. (1985) *Economic Institutions of Capitalism.*

WILTOX, F. (2000) Towards a relational view on industrial location theory. *Tijdschrift voor Economische en Sociale Geografie,* 91, 135-146.

WOHLGEMUTH, N.; MISSFELDT, F. (2000) The Kyoto Mechanisms and the Prospects for Renewable Energy Technologies. *Solar Energy,* 69, 305-314.

WOLFF, F. (2001) Green Global Governance. Eine sozialwissenschaftliche Einführung in Aufgaben des Öko-Instituts. Unpublished concept paper in the research project "*Schnittstellenentwicklung zur Integration akademischer und praxisbezogener Forschung im Bereich Sozial-Ökologie".* Berlin, BMZ.

WOLMAN, H. (1981) The Determinants of Program Success and Failure. *Journal of Public Policy,* 1, 433-464.

WORRELL, E.; VAN BERKEL, R.; FENQI, Z.; MENKE, C.; SCHAEFFER, R.; WILLIAMS, R. O. (2001) Technology transfer of energy efficient technologies in industry: a review of trends and policy issues. *Energy Policy,* 29, 29-43.

YEUNG, H. W.-C. (2002) Towards a relational economic geography: Old Wine in New Bottles? *98th Annual Meeting of the Association of American Geographers.* Los Angeles, USA.

YIN, R. (2003) *Case Study Research. Design and Methods,* London, Sage.

YOUNG, O. (1997) *Global governance: drawing insights from the environmental experience,* London, Routledge

YOUNG, O. (1999) *The Emergence of Ecological Modernisation. Integrating the Environment and the Economy?* London, Routledge.

YOUNG, O. (2007) Institutions and Environmental Change: The scientific legacy of a decade of IDGEC research. Conference Paper presented at the *Amsterdam Conference in the Human Dimensions of Global Environmental Change.* Amsterdam, Vrije Universiteit

ZINGEL, W.-P. (1998) Umweltpolitik und wirtschaftlicher Strukturwandel in Indien. In: SCHUCHER, G. (Ed.) *Asien zwischen Ökonomie und Ökologie. Wirtschaftwunder ohne Grenzen?* Hamburg, Institut für Asienkunde.

ZUKIN, S.; DI MAGGIO, P. (Eds.) (1990) *Structure of Capital. The social organisation of the economy,* New York, Cambridge University Press.
ZÜRCHER, M. (2007) When governance meets troubled states. In: BEISHEIM, M.; SCHUPPERT, F (Eds.) *Staatszerfall und Governance.* Baden-Baden, Nomos.
ZÜRN, M. (1998) *Regieren jenseits des Nationalstaats. Globalisierung and Denationalisierung als Chance,* Frankfurt/Main, Suhrkamp.

Interview Sources

Chapter I:
INT 1: Interview with Director of Senergi Global on 12.09.2007 (Mr Chintan Shah)
INT 2: Interview with Unit Chief at C-WET on 2.11.2007 (Dr Sreevalsan)

Chapter II
INT I Interview with AGM Corporate Affairs, Suzlon on 25.10.2007 (Mr Suresh Kumar)

Chapter III
INT 1 Interview with Business Head South, Suzlon on 29.10.2007 (Mr Saibaba)
INT 2 Interview with Business Head South, Suzlon on 29.10.2007 (Mr Saibaba)
INT 3 Interview with AGM Corporate Affairs, Suzlon on 25.10.2007 (Mr Suresh Kumar)
INT 4 Interview with CEO & Secretary General, Indian Wind Energy Association (InWEA) on 09.01.2008 (Mr Rakesh Kacker)
INT 5 Interview with Ret Chairman, TENRC on 04.11.2007 (Mr Balraj)
INT 6 Interview with President, Indian Wind Turbine Manufacturers' Association (IWTMA) on 03.11.2007 (Mr DV Giri)
INT 7 Interview with CEO & Secretary General, Indian Wind Energy Association (InWEA) on 21.09.2007 (Mr Rakesh Kacker)
INT 8 Interview with CEO & General Manager WISE on 26.09.2007 (Mr GM Pillai)

Chapter IV
INT 1: Interview with GTZ official on 16.09.2007 (Ms Pamposh Bhat)
INT 2: Interview with AGM Corporate Affairs at Suzlon Kerala Office on 24.10. and on 25.10.2007 (Mr Suresh Kumar)

Chapter V
INT 1 Interview with Principal Secretary of Tamil Nadu Energy Minister on 30.10.2007 (Mr Satapathy)
INT 2 Interview with Professor at JNU in New Delhi on 10.01.2008 (Dr Navroz Dubash)
INT 3 Interview with Principal Secretary of Tamil Nadu Energy Minister on 30.10.2007 (Mr Satapathy)
INT 4 Interview with Director of ABI Energy Consultancy on 29.10.2007 (Mr Vijayarajan)
INT 5 Interview with Director of InWEA on 17.09.2007 (Mr Rakesh Kacker)
INT 6 Interview with Director at MNRE on 18.09.2007 (Mr Ajit Gupta)
INT 7 Interview with Business Head-South of Suzlon on 29.10.2007 (Mr Saibaba Vutukuri)
INT 8 Interview with Managing Director of NEG Micon on 02.11.2007 (Mr Ramesh Kymal)
INT 9 interview with Consultant at ANERT on 26.10.2007 (Mr Harikumar)

INT 10 Interview with Assistant Executive Engineer at TNEB on 26.10.2007 (Mr R. Ramesh)
INT 11 Interview with Director of ANERT on 25.10.2007 (Mr Unnithan)
INT 12 Interview with Vice President of Indowind on 29.10.2007 (Mr Raja Sukumar)
INT 13 Interview with Director of MNRE on 12.09.2007 (Dr PC Maithani)
INT 14 Interview with Advisor at MNRE on 08.10.2007 (Dr TC Tripathi)
INT 15 Interview with Vice President of Indowind on 29.10.2007 (Mr Raja Sukumar)
INT 16 Interview with Director of Zenith Energy on 3.10.2007 (Mr. A. Mohan Reddy)
INT 17 Interview with Assistant Executive Engineer at TNEB on 26.10.2007 (Mr R. Ramesh)
INT 18 Interview with Consultant of ANERT on 26.10.2007 (Mr Harikumar)
INT 19 Interview with Deputy Director of ANERT on 25.10.2007 (Mr Unnithan)
INT 20 Interview with Member at KSERC on 25.10.2007 (Shri C Balakrishnan)
INT 21 Interview with Deputy Director and Senior Advisor at USAID on 20.11.2007 (Mr Padmanaban)
INT 22 Interview with Minister for Power and Electricity, Kerala State Government on 25.10.2007 (Mr CP Bramod)
INT 23 Interview with AGM Corporate Affairs on 25.10.2007 (Mr Suresh Kumar)
INT 24 Interview with Ret. Chairman at TSERC on 04.11.2007 (Mr A Balraj)
INT 25 Interview with Unit Chief at C-WET on 2.11.2007 (Dr Sreevalsan)
INT 26 Interview with CEO & General Manager WISE on 26.09.2007 (Mr GM Pillai)

Chapter VI

INT 1: Interview with Principal Secretary. Energy Ministry at Government of Tamil Nadu on 30.10.2007 (Mr Satapathy)
INT 2: Interview with Indian government official at the COP14 2008 in Poznan (wished to remain anonymous)
INT 3: Interview with Principal Secretary of Tamil Nadu Energy Minister on 30.10.2007 (Mr Satapathy)
INT 4: Interview with Assistant Executive Engineer at KSEB on 26.10.2007 (Mr R. Ramesh)
INT 5: Interview with AGM Corporate Affairs, Suzlon on 25.10.2007 (Mr Suresh Kumar)

Secondary Sources

AGENCY OF NON CONVENTIONAL ENERGY AND RURAL TECHNOLOGIES (2007) Budget Proposal 2006-07. Trivandrum, ANERT.

ANDHRA PRADESH STATE GOVERNMENT (2004): *Annual Plan 2004-05*. Hyderabad, Department for Statistics, Andhra Pradesh State Government

ASHOK LEYLAND (2007) Annual Report 2006/07. Chennai, Ashok Leyland.

BUNDESAGENTUR FUER AUSSENWIRTSCHAFT (bfai) (2006): Energiewirtschaft Indien 2004/05. bfai, Köln

CENSUS OF INDIA (2001): Census of India 2001. Vital Statistics. Office of Registrar General, New Delhi. Available at: http://censusindia.net, Accessed: 12.03.2007

CENTRAL ELECTRICITY AUTHORITY (2007) Power Sector Fundamentals. Ministry of Power, CEA.

CENTRAL ELECTRICITY AUTHORITY (2007) Power Sector Reports. Available at: http//www.cea.nic.in/power_sector_reports/general_review/0405/index.pdf, Accessed: 17.09.2007

CENTRE FOR WIND ENERGY TECHNOLOGY (C-WET) (2008) Annual Report 2007/2008. Chennai, C-WET.

CENTRE FOR WIND ENERGY TECHNOLOGY (C-WET) (2007): Wind Power Density Map. Available at: http://www.cwet.tn.nic.in/Docu/WPDmap.pdf, Accessed: 18.07.2007

GOVERNMENT OF INDIA (2006) Five-Year-Plan 2007-2011. In PLANNING COMMISSION, New Delhi,Government of India.

GOVERNMENT OF INDIA (2006) Power Sector Rating. In MINISTRY OF POWER; New Delhi,Government of India

GOVERNMENT OF INDIA (2007) Installed Power Capacity. In MINISTRY OF POWER, New Delhi,Government of India

GOVERNMENT OF INDIA (2001): Poverty Estimates 1999-2001. Government of India, Press Information Bureau, New Delhi

INDIAN WIND ENERGY ASSOCIATION (InWEA) (2007) Wind Energy Statistics. New Delhi, InWEA.

IndiaStat (2007): Power Sector Statistics. Available at: http://www.indiastat.com, Accessed: 18.09.2007

INTERNATIONAL ATOMIC ENERGY AGENCY (IAEA) (2005): Energy Indicators for Sustainable Development: Guidelines and Methodology, Geneva, IAEA

INTERNATIONAL ENERGY AGENCY (IEA) (1999) The Evolving Renewable Energy Market. Paris, IEA

INTERNATIONAL ENERGY AGENCY (IEA) (2006) World Energy Outlook 2006. Geneva, IEA

INTERNATIONAL ENERGY AGENCY (IEA) (2007) World Energy Outlook 2007. Geneva, IEA.

INTERNATIONAL PANEL ON CLIMATE CHANGE (IPCC) (2002): Climate Change 2001: Synthesis Report. Cambridge, New York, Cambridge University Press

KERALA STATE PLANNING BOARD (2002) Tenth Five-Year-Plan 2002-2007. Trivandrum, Kerala State Government

KERALA STATE PLANNING BOARD (2009) Eleventh Five-Year-Plan and Annual Plan 2007-08. Trivandrum, Kerala State Government.

KERALA STATE ELECTRICITY REGULATORY COMMISSION (2007) Annual Report 2006-2007. Trivandrum, Kerala State Electricity Regulatory Commission.

Maps of India – Available at: http://www.mapsofindia.com, Accessed: 15.05.2007

MINISTRY OF COMMERCE AND INDUSTRIES (2002): Foreign Direct Investment. SIA Newsletter 2002, Government of India, Secretariat for Industry Assistance, New Delhi

MINISTRY OF NEW AND RENEWABLE ENERGY SOURCES (MNRE) (2008) Renewable Energy Programme MNRE, Annual Report. New Delhi, MNRE.

MINISTRY OF NEW AND RENEWABLE ENERGY SOURCES (MNRE) (2008): Installed Wind Power. Available at: http://www.mnes.nic.in/wp-installed.htm, Accessed: 03.04.2008

MINISTRY OF POWER (2007): Power Sector Statistics. Available at: http://www.powermin.nic.in/JSP_SERVLETS/internal.jsp, Accessed: 15.05.2007

MINISTRY OF POWER (2006): State Power sector Performance Rating. ICRA Limited & Crisil Ratings, Ministry of Power, New Delhi

POWERLINE (2006) In Focus: Wind Energy. POWERLINE (Ed.) *India's First Power Magazine.*

SUZLON (2007) Annual Report 2006/2007. Pune, Suzlon.

TAMIL NAUD ELECTRICITY BOARD (2006) Annual Administrative Report 2005-06. Chennai,Tamil Nadu Electricity Board Press.

TAMIL NADU ELECTRICITY REGULATORY COMMISSION (2006) Annual Report and Statement of Accounts 2005/06. Chennai, Tamil Nadu State Regulatory Commission.

TAMIL NADU STATE PLANNING COMMISSION (2002) Tamil Nadu Tenth Five-Year Plan. IN STATE PLANNING COMMISSION (Ed.), Chennai, Tamil Nadu State Government.

TAMIL NADU ENERGY DEVELOPMENT AGENCY (2006) Annual Report 2005/06. Chennai, Tamil Nadu Energy Development Agency.

STATE ELECTRICITY REGULATORY COMMISSIONS (2007): Tariff Orders and Regulations. Accessed: 16.09.2007, Available at:

- Andhra Pradesh: http://www.ercap.org
- Karnataka: http://www.kerc.org
- West Bengal: http://www.wberc.net
- Madhya Pradesh: http://www.mperc.org
- Maharahstra: http://www.mercindia.org.in
- Rajasthan: http://www.rerc.gov.in
- Tamil Nadu: http://www.tnerc.tn.nic.in

- Gujarat: http://www.gercin.org
- Kerala: http://www.erckerala.org.in

THE ENERGY AND RESOURCES INSTITUTE (TERI) (2004): TEDDY – The Energy Data Directory and Yearbook 2003/04, New Delhi, TERI

TRANSPARENCY INTERNATIONAL INDIA (2005): India Corruption Study to Improve Governance. Transparency International India & Centre for Media Studies, New Delhi

UNDP (2001): Human Development Report India 2001. OUP, New Delhi

UNEP Risoe (2007): CDM Pipeline August 2007. Available at: http://cdmpipeline.org Accessed: 17.08.2007

UNEP Risoe (2008a): CDM Pipeline September 2008. Available at: http://cdmpipeline.org, Accessed: 23.09.2008

UNEP Risoe (2008b): CDM Pipeline November 2008. Available at: http://cdmpipeline.org, Accessed: 13.11.2008

WORLD BANK (2009): World Development Report 2009 "Reshaping Economic Geography". Washington DC, World Bank

WORLD COMMISSION ON ENVIRONMENT AND DEVELOPMENT (WCED) (1987) Our Common Future. Brundtland Report. New York, United Nations.

WORLD WIND ENERGY ASSOCIATION (WWEA) (2006): Global Wind Energy Statistics and Yearbook 2006. Vienna, WWEA

FSC
www.fsc.org
MIX
Papier | Fördert gute Waldnutzung
FSC® C083411

Kolibri 360 GmbH
Ferdinand-Jühlke-Straße 7
99095 Erfurt, Deutschland
produktsicherheit@kolibri360.de